Computational Aspects of Commutative Algebra

Computational Aspects of Commutative Algebra

from a special issue of the Journal of Symbolic Computation

Edited by

Dr Lorenzo Robbiano
Dipartimento di Matematica,
Università di Genova,
Italy

ACADEMIC PRESS
Harcourt Brace Jovanovich, Publishers
London San Diego New York Berkeley
Boston Sydney Tokyo Toronto

ACADEMIC PRESS LIMITED
24/28 Oval Road,
London NW1 7DX

United States Edition published by
ACADEMIC PRESS INC.
San Diego, CA 92101

British Library Cataloguing in Publication Data
is available

ISBN 0–12–589590–9

Printed in Great Britain at the University Press, Cambridge

Contributors

J. Apel, Naturwiss.-Theoretisches Zentrum und Sektion Mathematik, Karl-Marx-Universität, DDR-7010 Leipzig, DDR

D. Bayer, Department of Mathematics, Columbia University, New York, NY 10027, USA

M. Brundu, Dip. di Scienze Matematiche, Piazzale Europa 1, 34100 Trieste, Italy

G. Carrà Ferro, Dipartimento di Matematica, Università di Catania Viale A. Doria 6, 95125 Catania, Italy

T. Gateva-Ivanova, Institute of Mathematics, Bulgarian Academy of Sciences, P.O.B. 373, 1000 Sofia, Bulgaria

R. Gebauer, Springer-Verlag, New York, 175 Fifth Avenue, New York, NY 10010, USA

P. Gianni, IBM Research and University of Pisa, Pisa, Italy

M. Giusti, Centre de Mathématiques, Ecole Polytechnique, 91128 Palaiseau Cedex, Unité associée au CNRS No. 169, France

H. Kredel, Universität Passau, D-8390 Passau, FRG

W. Lassner, Naturwiss.-Theoretisches Zentrum und Sektion Mathematik, Karl-Marx-Universität, DDR-7010 Leipzig, DDR

V. Latyshev, Department of Mechanics and Mathematics, Moscow State University, 117 234 Moscow, USSR

A. Miola, Dipartimento di Informatica e Sistemistica dell'Università degli Studi di Roma "La Sapienza", Italy

H. M. Möller, FB Mathematik und Informatik, Fern Universität, Lützowstr., D-5800 Hagen, FRG

T. Mora, Dipartimento di Matematica dell'Università di Genova, Italy

I. Morrison, Department of Mathematics, Columbia University, New York, NY 10027, USA

L. Robbiano, Dipartimento di Matematica, Università di Genova, Via L.B. Alberti 4, I-16132 Genova, Italy

F. Rossi, Dip. di Scienze Matematiche, Piazzala Europa 1, 34100 Trieste, Italy

D. Shannon, Department of Mathematics, Transylvania University, Lexington, KY 40508, USA

R. Shtokhamer, Department of Computer and Information Sciences, University of Delaware, Newark, DE 19716, USA

M. Stillman, Department of Mathematics, Cornell University, Ithaca, NY 14853, USA

M. Sweedler, Department of Mathematics, Cornell University, Ithaca, NY 14853, USA

B. Trager, IBM Research, P.O. Box 218, Yorktown Heights, NY 10598, USA

V. Weipfenning, Universität Passau, D-8390 Passau, FRG

F. Winkler, Institut für Mathematik and Research Institute for Symbolic Computation, Johannes Kepler Universität, A-4040 Linz, Austria

G. Zacharias, MIT Al Laboratory, 545 Technology Square, Cambridge, MA 02139, USA

Preface

This volume features some recent results about computational problems in commutative algebra and related fields. The idea arose during the first meeting COCOA I (*Co*mputer and *Co*mmutative *A*lgebra), which was held in Genoa in May 1986; its success showed that the new ideas and techniques from computer science could strongly influence such a traditional section of mathematics as commutative algebra. Many new results were presented at the meeting, which afterwards were written and submitted as papers; in addition, we also obtained many other contributions. All the papers deal more or less directly with the fundamental notion of Gröbner basis (which is also called standard basis, so that the two words "Gröbner" and "standard" are usually interchangeable).

Let me now briefly describe the content of the volume.

The first three papers have in common a long history.

The core of "On the Complexity of Computing Syzygies" by D. Bayer and M. Stillman is formed by a presentation of the result by Mayr–Meyer on the double exponential nature of solving the ideal membership problem in polynomial rings over fields. In addition, a similar result is proven for the syzygies problem. By personal communication, this result is already part of the "folklore" on the subject, but originally it is due to the authors.

The paper "Gröbner Bases and Primary Decomposition of Polynomial Ideals" by P. Gianni, B. Trager, G. Zacharias is "well-known" in the sense that early versions of it were already available and quoted several times in the literature.

"Lifting Canonical Algorithms from a Ring R to the Ring $R[X]$" by R. Shtokhamer is an elaborated version of an early (1975) report on constructing canonical bases for polynomial ideals. Although results similar to those presented here meanwhile appeared in the literature, we felt that this early paper should be made available to a broader readership.

Then two related papers follow, "The Gröbner Fan of an Ideal" by myself and Mora, and "Standard Bases and Geometric Invariant Theory. I. Initial Ideals and State Polytopes" by D. Bayer and J. Morrison. They deal with the problem of understanding the effect of changing the ordering on the computation of a Gröbner basis.

Then "Gröbner Bases and Hilbert Schemes. I" by G. Carrà Ferro investigates the family of all the ideals that have the same associated monomial ideal with respect to a given ordering.

Two papers, "Computing Dimension and Independent Sets for Polynomial Ideals" by H. Kredel and V. Weispfenning, and "Combinatorial Dimension Theory of Algebraic Varieties" by M. Giusti deal with the problem of computing one of the most important invariant of ideals; while "Using Gröbner Bases to Determine Algebra Membership, Split Surjective Algebra Homomorphisms Determine Birational Equivalence" by D. Shannon and M. Sweedler gives another application of the theory of Gröbner Bases to the solution of some computational problems in commutative algebra.

The paper "On an Installation of Buchberger's Algorithm" by R. Gebauer and H. M. Möller is the source of some ideas which have already been used by the authors and other people for efficient implementations of the most important algorithm in this field, namely, Buchberger's algorithm. The problem of computing Gröbner bases modulo p, to avoid bad behaviour of integral coefficients is treated in "A p-adic Approach to the Computation of Gröbner Bases" by F. Winkler.

Then three papers, "Constructive Lifting in Graded Structures: A Unified View of Buchberger and Hensel Methods" by A. Miola and T. Mora, "On the Computation of

Generalised Standard Bases" by M. Brundu and F. Rossi, and "On the Construction of Gröbner Bases Using Syzygies" by H. M. Möller are mostly concerned with the problem of extending standard techniques for the computation of Gröbner bases to more general situations, mainly in the framework of graded structures.

The last two papers, "An Extension of Buchberger's Algorithm and Calculations in Enveloping Fields of Lie Algebras" by J. Apel and W. Lassner, and "On Recognisable Properties of Associative Algebras" by T. Gateva-Ivanova and V. Latyshev, do not properly belong to the domain of commutative algebra, but they were included as well, since they show how some non-commutative problems can be handled in a similar way.

It is my intention to express my thanks to the authors, who contributed to the success of this initiative, and my gratitude to Bruno Buchberger and Teo Mora.

The first gave me the opportunity of being the editor of this volume and the second worked with me in such a way that, without his help, I could hardly achieve the goal of managing the whole subject.

Let me finally express my personal hope that this volume can become a standard (here it would be difficult to use the word "Gröbner") reference and can serve as a good companion for researchers in this new and fascinating field.

Lorenzo Robbiano

Contents

On the Complexity of Computing Syzygies

DAVID BAYER AND MICHAEL STILLMAN‡

Department of Mathematics,
Columbia University, New York, NY 10027, U.S.A.

Department of Mathematics,
Cornell University, Ithaca, NY 14853, U.S.A.

We give a self-contained exposition of Mayr & Meyer's example of a polynomial ideal exhibiting double exponential degrees for the ideal membership problem, and generalise this example to exhibit minimal syzygies of double exponential degree. This demonstrates the existence of subschemes of projective space of double exponential regularity.

Introduction

Let k be a field, and let $R = k[x_1, \ldots, x_n]$ be a polynomial ring. Let $I \subset R$ be an ideal generated by $h_1, \ldots, h_s$ of degree $\leqslant d$. Many important operations in computational ring theory rely on the basic operations of constructing a standard or Gröbner basis for the ideal I, and constructing a basis of syzygies of I. Corresponding to these two operations are the following two problems, which are closely related:

(*IM*) *Ideal membership.* If $h \in I$, what degrees can occur for $g_1, \ldots, g_s \in R$ of minimal degree so $h = \Sigma_{i=1,s} g_i h_i$?

(*SYZ*) *Syzygies.* If I is homogeneous, what degrees can occur for the minimal (first) syzygies of I?

Once a standard or Gröbner basis has been constructed for I, the membership of $h \in I$ can be easily determined, and the $g_1, \ldots, g_s$ found, for any h (Buchberger, 1976). Thus, an answer to the problem (IM) gives a lower bound to the complexity of computing standard or Gröbner bases. Similarly, the complexity of computing a basis of syzygies of I is controlled by the degrees in which these syzygies occur.

Hermann (1926) (see also Seidenberg (1974) and Masser & Wüstholz (1983)) gave an upper bound, double exponential in the number of variables n, which applies equally to both of the above problems (IM) and (SYZ). More recently, Mayr & Meyer (1982) proved using methods of complexity theory that for (IM), the double exponential form of Hermann's bound for (IM) cannot be improved. This came as some surprise, since the rates of growth familiar to algebraic geometers had all been single exponential, such as Bezout's theorem on the number of points in the intersection of n hypersurfaces in n variables.

A careful look at Mayr & Meyer's construction yields the corresponding statement for (SYZ): The double exponential form of Hermann's bound for (SYZ) cannot be improved.

‡ Partially supported by N.S.F. grant DMS 8403168.

This is not apparent from the result of Mayr & Meyer (1982) alone, and in fact it was conjectured for some time after their work that Hermann's bound for (SYZ) could be substantially improved.

On a different front, Mayr & Meyer's result created widespread pessimism in the field of computer algebra about the viability of computing with standard or Gröbner bases, even as others were successfully solving large, naturally occurring problems using these bases.

These different points of view can be reconciled with the aid of the concept of regularity. Recall that a homogeneous ideal $I \subset R$ is defined to be m-regular if for $j \geqslant 0$, the jth syzygies of I are of degree $\leqslant m+j$ (Mumford, 1966; Eisenbud & Goto, 1984); the regularity of I is defined to be the least m for which I is m-regular.

There is considerable interest in finding sharp bounds for the regularity of ideals I with good geometric properties; see Eisenbud & Goto (1984). Mumford has shown (unpublished) that if I is the ideal of a characteristic zero non-singular variety of dimension p, degree d, then I is m-regular for $m=(p+1)(d-2)+2$. In Eisenbud & Goto (1984), it is conjectured that if I is a prime ideal defining a variety of codimension r, degree d in $\mathbf{P}^{n-1}$, then I is $(d+1-r)$-regular. In Gruson, Lazarsfeld & Peskine (1983), this statement is proved for reduced, irreducible curves. In Pinkham (1988), non-singular surfaces in $\mathbf{P}^5$ are shown to be $(d-1)$-regular. We conjecture that any reduced subscheme of total degree d in $\mathbf{P}^{n-1}$ is d-regular.

Where does the work of Mayr & Meyer (1982) fit into this picture? Their construction can be used to demonstrate that some ideals have regularity double exponential in the number of variables n. Thus, problem instances in this domain can be highly intractable. On the other hand, the study of the regularity of ideals with good geometric properties shows that many syzygy problem instances which arise naturally in algebraic geometry are tractable; this agrees with practical experience in finding syzygies by computer. Thus, regularity provides a framework for grading problem instances according to their computational complexity, and for resolving the conflict between theory and practice observed above.

It is imperative that the boundary between "nice" ideals with low regularity, and "wild" ideals such as the example of Mayr & Meyer (1982), be better understood; this paper is concerned with the wild side of the above boundary.

In this paper, we give a self-contained exposition of the key construction of Mayr & Meyer (1982). In section 1, we give a sufficient condition for a syzygy of a homogeneous ideal generated by differences of monomials to be minimal. In section 2, we construct an ideal $J_n \subset A$ which exhibits double exponential degrees for the ideal membership problem, and an ideal $K_n \subset A[z]$ having a minimal syzygy of double exponential degree. In section 3, we examine the underlying geometry of the ideal membership problem, and consider the related question of the complexity of ideal membership of 1. An exciting recent result of D. Brownawell (1986) establishes that this restricted problem is computationally more tractable than ideal membership in general.

The bound given by Hermann (1926), Seidenberg (1974), Mayr & Meyer (1982) and Masser & Wüstholz (1983) is as follows: Assume that k is an infinite field, and let $b_i = \Sigma_{j=1,s} g_j h_{ij}$, $i=1,\ldots,t$, be a system of linear equations in $g_1,\ldots,g_s$, with each $b_i, h_{ij} \in R$. Let $d=\max_{i,j}\{\deg(h_{ij})\}$ and $B=\max_i\{\deg(b_i)\}$, taking $\deg(0)=0$. If these equations have a solution, then there exists a solution $(g_1,\ldots,g_s)$ where $\deg(g_i) \leqslant B+2(sd)2^{n-1}$ for each i. Furthermore, when each $b_i=0$ the R-module of solutions is generated by elements $(g_1,\ldots,g_s)$ where $\deg(g_i) \leqslant 2(sd)2^{n-1}$. It follows for (IM), the degree of each g_i is bounded by $\deg(h)+2(sd)2^{n-1}$; the corresponding degree bound for (SYZ) is $d+2(sd)2^{n-1}$. In Masser & Wüstholz (1983), a sharper bound is given for ideal membership of h: $\deg(g_i) \leqslant 2(2d)2^{n-1}$.

Let $d \geqslant 2$, and define $e_n = d^{2^n}$. Mayr & Meyer (1982) construct a polynomial ring A in $10n$ variables, and an ideal $I_n \subset A$ which can in effect count to e_n: Included among the variables of A are a "start" variable S, a "finish" variable F, four "counter" variables $B_1, \ldots, B_4$, and four "catalyst" variables $C_1, \ldots, C_4$. The ideal I_n is generated by differences of monomials, and contains the four differences $S C_i - F C_i B_i^{e_n}$. I_n is defined recursively in terms of I_{n-1}; its construction relies on the identity $e_n = (e_{n-1})^2$.

The relations $S C_i - F C_i B_i^{e_n} \in I_n$ are used in Mayr & Meyer (1982) as part of a construction which realises the halting problem for a bounded 3-counter machine as an instance of the decision problem for ideal membership: Given h, and $h_1, \ldots, h_s \in R$, does $h \in (h_1, \ldots, h_s)$? Thus, the decision problem for ideal membership is seen to be exponential space complete; the argument in Mayr & Meyer (1982) is valid over any field k. Since there are problems solvable in exponential space which are known to require exponential space, a degree bound for ideal membership which grows double exponentially in the maximum of the number of variables and the number of generators is seen to be inevitable.

By setting $B_1, \ldots, B_4, C_1, \ldots, C_4 = 1$ at the top level of recursion in the above construction, we obtain an ideal $J_n \subset A$ so $S - F \in J_n$; this instance of ideal membership directly exhibits double exponential degrees for any $g_1, \ldots, g_s$ so $S - F = \Sigma_{i=1,s} g_i h_i$, where $h_1, \ldots, h_s$ denote the generators of J_n. After homogenising with z, and adjoining $S - F$, we obtain an ideal $K_n \subset A[z]$ exhibiting a minimal syzygy of double exponential degree. K_n is thus of double exponential regularity.

We would like to thank David Mumford for many helpful conversations.

1. Syzygies Formed by Differences of Monomials

Let $I \subset R$ be an ideal generated by $h_1, \ldots, h_s$, where each $h_i = x^{A_i} - x^{B_i}$, and the differences $A_i - B_i \in \mathbf{Z}^n$, $i = 1, \ldots, s$ are all distinct.

1.1. Definition. In the above situation, define $G = G(h_1, \ldots, h_s)$ to be the directed graph having as vertex set the monomials of R, and having as edge set all directed edges (A, B) from x^A to x^B, so $A - B = A_i - B_i$ for some (unique) i.

1.2. Example. *Let* $R = k[x, y, z]$, and let

$$h_1 = xy^3z - y^2z^3, \quad h_2 = xyz^3 - x^3z^2, \quad h_3 = x^3yz - x^2y^3.$$

The ideal $I = (h_1, h_2, h_3)$ is homogeneous, and has $xh_1 + yh_2 + zh_3 = 0$ as a minimal syzygy.

A portion of the graph $G = G(h_1, h_2, h_3)$, with vertex set consisting of the monomials in R_5 and R_6, is shown in Fig. 1. The monomials are arranged as their exponents appear on antidiagonal slices of $\mathbf{N}^3 \subset \mathbf{R}^3$; the monomial nearest the label x is x^5, for example. The three edges shown on the monomials for R_5 correspond to the generators h_1, h_2, h_3 of I. The edges for R_6 all correspond to monomial multiples of generators; the closed triangle formed by three of the edges corresponds to the syzygy $xh_1 + yh_2 + zh_3 = 0$.

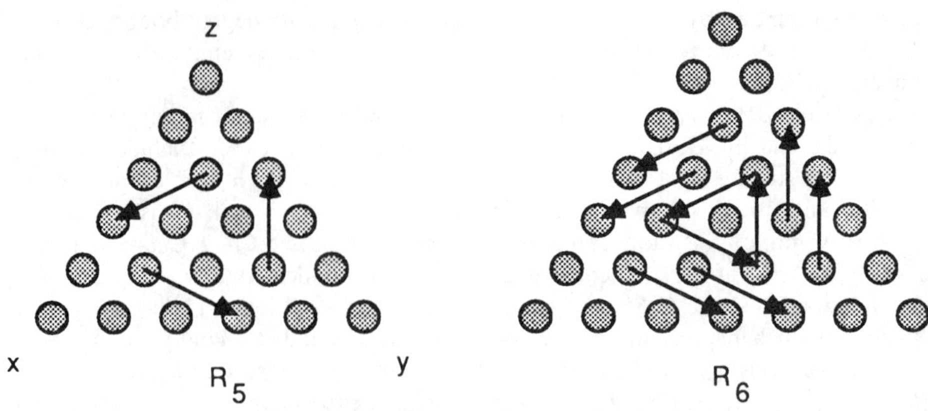

Fig. 1. A syzygy formed by differences of monomials.

1.3. DEFINITION. A chain C in $G=G(h_1, \ldots, h_s)$ is a formal sum $\Sigma c_{A,B}(A, B)$ of edges (A, B) in G with coefficients $c_{A,B} \in k$; $C(G)$ will denote the set of chains in G. The value $|C| \in R$ of a chain C is the sum $\Sigma(c_{A,B}x^A - c_{A,B}x^B)$. C is a cycle if $|C|=0$; $Z(G)$ will denote the set of cycles in G.

1.4. EXAMPLE. One can see from the diagram of example (1.2) that

$$((2, 3, 1), (1, 2, 3))+((1, 2, 3), (3, 1, 2))+((3, 1, 2), (2, 3, 1))$$

is a cycle in G, corresponding to the closed triangle of edges on R_6.

For a chain $C=\Sigma c_{A,B}(A, B)$, and a monomial $x^D \in R$, define

$$x^D C = \Sigma c_{A,B}(A+D, B+D).$$

This gives the set of chains $C(G)$ an R-module structure. The value map $C \to |C|$ is seen to be an R-module homomorphism. Thus its kernel, the set of cycles $Z(G)$, is an R-module.

1.5. LEMMA. *The module of syzygies, $SYZ(I)$, of $I=(h_1, \ldots, h_s)$ is isomorphic to the module of cycles $Z(G)$.*

PROOF. The map from the module $M=Rh_1 \oplus \ldots \oplus Rh_s$ to the module of chains $C(G)$ obtained by sending $(g_1, \ldots, g_s) \in M$ to the chain $\Sigma g_i \cdot (A_i, B_i)$ is an R-module isomorphism. Since $\Sigma g_i h_i = 0$ iff the chain $\Sigma g_i \cdot (A_i, B_i)$ is a cycle, the submodule $\mathrm{SYZ}(I) \subset M$ is mapped isomorphically onto $Z(G) \subset C(G)$. ■

In fact, we have shown that the two exact sequences

$$0 \to \mathrm{SYZ}(I) \to Rh_1 \oplus \ldots \oplus Rh_s \to R$$

and

$$0 \to Z(G) \to C(G) \to R$$

are isomorphic, where the right-hand map of the second sequence is the value map for chains. The formalism of chains and cycles on G provides a graph-theoretic language for discussing the syzygy exact sequence; we shall make particular use of the ability to consider connected components of G.

Define the monomials of a chain $C = \Sigma c_{A,B} \cdot (A, B)$ to be those monomials x^A so $c_{A,B} \neq 0$ or $c_{B,A} \neq 0$ for some monomial x^B; this set of monomials will be called the support of C. Define

$$gcd(C) = gcd\{x^A | x^A \text{ is a monomial of } C\}.$$

The following is a sufficient criterion for a syzygy of I to be minimal, when I is homogeneous:

1.6. LEMMA. *Let I be a homogeneous ideal, and let $G = G(h_1, \ldots, h_s)$. Let G_A be the (connected) component of G containing the monomial x^A. Suppose that*

(i) *the set of cycles with support in G_A is a one-dimensional vector space $\{aC | a \in k\}$ for some cycle C, and*

(ii) *$gcd(C) = 1$.*

Then the syzygy corresponding to C is a minimal syzygy of I.

PROOF. We show that in the module of cycles $Z(G)$, C cannot be expressed as $C = \Sigma \alpha_i C_i$ for cycles C_i, where each $\alpha_i = ax^D$ for some $a \in k$, $D \in \mathbf{N}^n$, and each $\deg(\alpha_i) > 0$. The result follows by the isomorphism of lemma 1.5.

Suppose on the contrary that C can be so expressed. Write each C_i as $C_i' + C_i''$, where C_i' is a cycle so $\alpha_i C_i'$ has support in G_A, and C_i'' is a cycle so $\alpha_i C_i''$ has support in $G - G_A$. Then $C = \Sigma \alpha_i C_i'$. However, each $\alpha_i C_i' = a_i C$ for some $a_i \in k$, by (i). At least one $a_i \neq 0$; for this i, we have $\alpha_i | gcd(C)$. Since $\deg(\alpha_i) > 0$, this contradicts (ii). ∎

2. The Example of Mayr and Meyer

In this section, we describe the construction of Mayr & Meyer (1982), and modify it to give examples for ideal membership and syzygies.

Let $n \geqslant 0$ and $d \geqslant 2$ be integers, and define $e_n = d^{2^n}$. Let V_j denote the set of variables $\{s_j, f_j, b_{j1}, \ldots, b_{j4}, c_{j1}, \ldots, c_{j4}\}$, for $j = 0, \ldots, n$; the variables in V_j will be said to be of level j. Let $A = k[V_0, \ldots, V_n]$.

As a notational convenience, when an integer r, $0 \leqslant r \leqslant n$, is fixed, let upper-case letters denote level r variables, and let lower-case letters denote level $r-1$ variables: Let $S = s_r$, $F = f_r$, $B_i = b_{ri}$, $C_i = c_{ri}$, $s = s_{r-1}$, $f = f_{r-1}$, $b_i = b_{r-1,i}$, and $c_i = c_{r-1,i}$.

Define $I_0 \subset A$ to be the ideal generated by

$$s_0 c_{0i} - f_0 c_{0i} b_{0i}^d \quad \text{for } i = 1, \ldots, 4;$$

these generators will be said to be of level 0. Given the ideal I_{r-1}, for $1 \leqslant r \leqslant n$, define I_r to be the ideal generated by I_{r-1} and the new generators of level r,

$$\begin{array}{ll} S - sc_1, & sc_4 - F, \\ fc_1 - sc_2, & sc_3 - fc_4, \\ fc_2 b_1 - fc_3 b_4, & sc_3 - sc_2, \\ fc_2 C_i b_2 - fc_2 C_i B_i b_3 & \text{for } i = 1, \ldots, 4. \end{array}$$

A monomial $x^D \in A$ will be said to be of level j if

(i) x^D involves only variables of level $\geqslant j$, and

(ii) x^D is linear in $\{s_j, f_j\}$, and in $\{c_{j1}, \ldots, c_{j4}\}$, and is not divisible by any of $s_{j+1}, \ldots, s_n$ or $f_{j+1}, \ldots, f_n$.

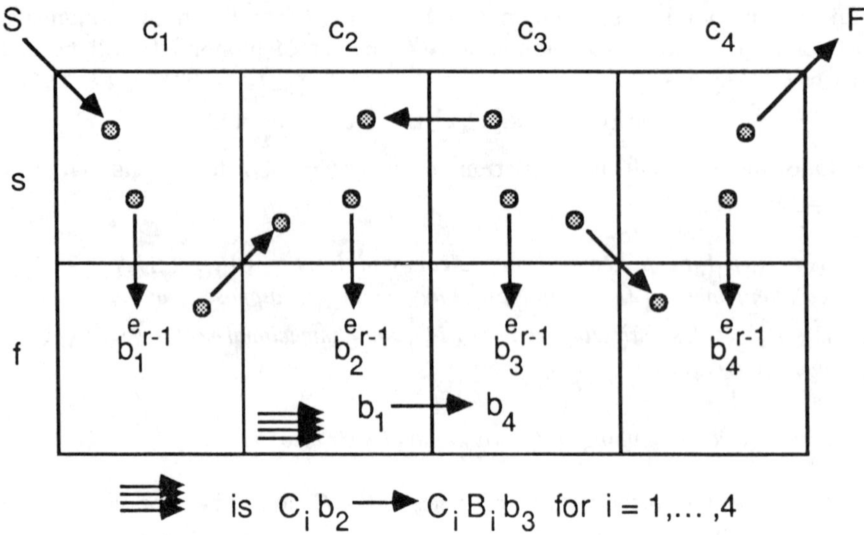

Fig. 2. The example of Mayr and Meyer.

Note that some monomials of A have no level; we shall be studying chains whose support consists of monomials which are of some level.

In abuse of notation, write $G(I_r)$ for the graph of definition 1.1 corresponding to the given generators of I_r. Figure 2 can be used as an aid in visualising $G(I_r)$. The eight boxes shown have row labels s, f, and column labels $c_1, \ldots, c_4$. Each monomial xD of level $r-1$ in A is divisible by unique row and column labels; assign x^D to the box with these labels. Assign the monomials of level r in A to the region outside of all eight boxes. The other monomials of A are not represented in Fig. 2.

The level r generators for I_r induce a multitude of directed edges on these monomials; each solid arrow is meant to denote all edges induced on diagram monomials by a given level r generator $x^D - x^E$. Each such x^D (or x^E) divides the monomials in exactly one diagram region; the corresponding arrow end is located in this region, and labelled with x^D (or x^E). If the region is a box, the variables s, f, and $c_1, \ldots, c_4$ have been suppressed from this label.

We shall see in lemma 2.2 that $sc_i - fc_i b_i^{e_{r-1}} \in I_{r-1}$ for $i = 1, \ldots, 4$; the vertical arrows represent these relations. In the graph $G(I_r)$, these relations can be obtained when $r \geqslant 2$ via multistep chains through vertices corresponding to monomials of level $<r-1$.

2.1. Example. Let $r = 1$, and $d = 3$. Then the ideal I_r contains the relations $S C_i - F C_i B_i^9$ for $i = 1, \ldots, 4$.

Figure 3 illustrates a chain whose value is this relation for a fixed i; each arrow which lies entirely inside box fc_2 is meant to represent a sequence of three edges. To interpret the path shown as a chain, follow it from left to right. Assign coefficient $+1$ to those edges that point in the direction of the path, and coefficient -1 to the remaining edges.

The relation $S C_i - F C_i B_i^9$ can be best understood by keeping track of the variables $b_1, \ldots, b_4$, which act as counters, as one moves along the chain. The chain obtains e_{r-1} copies of the counter b_1 on entering the box fc_1. It crosses from the box fc_2 to the box fc_3 a total of e_{r-1} times, as it laps around in the middle four boxes; a copy of b_1 is

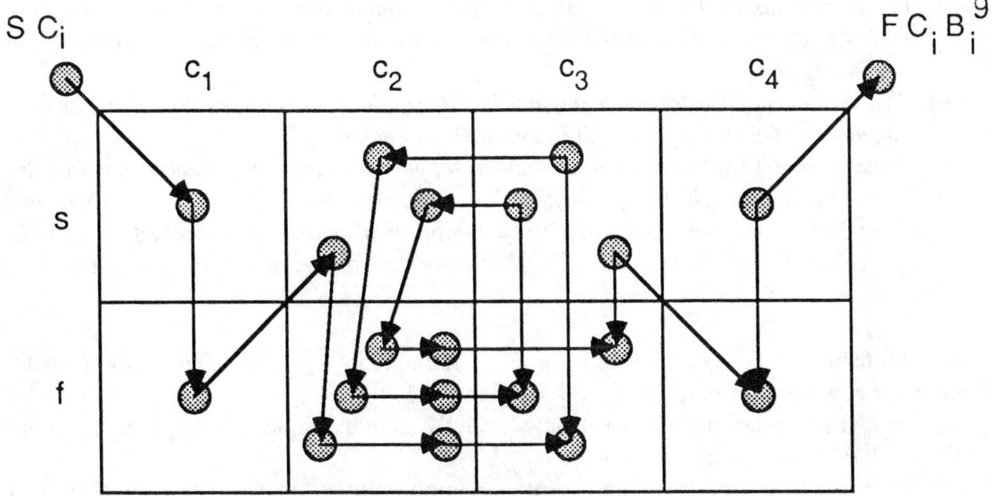

Fig. 3. A chain of equivalences.

converted to b_4 to count each crossing. The e_{r-1} copies of b_2 obtained on entering box fc_2 are each converted there into $B_i b_3$ in the presence of the catalyst C_i. The copies of b_3 are all consumed on exiting box fc_3. In this way, a total of $e_r = (e_{r-1})^2$ copies of B_i are obtained. The e_{r-1} copies of b_4 accumulated by the chain are consumed on exiting box fc_4.

This chain corresponds to the following step by step computation in A/I_r, using generators of I_r:

$$\begin{aligned}
SC_i &= sc_1 C_i = fc_1 C_i b_1^3 \\
&= sc_2 C_i b_1^3 = fc_2 C_i b_1^3 b_2^3 \\
&= \ldots = fc_2 C_i B_i^3 b_1^3 b_3^3 \\
&= fc_3 C_i B_i^3 b_1^2 b_3^3 b_4 = sc_3 C_i B_i^3 b_1^2 b_4 \\
&= sc_2 C_i B_i^3 b_1^2 b_4 = fc_2 C_i B_i^3 b_1^2 b_2^3 b_4 \\
&= \ldots = fc_2 C_i B_i^6 b_1^2 b_3^3 b_4 \\
&= fc_3 C_i B_i^6 b_1 b_3^3 b_4^2 = sc_3 C_i B_i^6 b_1 b_4^2 \\
&= sc_2 C_i B_i^6 b_1 b_4^2 = fc_2 C_i B_i^6 b_1 b_2^3 b_4^2 \\
&= \ldots = fc_2 C_i B_i^9 b_1 b_3^3 b_4^2 \\
&= fc_3 C_i B_i^9 b_3^3 b_4^3 = sc_3 C_i B_i^9 b_4^3 \\
&= fc_4 C_i B_i^9 b_4^3 = sc_4 C_i B_i^9 = FC_i B_i^9.
\end{aligned}$$

Define $p_r: A \to A$ by $p_r(v) = v$ for all variables v of level $<r$ and for $v = s_r$ or f_r, and $p_r(v) = 1$ for all other variables v. For $r \geqslant 1$, define J_r to be the ideal generated by $p_r(I_r)$. J_r differs from I_r only in that the four level r generators $fc_2 C_i b_2 - fc_2 C_i B_i b_3$, $i = 1, \ldots, 4$, are replaced by the single generator $fc_2 b_2 - fc_2 b_3$.

2.2. Lemma. *The following statements hold for I_r, $r \geqslant 0$, and for J_r, $r \geqslant 1$.*

(i) *I_r contains the four relations*

$$SC_i - FC_i B_i^{e_r}, \quad i = 1, \ldots, 4,$$

and J_r contains the relation

$$S - F.$$

(ii) *The monomials in the component of a level r monomial in the graph $G(I_r)$ are all of level $\leqslant r$; the monomials in the component of S and F in the graph $G(J_r)$ are S, F, or monomials of level $<r$.*

(iii) *The component of a level r monomial in the graph $G(I_r)$ contains no cycles, and the component of S in the graph $G(J_r)$ contains no cycles.*

(iv) *In the graph $G(J_r)$, there is a unique chain supported in the component of S with value $S-F$. In the graph $G(I_r)$, if x^D, x^E are distinct monomials of level $\geqslant r$, and $x^D - x^E \in I_r$, then there is a unique chain supported in the component of x^D in $G(I_r)$ with value $x^D - x^E$. In this case $x^D - x^E$ is a multiple of one of the relations given in (i) for I_r.*

PROOF. Statements (i)–(iv) are evident for $r=0$; inductively assuming these statements for I_j when $j<r$, we prove them for I_r and J_r when $r \geqslant 1$.

(i) For I_r, these relations are constructed exactly as in example 2.1. Applying p_r to one of these relations, we obtain $S-F \in J_r$.

(ii) We are considering the component of a monomial x^D in $G=G(I_r)$, or $G=G(J_r)$, where x^D is of level r, or $x^D=S$, respectively. One edge of G is incident on x^D; it connects x^D to a monomial of level $r-1$. From statement (ii) for I_{r-1}, all paths in G from monomials of level $r-1$, consisting of edges arising from I_{r-1}, lead to monomials of level $\leqslant r-1$. Edges in G arising from level r generators are not incident on any monomials of level $<r-1$; starting from a level $r-1$ monomial, such an edge leads either to

(a) another level $r-1$ monomial, or to

(b) a monomial which is divisible by S or F, and not divisible by any variables of level $<r$, with the possible exception of $b_1, \ldots, b_4$. In this case, this edge is the only edge incident on the monomial described.

We need to show that a monomial in case (b) is not divisible by $b_1, \ldots, b_4$. For $G(I_r)$, this monomial is then of level r; for $G(J_r)$, this monomial is then either S or F.

The degrees in $\{b_1, b_4\}$ and in $\{b_2, b_3\}$ of monomials of level $r-1$ on these connected components can be given as a function of which box they occupy in Fig. 2: Monomials in boxes sc_1 and sc_4 are of degree 0 in $\{b_1, b_4\}$, and monomials in boxes $sc_2, sc_3, fc_1, fc_2, fc_3$, and fc_4 are of degree e_{r-1} in $\{b_1, b_4\}$. Monomials in boxes fc_2 and fc_3 are of degree e_{r-1} in $\{b_2, b_3\}$, and monomials in boxes $sc_1, sc_2, sc_3, sc_4, fc_1$, and fc_4 are of degree 0 in $\{b_2, b_3\}$. One verifies this by observing that the edges arising from level r generators of I_r or J_r, and the chains of statement (iv) for I_{r-1}, preserve this grading. Thus the monomials of case (b) are not divisible by $b_1, \ldots, b_4$.

(iii) Suppose that C is a cycle in the connected component of S in the graph $G(J_r)$. Assume without loss of generality that exactly two edges of C are incident on each monomial of C; any cycle C can be decomposed into simple cycles of this form. S or F cannot be monomials of C, since only one generator of J_r applies to each of S or F. Thus by statement (ii), the remaining monomials in this component are all of level $\leqslant r-1$.

Suppose that the highest level of a monomial of C is $j-1$, for some $j<r$. Then no edges corresponding to generators of level $>j$ are involved in C, so C is already a cycle in the connected component of a level j monomial in the graph $G(I_j)$, which contradicts statement (iii) for j.

Thus the support of C must include monomials of level $r-1$. Choose a traversal of C, and call two level $r-1$ monomials of C adjacent if no other level $r-1$ monomials occur between them in this traversal. Such adjacent monomials are either joined in C by an edge corresponding to a level r generator of J_r, or by a chain corresponding to one of the relations $s\,c_i - f c_i b_i^{e_{r-1}}$; this follows from statement (iv) for I_{r-1}. From this, we see that no level $r-1$ monomials of C can lie in the outer boxes $s\,c_1$, $f c_1$, $s\,c_4$, or $f c_4$ of Fig. 2: A monomial in boxes $s\,c_1$ or $s\,c_4$ cannot have two adjacent monomials, and once these boxes are excluded, the same claim can be made for boxes $f c_1$ and $f c_4$. Furthermore, a monomial in boxes $s\,c_2$, $s\,c_3$, or $f c_3$, and its two adjacent monomials, lie in three distinct boxes. A monomial in box $f c_2$ is of degree e_{r-1} in $\{b_2, b_3\}$, by the above proof of statement (ii). Thus it is part of a sequence of e_{r-1} adjacent monomials in box $f c_2$, joined by edges corresponding to the generator $b_2 - b_3$, with the monomials on each end adjacent to monomials in boxes $s\,c_1$ and $f c_3$, respectively.

From these local considerations along the cycle C, we see that a traversal of C must loop some non-zero number of times through the four boxes $s\,c_2$, $f c_2$, $f c_3$, and $s\,c_3$, always in the same direction. In particular, edges arising from the generator $f c_2 b_1 - f c_3 b_4$ are used in the cycle C, and are always traversed in the same direction. Since no other edges in C affect the degree of b_1 in monomials of C, the degree of b_1 must decrease or increase along a traversal of C. This is impossible, so statement (iii) holds for $G(J_r)$.

Applying the map p_r to a cycle in $G(I_r)$ produces a cycle in $G(J_r)$. Since p_r maps level r monomials of A to either S or F, statement (iii) follows for $G(I_r)$.

(iv) The existence of a chain in the component of S in $G(J_r)$ with value $S-F$ is guaranteed by statement (i). This chain is unique, since the existence of more than one such chain would permit the construction of a cycle contradicting statement (iii) for $G(J_r)$.

Let C be a chain in the component of x^D in $G(I_r)$ with value $x^D - x^E$. Let C' be the chain in the compoment of $p_r(x^D)$ in $G(J_r)$ which is the image under p_r of C; C' has value $p_r(x^D - x^E) \in J_r$.

x^D or x^E must both be of level r, for otherwise we would have $p_r(x^D) = 1$ or $p_r(x^E) = 1$, contradicting the fact that no relations in J_r involve the monomial 1. Thus p_r maps x^D to either S or F, and x^E to either S or F.

p_r cannot map x^D and x^E both to S, or both to F, for in these cases $p_r(x^D - x^E) = 0$, so C' is a cycle, contradicting statement (iii) for $G(J_r)$. Thus, we can assume without loss of generality that $p_r(x^D) = S$ and $p_r(x^E) = F$. C' is then the unique chain in the component of S in $G(J_r)$ with value $S-F$.

Since x^D and x^E are of level r, and no relation of I_r changes the variables $C_1, \ldots, C_4$, every monomial of C is divisible by C_i for some fixed i. C is determined by the chain C' and i, and is thus unique: C differs from C' only in that edges corresponding to the generator $f c_2 b_2 - f c_2 b_3 \in J_r$ are replaced by edges corresponding to the generator $f c_2 C_i b_2 - f c_2 C_i B_i b_3 \in I_r$. There are a total of e_r such edges, so for some monomial α, $x^D = \alpha S\, C_i$, and $x^E = \alpha F\, C_i B_i^{e_r}$. Thus, $x^D - x^E$ is a multiple of one of the relations given in (i) for I_r. ■

2.3. Lemma. *Let $h = S - F$, and let $h_1, \ldots, h_s$ be the generators of J_r. For any expression $h = \Sigma_{i=1,s} g_i h_i$, some g_i will have degree $\geqslant r - 1 + 2e_0 + \ldots + 2e_{r-1}$.*

Proof. One of the monomials occurring in the unique chain of lemma 2.2(iv) from S to F in the compoment of S in $G(J_r)$ is

$$f_0\, c_{03}\, b_{03}^{e_0}\, b_{04}^{e_0} \ldots c_{r-1,3}\, b_{r-1,3}^{e_{r-1}}\, b_{r-1,4}^{e_{r-1}}.$$

One of the two edges of the chain which are incident on this monomial is a degree $r - 1 + 2e_0 + \ldots + 2e_{r-1}$ multiple of the generator $s_0 c_{03} - f_0 c_{04}$ of J_r. Any expression

$\Sigma_{i=1,s} g_i h_i$ for $S-F$ corresponds to a chain which differs from this chain only by the addition of cycles supported in other components of $G(J_r)$, so some g_i has at least the stated degree. ■

2.4. THEOREM (*Mayr & Meyer*, 1982). *Any degree bound for the ideal membership problem must grow double exponentially in the maximum of the number of variables and the number of generators.*

PROOF. Taking $r=n$ in lemma 2.3, the polynomial ring A has $10n$ variables, and the ideal J_n has $10n+1$ generators of degree $\max\{5, d+2\}$. Any set of g_i's for the instance $S-F \in J_n$ of ideal membership have maximum degree exceeding $e_{n-1} = d2^{n-1}$, which grows double exponentially in $10n+1$. ■

Let $A' = A[z]$, where z is a homogenising variable. Define the projection $q: A' \to A$ to be the identity on $A \subset A'$, and to map z to 1. Let $r=n$, and let $J'_n \subset A'$ be the ideal generated by the homogenisations, using z, of the generators of J_n. Let $K_n \subset A'$ be the ideal generated by J'_n and $S-F$. J'_n and K_n are homogeneous ideals of A'.

2.5. LEMMA. *K_n has a minimal syzygy of degree $m+1$, where*

$$m = n + 2e_0 + \dots + 2e_{n-1}.$$

PROOF. Let H denote the component of $z^m S$ in $G(J'_n)$. All of the monomials of H have the same degree, so q maps this set of monomials injectively into A. This induces an embedding of the graph H into the component of S in $G(J_n)$. Thus by lemma 2.2(iii) for J_n, H has no cycles.

The component of $z^m S$ in $G(K_n)$ has the same monomials as H, and in addition to the edges of H, one new edge e corresponding to $z^m(S-F)$. There are no other monomials or edges, since $z^m S$ and $z^m F$ are the only monomials of H which are divisible by S or F, so $z^m(S-F)$ is the only multiple of $S-F$ giving an edge incident on monomials of H. Denote this component of $G(K_n)$ by $H \cup \{e\}$.

Let C be the unique chain with value $S-F$ supported in the component of S in $G(J_n)$, given by lemma 2.2(iv). One verifies that every monomial of C has degree $\leqslant m+1$. The chain C can be homogenised to a chain C' in $G(J'_n)$ all of whose monomials are of degree $m+1$, by multiplying the monomials of C by appropriate powers of z. Each edge of C' is then a multiple, by some power of z, of the corresponding edge in C. C' is supported on $H \cup \{e\}$, and has value $z^m(S-F)$.

$C' - \{e\}$ is then a cycle supported on $H \cup \{e\}$. $gcd(C' - \{e\}) = 1$, since

$$f_0 c_{03} b_{03}^{e_0} b_{04}^{e_0} \dots c_{r-1,3} b_{r-1,3}^{e_{r-1}} b_{r-1,4}^{e_{r-1}},$$

$z^m S$ and $z^m F$ are all monomials of this cycle. Since H has no cycles, every cycle supported on $H \cup \{e\}$ is a multiple $a(C' - \{e\})$, $a \in k$, of $C' - \{e\}$. Thus, by lemma 1.6, this cycle corresponds to a minimal syzygy of K_n, of degree $m+1$. ■

2.6. THEOREM. *Any bound for the regularity of homogeneous ideals must grow double exponentially in the maximum of the number of variables and the number of generators.*

PROOF. The polynomial ring A' has $10n+1$ variables, and the ideal K_n has $10n+1$ generators of degree $\max\{5, d+2\}$. By lemma 2.5, K_n has a minimal (first) syzygy of degree $m+1$, where $m \geqslant e_{n-1} = d2^{n-1}$. Thus, the regularity of K_n grows double exponentially in $10n+1$. ■

In theorems 2.4 and 2.6, we would like to be able to assert simply that double exponential growth in the number of variables is inevitable, to agree with the form of Hermann's bound (1926). However, the number of variables and the number of generators grow together in the construction of Mayr & Mayer (1982). It remains an open question whether or not these bounds must grow double exponentially in the number of variables, when both the degree of the generators and the number of generators are held fixed.

3. The Geometry of the Ideal Membership Problem

In this section, we discuss the ideal membership problem from a geometric point of view, and consider the related question of the complexity of ideal membership of 1.

Given an ideal I generated by $h_1, \ldots, h_s \in R$, homogenise each h_i with z, to produce $h'_1, \ldots, h'_s$ generating a homogeneous ideal $I' \subset R[z]$. I defines a subscheme X of the affine space $\mathbf{A}^n$, and I' defines a subscheme X' of the projective space $\mathbf{P}^n$. In general, I' is not the homogenisation of I; there will exist $h \in I$ whose homogenisations h' will fail to belong to I'. Corresponding to this, X' is not in general the projective closure of X. Let $H \subset \mathbf{P}^n$ be the hyperplane at infinity defined by $z=0$, so $\mathbf{P}^n = \mathbf{A}^n \cup H$; X' in general has primary components supported on H.

Let $h \in I$, so $h - \Sigma_{i=1,s} g_i h_i = 0$ for some $g_1, \ldots, g_s \in R$. Homogenising with z, we obtain $z^m h' - \Sigma_{i=1,s} z^{m_i} g'_i h'_i = 0$ for integers $m, m_1, \ldots, m_s$, where $g'_1, \ldots, g'_s$ are the homogenisations of $g_1, \ldots, g_s$, and at least one $m_i = 0$. In particular, $z^m h' \in I'$. Assume for simplicity that $\deg(h_i) = d$ for each h_i; we have $\max\{deg(g_i)\} = m + \deg(h) - d$. Conversely, if $z^j h' \in I'$ for some $j \geqslant 0$, then we can find $g_1, \ldots, g_s \in R$ so $h - \Sigma_{i=1,s} g_i h_i = 0$, with $\max\{\deg(g_i)\} = j + \deg(h) - d$. Thus, the degrees that can occur for $g_1, \ldots, g_s$ of minimal degree in the ideal membership problem are determined by the minimum m so $z^m h' \in I'$.

There is a relationship between the problem of ideal membership of 1, the ideal membership problem, and the syzygy problem: Consider the homogeneous versions of these problems. If $z^m h' \in I'$, then $z^m \in (I' : h')$, where $(I' : h') = \{f \in R[z] \mid f h' \in I'\}$. The generators of $(I' : h')$ can be computed from the generators of I', and the minimal syzygies of $I + (h)$. Thus, the ideal membership problem can be reduced to a combination of the syzygy problem and ideal membership of 1.

Since $h \in I$, h vanishes on X, so h' vanishes on the projective closure of X in $\mathbf{P}^n$. Thus, $z^m h' \in I'$ if z^m vanishes on the primary components of X' supported on H. This motivates the following definition.

3.1. Definition. Let $Y \subset \mathbf{A}^{n+1}$ be a scheme defined by the ideal $J \subset R[z]$, let $Z \subset \mathbf{A}^{n+1}$ be a reduced scheme defined by the ideal $K \subset R[z]$, and suppose that $\mathrm{Supp}(Y) \subset Z$. Define the thickness of Y relative to Z to be the least integer m so $K^m \subset J$.

Intuitively, the thickness m measures how much the structure of Y extends in directions normal to Z.

Let Y denote the union of those primary components of X' which are supported on the hyperplane H. It follows from the preceding discussion that if the thickness of Y relative to H is m, then $z^m h' \in I'$.

In this setting, theorem 2.4 asserts the existence of subschemes $X' \subset \mathbf{P}^n$ having primary components whose thickness relative to H is double exponential in the maximum of the number of variables and the number of ideal generators. This is startling when compared to other known bounds in algebraic geometry. For example, Bezout's theorem asserts that a complete intersection of n hypersurfaces of degree d in $\mathbf{P}^n$ has degree d^n, which is single exponential in n.

In considering the above geometric picture, one is lead to believe that the primary components of X' which cause this double exponential thickness relative to H are embedded components of X', and that these components have at most single exponential thickness relative to the support of X'. This, in fact, has been confirmed in characteristic zero by the following recent results of Brownawall (1986).

3.2. THEOREM (IDEAL MEMBERSHIP OF 1). (*Brownawell*, 1986). *Let*

$$I = (h_1, \ldots, h_s) \subset R = k[x_1, \ldots, x_n],$$

where k is of characteristic zero, and $\deg(h_i) \leqslant d$. *If* $1 \in I$, *then there exist* $g_1, \ldots, g_s \in R$ *so* $1 = \Sigma_{i=1,s} g_i h_i$, *where* $\deg(g_i h_i) \leqslant 3\mu n d^\mu$, *and* $\mu = \min(n, s)$.

3.3. COROLLARY (RADICAL IDEAL MEMBERSHIP). (*Brownawell*, 1986). *Following the same notation as in theorem* 3.2, *if f belongs to the radical rad*(I) *of I, then*

$$f^e \in I \text{ for some } e \leqslant e' = 3(\mu+1)(n+1)(d+1)^{\mu+1}.$$

The following corollary follows easily from these results:

3.4. COROLLARY. *Following the same notation as in theorem* 3.2, *let* $X \subset \mathbf{A}^n$ *be the scheme defined by the ideal* $I \subset R$, *let* $Y \subset \mathbf{A}^n$ *be a primary component of X defined by the ideal* $J \subset R$, *and let* $Z \subset \mathbf{A}^n$ *be the scheme defined by the ideal rad*$(I) \subset R$, *so Supp*$(Y) \subset Z$. *Then the thickness of Y relative to Z is given by some integer* $e \leqslant e' = 3(\mu+1)(n+1)(d+1)^{\mu+1}$.

PROOF. We need to show that $\mathrm{Rad}(I)^{e'} \subset J$. From corollary 3.3, $\mathrm{Rad}(I)^{e'} \subset I$. Since J is a primary component of I, $I \subset J$, and the corollary follows. ■

Conjecturally, theorem 3.2 and corollaries 3.3, 3.4 hold also in positive characteristic, but these statements are not yet known; they await an algebraic proof. The construction of Mayr & Meyer (1982) does not generalise to produce a counterexample, for the following reason: The variables s_j, f_j, and $c_{j1}, \ldots, c_{j4}$ limit the number of generators of I_n that apply at once to a monomial of level $\leqslant n$. If one introduces a new generator of the form $1 - x^D$, where any of these variables divide x^D, then one loses control of the degree of these variables. For example, the linearity in $\{s_0, \ldots, s_n, f_0, \ldots, f_n\}$ for monomials of level $\leqslant n$ cannot be preserved. Intuitively, one is no longer restricted to visiting a single box at a time in Fig. 2.

We have seen that the ideals exhibiting double exponential behaviour for the ideal membership problem are characterised by pathological primary components on the hyperplane at infinity; it can be shown that these ideals are of double exponential regularity. We have also seen that these primary components have double exponential thickness relative to the hyperplane at infinity, but that in characteristic zero they have only single exponential thickness relative to the support of the ideal I itself. In contrast, as discussed in the introduction, when geometric conditions are imposed that preclude such primary components, investigators have proven or conjectured regularity bounds exhibiting polynomial growth. For computer algebra systems to be able to safely navigate this problem domain, it is necessary both that the cases which exhibit polynomial behaviour be better understood, and that algorithms be able to recognise these cases.

Note added in proof

J. Kollàr (sharp effective Nullstellensatz, preprint 1988) has improved theorem 3.2 to include all characteristics. He obtains sharp bounds for $\deg(g_i h_i)$, if each $\deg h_i > 2$.

References

Brownawell, W. D. (1987). Bounds for the degrees in the Nullstellensatz. *Annals of Mathematics*, **126,** 577–591.

Buchberger, B. (1976). A theoretical basis for the reduction of polynomials to canonical forms. *ACM SIGSAM Bull.* 39, 19–29.

Eisenbud, D., Goto, S. (1984). Linear free resolutions and minimal multiplicity. *J. Algebra* **88,** 89–133.

Gruson, L., Lazarsfeld, R., Peskine, C. (1983). On a theorem of Castelnuovo, and the equations defining space curves. *Inventiones Math.* **72,** 491–506.

Hermann, G. (1926). Die Frage der endlich vielen Schritte in der Theorie der Polynomideale. *Math. Ann.* **95,** 736–788.

Masser, D. W., Wüsthoïz, G. (1983). Fields of large transcendence degree generated by values of elliptic functions. *Inventiones Math.* **72,** 407–464.

Mayr, E., Meyer, A. (1982). The complexity of the word problem for commutative semigroups and polynomial ideals. *Adv. Math.* **46,** 305–329.

Mumford, D. (1966). *Lectures on Curves on an Algebraic Surface*. Princeton University Press, Princeton, New Jersey.

Pinkham, H. A. (1986). Castelnuovo bound for smooth surfaces. *Invent. math.* **83,** 321–332.

Seidenberg, A. (1974). Constructions in algebra. *Trans. Amer. Math. Soc.* **197,** 273–313.

Gröbner Bases and Primary Decomposition of Polynomial Ideals

PATRIZIA GIANNI, BARRY TRAGER* AND GAIL ZACHARIAS†

IBM Research and University of Pisa, Pisa, Italy
**IBM Research, P.O. Box 218, Yorktown Heights, NY 10598, USA*
†MIT AI Laboratory, 545 Technology Square, Cambridge, MA 02139, USA

We present an algorithm to compute the primary decomposition of any ideal in a polynomial ring over a factorially closed algorithmic principal ideal domain R. This means that the ring R is a constructive PID and that we are given an algorithm to factor polynomials over fields which are finitely generated over R or residue fields of R. We show how basic ideal theoretic operations can be performed using Gröbner bases and we exploit these constructions to inductively reduce the problem to zero dimensional ideals. Here we again exploit the structure of Gröbner bases to directly compute the primary decomposition using polynomial factorization. We also show how the reduction process can be applied to computing radicals and testing ideals for primality.

1. Introduction

A fundamental construction in commutative algebra is the primary decomposition of ideals. From an algebraic point of view this generalizes the operation of factorization into products of irreducible elements, while it is connected from a geometric viewpoint with the decomposition of a variety into its irreducible components. In this paper we present an algorithm to compute the primary decomposition of any ideal in a polynomial ring over a factorially closed algorithmic principal ideal domain (Ayoub, 1982). The rational integers **Z** form such a ring, but more generally this means that the ring R is a constructive PID and that we are given an algorithm to factor polynomials over fields which are finitely generated over R or residue fields of R. In particular this is true if R is any prime ring (Seidenberg, 1974; Davenport & Trager, 1981).

Algorithms for primary decomposition in polynomial rings over **Z** have been presented by Seidenberg (1978) and Ayoub (1982). Seidenberg was able to present a simplified construction when the base ring was a field by reducing the problem to zero-dimensional ideals (Seidenberg, 1978, Theorem 9). In the more general case when the base ring was the integers, he was forced to give a more indirect construction involving first computing all the associated primes, and then isolating the primary component associated with each prime. Ayoub attempted to generalize his construction for fields to principal ideal domains. She presented an algorithm which proceeded by induction on the number of variables in the polynomial ring, rather than on the dimension of auxiliary ideals at each stage of the process. Subsequently Seidenberg (1984) investigated more general rings R and presented conditions on R which are sufficient to allow the computation of primary decompositions in polynomial rings over R.

We base our construction on the Gröbner basis algorithm, a very powerful tool in computational ring theory (Buchberger, 1983). This method was introduced in 1965 by

Buchberger to solve systems of polynomial equations (Buchberger, 1965). It provides a canonical (relative to a monomial ordering) set of generators for an ideal which facilitates testing ideal membership and contraction of ideals of subrings. Lazard (1985) has also exploited the structure of a Gröbner basis to give a very efficient primary decomposition algorithm for the special case of polynomial rings in two variables over fields.

Our construction of the primary decomposition is based on an induction on dimension which generalizes the one presented by Seidenberg for the field case. We use localization at principal primes in place of quotient field formation to decrease the dimension and we present in Proposition 3.7 the fundamental construction which enables us to reduce the primary decomposition computation to its zero-dimensional counterpart. We show how the zero-dimensional problem can be solved by exploiting the structure of Gröbner bases.

In the first section we introduce our notations and recall the known properties of Gröbner bases which we will use. The next section shows how basic ideal-theoretic operations such as contractions, intersections, and ideal quotients can be directly computed using Gröbner bases. Next we present the properties of Gröbner bases for zero-dimensional ideals. This is used to develop primary decomposition algorithms, first for general zero-dimensional ideals over PID's, and later a simpler and more efficient one when the coefficient ring is a field and the ideal is in general position. Finally we develop our fundamental construction which enables us to reduce the decomposition of general ideals to the zero-dimensional case. We also show how the reduction process can be applied to other problems, specifically computing radicals and testing ideals for primality.

2. Definitions

Let R be a Noetherian commutative ring with identity.

We use the following standard notation:

If S is a multiplicatively closed subset of R, then $S^{-1}R = \{r/s \mid s \in S\}$ denotes the ring of fractions of R with respect to S.

If $f \in R$ then $R_f = S^{-1}R$, where $S = \{f^n\}$, is the localization of R at f.

If $P \subset R$ is a prime ideal then $RP = S^{-1}R$, where $S - R - P$, is the localisation of R at P.

If I, J are ideals in R then $I\colon J = \{a \mid aJ \subset I\}$ is the ideal quotient of I and J.

If $I \subset R$ is an ideal then $\sqrt{I} = \{a \mid a^m \in I \text{ for some } m\}$ is the radical of I.

We will say that an ideal I is "given" if we are given a finite set of generators for I.

DEFINITION 2.1. We say that linear equations are solvable in R if:

(a) (ideal membership) Given $a, a_1, \ldots, a_m \in R$ it is possible to decide whether a is in the ideal $(a_1, \ldots, a_m)R$ and if so, find $b_1, \ldots, b_m$ such that $a = \Sigma b_i a_i$.

(b) (syzygies) Given $a_1, \ldots, a_m \in R$ one can find a finite set of generators for the R-module $\{(b_1, \ldots, b_m) \in R^m \mid \Sigma b_i a_i = 0\}$.

In all that follows we assume that R is a ring in which linear equations are solvable. We now review the definitions and basic properties of Gröbner bases and associated concepts.

DEFINITION 2.2. A total order $>$ on N^n is compatible with the semi-group structure if

(i) $A \geqslant 0$ for all $A \in N^n$.

(ii) $A > B \Rightarrow A + C > B + C$ for all $A, B, C \in N^n$.

We now fix a compatible order $>$. Note that such an order is necessarily a well-ordering (Zacharias, 1978).

DEFINITION 2.3. For any non-zero $f \in R[x] = R[x_1, \ldots, x_n]$, write

$$f = cx^A + f'$$

with $c \in R$, $c \neq 0$, and $A > A'$ for every non-zero term $c'x^{A'}$ of f'. With this notation we set

$$\text{lt}(f) = cx^A, \text{ the leading term of } f.$$

$$\text{lc}(f) = c, \text{ the leading coefficient of } f.$$

$$\deg(f) = A, \text{ the degree of } f.$$

For a subset G of $R[x]$, we define

$\text{Lt}(G)$ = the ideal generated by $\{\text{lt}(g) \mid g \in G\}$, the leading term ideal of G.

By convention we let $\text{lt}(0) = \text{lc}(0) = 0$ and $\deg(0) = -\infty$.

DEFINITION 2.4. $f \in R[x]$ is reducible modulo $G \subset R[x]$ if f is non-zero and $\text{lt}(f) \in \text{Lt}(G)$. Otherwise f is reduced modulo G.

PROPOSITION 2.5. (Reduction algorithm) *Given f and $G = \{g_1, \ldots, g_m\}$ in $R[x]$, it is possible to construct f' such that $f = f'$* $\text{mod}(g_1, \ldots, g_m)R[x]$ *and f' is reduced modulo G.*

PROOF. The ideal membership condition on R insures that we can decide whether f is reducible modulo G, and if so, find terms t_i such that $\text{lt}(f) = \Sigma t_i \, \text{lt}(g_i)$. If f is not reducible, then $f' = f$ will do. Otherwise let $f_1 = f - \Sigma t_i g_i$. By construction, the leading term of $\Sigma t_i g_i$ cancels the leading term of f, so $\deg(f_1) < \deg(f)$. Thus by induction on the well-ordering $<$ we can find a reduced f' with $f' \equiv f_1 \,\text{mod}(g_1, \ldots, g_m)$. But $f \equiv f_1$ so $f \equiv f'$ as required. □

REMARK. It is clear from the proof that the non-constructive version of Proposition 2.5 also holds: For any f and G (whether explicitly given or not), there exists a reduced f' with $f \equiv f'$ modulo the ideal generated by G.

DEFINITION 2.6. A subset G of an ideal $I \subset R[x]$ is a Gröbner basis for I if $\text{Lt}(G) = \text{Lt}(I)$, that is if every non-zero element of I is reducible modulo G. G is a minimal Gröbner basis if additionally every $g \in G$ is non-zero and reduced modulo $G - \{g\}$.

If g is reducible modulo $G - \{g\}$, i.e. $\text{lt}(g) \in \text{Lt}(G \neq \{g\})$, then $\text{Lt}(G - \{g\}) = \text{Lt}(G)$. Thus $G - \{g\}$ is a Gröbner basis for I if G is. In particular any given Gröbner basis can be made minimal by simply removing those elements which are reducible modulo the others.

The following proposition describes the fundamental property of Gröbner bases.

PROPOSITION 2.7. *Let G be a Gröbner basis for $I \subset R[x]$. Then $f \in I$ if and only if applying the reduction algorithm* (Proposition 2.5) *to f returns* 0.

PROOF. Let f be a non-zero element of $R[x]$ and let f' be as in Proposition 2.5. Since $G \subset I$, $f \equiv f'$ mod I. Thus if $f' = 0$ then $f \in I$ and we are done. Conversely, if $f \in I$ then $f' \in I$ and hence $\text{lt}(f') \in \text{Lt}(I) = \text{Lt}(G)$. But f' is reduced modulo G, so we must have $f' = 0$. □

Corollary 2.8. If G *is a given Gröbner basis for* I *then ideal membership in* I *is decidable.*

Corollary 2.9. *If* G *is a Gröbner basis for* I *then* G *generates* I.

We also record the following useful consequence of previous results:

Corollary 2.10. *If* $J \subset I$ *are ideals in* $R[\mathrm{x}]$ *and* $\mathrm{Lt}(I) = \mathrm{Lt}(J)$ *then* $I = J$.

Proof. J forms a (non-finite) Gröbner basis for I, so by the remark following Proposition 2.5, we may conclude that J generates I. But since J is an ideal, it only generates itself, so $J = I$. □

The importance of Gröbner bases in constructive algebra derives from the following fact:

Proposition 2.11. *One can compute a Gröbner basis for an ideal in* $R[\mathrm{x}]$ *from any given set of generators.*

Proof. See Trinks (1978) or Zacharias (1978). □

We remark that the Gröbner basis algorithm automatically produces a basis for the syzygy module of the generators. Thus it can be used to demonstrate that $R[\mathrm{x}]$ satisfies both the computability conditions of Definition 2.1 whenever R does.

The computation of Gröbner bases takes a particularly simple form when the coefficient ring R is a Principal Ideal Domain (PID) or a field. In fact the algorithm was originally discovered in the context of fields (Buchberger, 1965; 1970; 1976). See also (Buchberger, 1979) for some additional results which can be used to improve the efficiency of the algorithm.

Finally, we note that our definition of Gröbner bases is the least restrictive one possible, in the sense that there exist definitions in the literature which place additional conditions on the leading coefficients and/or non-leading terms of the basis elements. All the algorithms presented in this paper can of course be applied to any such stricter types of Gröbner bases, provided only that the condition of Definition 2.6 is satisfied.

3. Operations on Ideals

In this section we discuss the use of Gröbner bases to perform some basic ideal operations in $R[\mathrm{x}]$. Most of the constructions we describe are based on an observation by D. Spear (1977) that Gröbner bases computed with respect to the lexicographical order on monomials have the effect of eliminating the more "main" variables. The following proposition describes this property in more detail.

Proposition 3.1. *Let* I *be an ideal in* $R[\mathrm{y}, \mathrm{x}] = R[y_1, \ldots, y_n, x_1, \ldots, x_m]$. *Given any two orders* $>_1$ *and* $>_2$ *on monomials in* x *and* y *respectively, define an order* $>$ *by* $\mathrm{x}^A\mathrm{y}^B > \mathrm{x}^{A'}\mathrm{y}^{B'}$ *if* $\mathrm{x}^A >_1 \mathrm{x}^{A'}$, *or if* $\mathrm{x}^A = \mathrm{x}^{A'}$ *and* $\mathrm{y}^B >_2 \mathrm{y}^{B'}$. *If* $G \subset R[\mathrm{y}, \mathrm{x}]$ *is a Gröbner basis for* I *with respect to* $>$ *then*

(*i*) *G is a Gröbner basis for* I *with respect to the order* $>_1$ *on* $(R[\mathrm{y}])[\mathrm{x}]$, *the polynomial ring in* $x_1, \ldots, x_m$ *with coefficients in* $R[\mathrm{y}]$.
(*ii*) $G \cap R[\mathrm{y}]$ *is a Gröbner basis for* $I \cap R[\mathrm{y}]$ *with respect to the order* $>_2$.

PROOF. (i) By the definition of $>$, we have $\mathrm{lt}_>(\mathrm{lt}_{>1}(f)) = \mathrm{lt}_>(f)$ for any $f \in R[\mathrm{y}, \mathrm{x}]$. Hence

$$\mathrm{Lt}_>(\mathrm{Lt}_{>1}(G)) = \mathrm{Lt}_>(G) = \mathrm{Lt}_>(I) = \mathrm{Lt}_>(\mathrm{Lt}_{>1}(I))$$

Thus by Corollary 2.10, $\mathrm{Lt}_{>1}(G) = \mathrm{Lt}_{>1}(I)$.

(ii) Since no term involving any x_i can be $>$-larger than a term involving only the y_i, a polynomial whose leading term is in $R[\mathrm{y}]$ cannot involve any x_i in the remaining, smaller, terms. In other words, $\mathrm{lt}_>(g) \in R[\mathrm{y}] \Leftrightarrow g \in R[\mathrm{y}]$. Hence

$$\mathrm{Lt}_>(G \cap R[\mathrm{y}]) = \mathrm{Lt}_>(G) \cap R[\mathrm{y}] = \mathrm{Lt}_>(I) \cap R[\mathrm{y}] = \mathrm{Lt}_>(I \cap R[\mathrm{y}])$$

So $G \cap R[\mathrm{y}]$ is a Gröbner basis for $I \cap R[\mathrm{y}]$ with respect to $>$. But $>$ coincides with $>_2$ on $R[\mathrm{y}]$. □

Part (i) of the proposition shows that in order to compute Gröbner basis with coefficients in a ring $R[\mathrm{y}]$ which is itself a polynomial ring over a base ring R, we can instead group the coefficient variables y and ring variables x together using an appropriate order $>$ (extending the desired order $>_1$) and apply the algorithm over R itself. This is of great practical importance when the base ring R is a field or a PID, for in those cases the Gröbner basis computation over R is much simpler than the general algorithm which would have to be used if we were to work directly over the coefficient ring $R[\mathrm{y}]$.

In the remainder of this paper, we will often appear to require the calculation of Gröbner bases with coefficients in polynomial rings constructed from an initial base ring. It is a consequence of this proposition that in practice all our constructions can be performed using the simpler PID (respectively field) variant of the Gröbner basis algorithm, provided the original base ring is a PID (respectively a field).

Part (ii) of the proposition shows that we can compute the contraction of an ideal to a coordinate subring: We simply compute the Gröbner basis for the ideal, with respect to an order $>$ based on whatever order $>_2$ we want for the contraction. Then the elements of the Gröbner basis which involve only the subring variables give a Gröbner basis for the contraction.

EXAMPLE. Consider the ideal $I = (xy + y, xz + 1) \subset Q[z, y, x]$. Using the full lexicographical order with $x > y > z$, we can compute a minimal Gröbner basis $G = \{yx + y, zx + 1, yz - y\}$ for I. Since $G \cap Q[z, y] = \{yz - y\}$, Proposition 3.1(ii) implies that $I \cap Q[z, y] = (yz - y)Q[z, y]$. Proposition 3.1(i) states that G is also a Gröbner basis for I when it is considered as an ideal in the polynomial ring $Q[z, y][x]$ with variable x and coefficients in $Q[z, y]$. In other words, $\mathrm{Lt}_x(I) = \mathrm{Lt}_x(G)$ where Lt_x refers to leading terms taken with respect to the ordering of powers of x only. Note that in the interpretation, G is not a minimal Gröbner basis, for we have

$$\mathrm{lt}_x(xy + y) = xy = y(xz) - x(yz - y) = y\,\mathrm{lt}_x(xz + 1) - x\,\mathrm{lt}_x(yz - y)$$
$$\in (\mathrm{lt}_x(xz + 1), \mathrm{lt}_x(yz - y)).$$

Thus a minimal Gröbner basis for $I \subset (Q[z, y])[x]$ is $G' = \{xz + 1, yz - y\}$. Note that G' is not a Gröbner basis for I when all the variables are considered, because then $\mathrm{lt}(yz - y) = yz$ and $xy \notin (xz, yz)$. As this example shows, the price for computing the Gröbner bases over Q when we are only interested in the basis over $Q[z, y]$ is the construction of some unnecessary basis elements. The advantage is the ability to use simpler versions of the algorithms.

A number of useful ideal operations which can be expressed in terms of coordinate subring contractions are listed below.

COROLLARY 3.2. *Let I and J be given ideals in $R[\mathrm{x}]$. Then the following can be computed:*

(i) $I \cap J$.
(ii) $I: J$, *if the generators of J aren't zero divisors.*
(iii) *The kernel of a given homomorphism* $\varphi: R[\mathrm{y}] \mapsto R[\mathrm{x}]/I$.
(iv) *The ideal of polynomial relations among* $f_1, \ldots, f_m \in R[\mathrm{x}]$.
(v) $IR[\mathrm{x}]_f \cap R[\mathrm{x}]$ *for any nonzero divisor* $f \in R[\mathrm{x}]$.

PROOF. (i) Let t be a new indeterminate and observe that

$$I \cap J = (tI, (t-1)J)R[\mathrm{x}, t] \cap R[\mathrm{x}].$$

We note that we could compute the intersection directly by constructing the basis for an appropriate syzygy module, but the approach through subring contraction is simpler and more efficient. This is because the contraction computes the basis for the intersection of the ideals directly, without needing to first construct and store the explicit expression of each element as a linear combination of generators of both I and J.

(ii) Let $J = (f_1, \ldots, f_m)$. Then $I: J = \bigcap_i I: (f_i)$, so $I: J$ can be constructed provided each $I: (f_i)$ can. Now $I \cap (f_i)$ can be computed by (i), and if $\{g_1, \ldots, g_k\}$ is a basis for $I \cap (f_i)$ then $\{g_1/f_i, \ldots, g_k/f_i\}$ is a basis for $I: (f_i)$.
(iii) If the homomorphism φ is given by $\varphi(y_i) = f_i$, with $f_i \in R[\mathrm{x}]$, the kernel is the ideal $((y_i - f_i), I)R[\mathrm{x}, \mathrm{y}] \cap R[\mathrm{y}]$
(iv) Take $I = 0$ in (iii).
(v) $R[\mathrm{x}]_f \simeq R[\mathrm{x}, t]/(tf - 1)$, where t is a new indeterminate. Thus

$$IR[\mathrm{x}]_f \cap R[\mathrm{x}] = (I, tf - 1)R[\mathrm{x}, t] \cap R[\mathrm{x}]. \ \square$$

For example consider the ideal $I = (12, xy + 2) \subset \mathbf{Z}[y, z]$. In order to compute $I \cap (y)$ we first find the Gröbner basis for

$$(It, (t-1)y) = (12t, xyt + 2t, ty - y) \subset \mathbf{Z}[y, x, t]$$

using an order which puts t first. For example with total lexicographical order with $t > x > y$, the Gröbner basis is

$$G = (yxt - xy, yt - y, 2t + yx, 12y, xy^2 + 2y, 6xy).$$

Contracting this to $\mathbf{Z}[y, x]$ we find that

$$I \cap (y) = (12y, xy^2 + 2y, 6xy).$$

Dividing by y we see that $(I: y) = (6x, 12, xy + 2)$. In order to divide out all powers of y from I, we compute

$$I\mathbf{Z}[y, x]_y \cap \mathbf{Z}[y, x].$$

Proceeding as in part (v) above, we consider the ideal $(I, ty - 1) \subset \mathbf{Z}[y, x, t]$.

$$G = (ty - 1, 2t + x, 3x^2, 6x, 12, xy + 2)$$

is a Gröbner basis for it in lexicographical order, so by contracting the basis we find that

$$I\mathbf{Z}[y, x]_y \cap \mathbf{Z}[y, x] = (3x^2, 6x, 12, xy + 2).$$

Proposition 3.1(ii) shows in particular that if G is a Gröbner basis for $I \subset R[x]$ then $G \cap R$ generates $I \cap R$. We can in fact characterize the situation fully as follows:

PROPOSITION 3.3. *Let I be an ideal in $R[x]$ and let $\pi: R[x] \mapsto (R/I \cap R)[x]$ be the quotient map. Then for $G \subset I$ we have*

(i) *If G is a Gröbner basis for I then $G \cap R$ generates $I \cap R$ and $\pi(G)$ is a Gröbner basis for $\pi(I)$.*

(ii) *G is a minimal Gröbner basis for I if and only if $G \cap R$ is a minimal basis for $I \cap R$, $\pi(G - G \cap R)$ is a minimal Gröbner basis for $\pi(I)$, and $\pi(\mathrm{lt}(g)) \neq 0$ for all $g \in G - G \cap R$.*

PROOF. Since $\pi(\mathrm{lt}(f))$ is either 0 or $\mathrm{lt}(\pi(f))$, we have $\pi(\mathrm{Lt}(I)) \subset \mathrm{Lt}(\pi(I))$, and, conversely, given $f \in I$, we can write $f = f_0 + f_1$ where $\pi(f_0) = 0$ and $\pi(\mathrm{lc}(f_1)) \neq 0$. In particular, $f_0 \in I$ and so $f_1 \in I$ and $\mathrm{lt}(\pi(f)) = \mathrm{lt}(\pi(f_1)) = \pi(\mathrm{lt}(f_1)) \in \pi(\mathrm{Lt}(I))$. Thus we observe that $\pi(\mathrm{Lt}(I)) = \mathrm{Lt}(\pi(I))$. The result now follows from the definitions and Proposition 3.1(ii). □

Finally, we consider the construction of the ring of fractions of $R[x]$ with respect to multiplicative subsets of R. We first observe that Gröbner bases are well behaved under this operation.

PROPOSITION 3.4. *Let S be a multiplicatively closed subset of R. If G is a Gröbner basis for an ideal $I \subset R[x]$ then it is a Gröbner basis for $S^{-1}I \subset (S^{-1}R)[x]$.*

PROOF. We have $\mathrm{Lt}(S^{-1}I) = S^{-1}\mathrm{Lt}(I) = S^{-1}\mathrm{Lt}(G)$, i.e. the leading terms of elements of G generate $\mathrm{Lt}(S^{-1}I)$ in $S^{-1}R[x]$. □

Thus computing with Gröbner bases in $(S^{-1}R)[x]$ presents no special problems. An important construction which we need to consider now is the saturation $S^{-1}I \cap R[x]$ with respect to S of an ideal $I \subset R[x]$. We note that this operation can be determined from the behaviour of the leading term ideal, in the following sense:

LEMMA 3.5. *Let $T \subset S$ be multiplicatively closed subsets of R, and let I be an ideal in $R[x]$ if*

$$S^{-1}\mathrm{Lt}(I) \cap R[x] = T^{-1}\mathrm{Lt}(I) \cap R[x]$$

then

$$S^{-1}I \cap R[x] = T^{-1}I \cap R[x].$$

PROOF. We have

$$\begin{aligned} \mathrm{Lt}(S^{-1}I \cap T^{-1}R[x]) &\subseteq \mathrm{Lt}(S^{-1}I) \cap T^{-1}R[x] \\ &= S^{-1}\mathrm{Lt}(I) \cap T^{-1}R[x] \\ &= T^{-1}(S^{-1}\mathrm{Lt}(I) \cap R[x]) \quad \text{because } T \subset S \\ &= T^{-1}(T^{-1}\mathrm{Lt}(I) \cap R[x]) \quad \text{by assumption} \\ &= T^{-1}\mathrm{Lt}(I) \\ &= \mathrm{Lt}(T^{-1}I) \end{aligned}$$

Since $S^{-1}I \cap T^{-1}R[x] \supset T^{-1}I$, we obtain $\mathrm{Lt}(S^{-1}I \cap T^{-1}R[x]) = \mathrm{Lt}(T^{-1}I)$. Thus by Corollary 2.10 we have

$$S^{-1}I \cap T^{-1}R[x] = T^{-1}I.$$

Taking the intersection with $R[x]$ completes the proof. □

We remark that taking $T = \{1\}$ shows that I is saturated with respect to S if $\mathrm{Lt}(I)$ is. More generally, the lemma states that we may attempt to compute the saturation with respect to S using a smaller multiplicative set T, provided the change does not affect the behaviour of the leading term ideal.

PROPOSITION 3.6. *Let S be a multiplicatively closed subset of R, I an ideal in $R[x]$. If for some $s \in S$,*

$$S^{-1}\mathrm{Lt}(I) \cap R[x] = (\mathrm{Lt}(I)R_s[x]) \cap R[x] \qquad (*)$$

then

$$S^{-1}I \cap R[x] = IR_s[x] \cap R[x].$$

PROOF. Apply the lemma with $T = \{s^n\}$. □

Since we know how to compute $IR_s[x] \cap R[x]$ by Corollary 3.2(v), $S^{-1}I \cap R[x]$ can be computed if we can find an $s \in S$ satisfying condition (*). Thus Propostion 3.6 reduces the problem of computing the saturation $S^{-1}I \cap \mathrm{R}[x]$ of an arbitrary ideal I in $R[x]$ to an analogous problem for ideals generated by terms. This is equivalent to solving the problem for finite sets of ideals in R, and whether it can be done depends of course on the given R and S.

Of particular interest to us is the localization R_P at a prime ideal $P \subset R$, i.e. the case where $S = R - P$. While we do not know of any general algorithms to compute the saturations of ideals in R with respect to arbitrary prime ideals P, the problem can be solved in the case where P is a principal ideal. The following proposition will be central to a dimension reduction process which will be developed later in the paper.

PROPOSITION 3.7. *Let R be an integral domain, $(p) \subset R$ a principal prime ideal. For any given ideal $I \subset R[x]$ it is possible to find $s \in R - (p)$ such that*

$$IR_{(p)}[x] \cap R[x] = IR_s[x] \cap R[x].$$

In particular $IR_{(p)}[x] \cap R[x]$ can be computed

PROOF. Since R is a domain, $\cap\ (p^k) = 0$. Thus for any non-zero element r of R there exists a k such that $r \in (p^k)$, and $r \notin (p^{k+1})$. Thus $r = sp^k$ for some $s \notin (p)$. We can compute k and s by applying the ideal membership algorithm. Let $G = \{g_1, \ldots, g_m\}$ be a Gröbner basis for I. Express $\mathrm{lt}(g_i)$ as $\mathrm{lt}(g_i) = s_i p^{k_i} x^{A_i}$, where $s_i \notin (p)$ as described above. Then $\mathrm{Lt}(I) = (s_i p^{k_i} x^{A_i})$ while $\mathrm{Lt}(I)R_{(p)}[x] \cap R[x] = (p^{k_i} x^{A_i})$. Thus in order to apply Proposition 3.6, we only need to find an s such that every s_i is invertible in $R_s[x]$. The choice $s = \Pi\ s_i$ satisfies this condition. (In fact any common multiple of the radicals of s_i will be sufficient. Furthermore it is only necessary to consider those i with $(p^{k_i} x^{A_i})$ minimal.) □

Finally we note a very useful special case of Proposition 3.7.

COROLLARY 3.8. *Let R be an integral domain, K the quotient field of R. Then for any given ideal $I \subset R[\mathrm{x}]$ it is possible to compute $IK[\mathrm{x}] \cap R[\mathrm{x}]$.*

PROOF. Apply the proposition with $p = 0$. □

REMARK. When $p = 0$, the s of Proposition 3.7 is simply the product of the leading coefficients of a Gröbner basis for I.

4. Primality Test

As an application of the results developed in the previous section, we now present an algorithm for testing the primality of ideals in $R[\mathrm{x}]$. We first recall the following basic facts:

LEMMA 4.1. *An ideal $I \subset R[\mathrm{x}]$ is prime if and only if $I \cap R$ is prime and the image of I in $(R/I \cap R)[\mathrm{x}]$ is prime.*

PROOF. Zariski & Samuel (1975) Chapter III, Theorem 11. □

LEMMA 4.2. *Let R be an integral domain, K the quotient field of R. If I is an ideal of $R[\mathrm{x}]$ such that $I \cap R = (0)$ then I is prime if and only if $IK[\mathrm{x}]$ is prime and $I = IK[\mathrm{x}] \cap R[\mathrm{x}]$.*

PROOF. Zariski & Samuel (1975) Chapter IV, Corollary 1 to Theorem 16. □

We assume that we have a primality test for ideals in R and that we can test the irreducibility of univariate polynomials over quotient fields of residue rings of $R[\mathrm{x}]$ (this will be the case for instance if R is a prime field or Z). Then we obtain:

PROPOSITION 4.3. *It is possible to decide the primality of ideals in $R[\mathrm{x}]$*

PROOF. Proceeding by induction on the number of variables we may assume that we have an ideal I in $R[x_1]$. We can compute $I^c = I \cap R$ by Proposition 3.1(ii). If I^c is not prime then neither is I and we are done. Otherwise by Lemma 4.1, we need only to test the primality of the image of I in $R/I^c[x_1]$. Replacing R by R/I^c, we may assume R is an integral domain and $I \cap R = (0)$. Let K be the quotient field of R. Then $IK[x_1]$ is a principal ideal and hence we can test its primality by checking the irreducibility of its generator. We can compute $IK[x_1] \cap R[x_1]$ by Corollary 3.8. Thus we can test the primality of I by Lemma 4.2. □

ALGORITHM PT $(R; \mathrm{x}; I)$. Primality test

Input: Ring R; variables $\mathrm{x} = x_1, \ldots, x_n$; ideal $I \subset R[\mathrm{x}]$.
Assumptions: (none)
Output: TRUE if I is prime, otherwise FALSE.
Step 1: If $n = 0$ then if $I \subset R$ is prime the return TRUE otherwise return FALSE.
Step 2: Compute $J = I \cap R[x_2, \ldots, x_n]$. [Proposition 3.1(ii)]
Step 3: If PT $(R; x_2, \ldots, x_n; J)$ = FALSE then return to FALSE.
Step 4: Let $R' = R[x_2, \ldots, x_n]/J$, $I' = IR'[x_1]$, K' = the quotient field of R'.

Step 5: Compute $I'K'[x_1] = (f)$.
Step 6: If f is not irreducible over K' then return FALSE.
Step 7: Compute $I^{ec} = I'K'[x_1] \cap R'[x_1]$. [Corollary 3.8]
Step 8: If $I^{ec} \subset I'$ then return TRUE, otherwise return FALSE.

5. Zero-dimensional Ideals

We now begin a deeper study of the properties of Gröbner bases by examining the structure of zero-dimensional ideals. First we show that under certain conditions we can determine whether an ideal is zero-dimensional by simply inspecting its Gröbner basis.

LEMMA 5.1 *Let $I \subset R[\mathrm{x}]$ be an ideal such that $I \cap R$ is zero-dimensional. Then I is zero-dimensional if and only if $R[\mathrm{x}]/I$ is integral over R.*

PROOF. $\Leftarrow$: If $R[\mathrm{x}]/I$ is integral over R then it is also integral over the subring $R/I \cap R \subset R[\mathrm{x}]/I$. Thus $R/I \cap R$ and $R[\mathrm{x}]/I$ have the same dimension.

$\Rightarrow$: Let $I = \cap Q_k$ be a primary decomposition of I, and let $M_k = \sqrt{Q_k}$. By assumption, M_k is maximal. Since $M_k \cap R$ contains $I \cap R$, it is zero-dimensional and hence maximal. Therefore, by the Nullstellensatz, the field $R[\mathrm{x}]/M_k$ is a finite algebraic extension of the subfield $R/M_k \cap R$. In particular, for each i, M_k contains a monic polynomial $f_{i,k}(x_i)$. Then $f_{i,k}(x_i)^N \in Q_k$ for some N, and so $\prod_k f_{i,k}(x_i)^N \in I$ is an equation of integral dependence for x_i mod I. □

The requirement that $I \cap R$ be zero-dimensional cannot be omitted. For instance consider $R = Z_{(2)}$, the localization of Z at the prime ideal generated by $2 \in Z$. Then the ideal $I = (2x - 1)R$ in $R[\mathrm{x}]$ is maximal but contains no monic polynomials, so x mod I is not integral over R. And indeed $I \cap R = (0)$ is not zero-dimensional.

We note however that the condition that $I \cap R$ be zero-dimensional for every zero-dimensional ideal $I \subset R[\mathrm{x}]$ is satisfied in Hilbert rings (see Kaplansky, 1968). In particular the condition holds for polynomial rings with coefficients in a field. Furthermore, it follows from the lemma that if I and $I \cap R[\mathrm{x}]$ are zero-dimensional then so is $I \cap R[x_i, \ldots, x_n]$ for any i.

The following proposition gives an effective criterion for detecting integral extensions.

PROPOSITION 5.2. *$R[\mathrm{x}]/I$ is integral over R if and only if $(x_1, \ldots, x_n) \subset \sqrt{\mathrm{Lt}(I)}$.*

PROOF. $\Rightarrow$: Each $x_i + I \in R[\mathrm{x}]/I$ is integral over R, so for each i, I contains a monic polynomial $f(x_i) \in R[x_i]$. Then $\mathrm{lt}(f(x_i)) \in \mathrm{Lt}(I)$, but the leading term of $f(x_i)$ is just a power of x_i.

$\Leftarrow$: We will show that $R[\mathrm{x}]/I$ in finitely generated as an R-module, which implies that it is integral over R. Suppose $x_i^{m_i} \in \mathrm{Lt}(I)$, and consider the finitely generated R-module

$$K = \sum_{a_i < m_i} R x_1^{a_1} \ldots x_n^{a_n}.$$

We claim that the R-module map $\pi: K \mapsto R[\mathrm{x}]/I$, defined by $\pi(h) = h + I$, is surjective. Let f be an element of $R[\mathrm{x}]$ and consider $f + I \in R[\mathrm{x}]/I$. We may assume $f \notin I$, since $0 + I$ is clearly in the image of π. By the remark following Proposition 2.5, there exists an $f' \in f + I$ such that $\mathrm{lt}(f') \notin \mathrm{Lt}(I)$. In particular $\mathrm{lt}(f') \notin (x_1^{m_1}, \ldots, x_n^{m_n})$, so in fact $\mathrm{lt}(f') \in K$.

Furthermore, since $f-f'\in I$ and $\mathrm{lt}(f')\notin \mathrm{Lt}(I)$, we have $\mathrm{lt}(f-f')\neq \mathrm{lt}(f')$. It follows that $\deg(f')\leqslant \deg(f)$ and so $\deg(f'-\mathrm{lt}(f'))<\deg(f)$. By induction on the degree of f, we may assume that $(f'-\mathrm{lt}(f'))+I$ is in the image of π, say $(f'-\mathrm{lt}(f'))+I=\pi(h)$ for some $h\in K$. Then $\pi(\mathrm{lt}(f')+h)=\pi(\mathrm{lt}(f'))+\pi(h)=(\mathrm{lt}(f')+I)+(f'-\mathrm{lt}(f')+I)=f'+I=f+I$, showing that $f+I$ is in the image of π. □

If G is a Gröbner basis for I, let $G_i=\{g\in G\mid \mathrm{lt}(g)=cx_i^m \text{ for some } c\in R, m\geqslant 0\}$ and let $L_i\subset R$ be the ideal generated by the leading coefficients of elements of G_i. Clearly $\mathrm{Lt}(G_i)=\mathrm{Lt}(G)\cap R[x_i]$ so $x_i\in\sqrt{\mathrm{Lt}(I)}=\sqrt{\mathrm{Lt}(G)}$ if and only if $x_i\in\sqrt{\mathrm{Lt}(G_i)}$. This can happen if and only if $L_i=(1)$, a condition we can verify. Thus we can decide whether $x_i\in\sqrt{\mathrm{Lt}(I)}$ just by examining a Gröbner basis for I. Furthermore it follows from that first part of the proof that if $x_i\notin\sqrt{\mathrm{Lt}(I)}$ then x_i+I cannot be integral over R. Thus we have:

COROLLARY 5.3. *It is possible to decide whether $R[\mathrm{x}]/I$ is integral over R, and if not, to find an i such that x_i+I is not integral over R.*

Applying the lemma, we get

COROLLARY 5.4. *If $I\cap R$ is zero-dimensional then it is possible to decide whether I is zero-dimensional, and if not, to find an i such that $I\cap R[x_i]$ is not zero-dimensional.*

When $I\cap R$ is primary then we can further simplify the criterion described above.

PROPOSITION 5.5. *Let $I\subset R[\mathrm{x}]$ be an ideal such that $I\cap R$ is zero-dimensional primary. Let G be a Gröbner basis for I. Then I is zero-dimensional if and only if for each I there exists a $g_i\in G$ such that $\mathrm{lt}(g_i)=c_ix_i^{m_i}$, $c_i\in R$ a unit modulo $I\cap R$.*

PROOF. Let G_i and L_i be as in the discussion preceding Corollary 5.3. Note that G_i contains $G\cap R$ and hence L_i contains $I\cap R$. Since $\sqrt{R\cap I}$ is maximal, $L_i=(1)$ if and only if $L_i\not\subset\sqrt{R\cap I}$, which can occur if and only if there is some $g_i\in G_i$ such that $\mathrm{lc}(g_i)\notin\sqrt{R\cap I}$. But this is equivalent to the requirement that $(\mathrm{lc}(g_i), R\cap I)=(1)$. □

Note that, with notation as above, every element of I whose leading term is divisible by $x_i^{m_i}$ is reducible modulo $\{g_i\}\cup(G\cap R)$. In particular, if G is a minimal Gröbner basis then all elements of G_i other than g_i have degree in x_i strictly smaller than m_i. Thus to decide whether I is zero-dimensional using a minimal Gröbner basis, one needs only to check that G_i contains exactly one element of the maximal degree, and that its leading coefficient together with $G\cap R$ generates the unit ideal—it is not necessary to check the leading coefficients of any other elements of G_i. Conversely, if I is known to be zero-dimensional then g_i can be uniquely identified as the highest degree element of G_i—there is no need to verify the condition on its leading coefficient.

We now investigate the structure of zero-dimensional primary ideals. When we say a polynomial has some property modulo an ideal J in R, we mean that its image as a polynomial in R/J has that property. We first note the following univariate results.

LEMMA 5.6. *Let $I\subset R[\mathrm{x}_1]$ be an ideal such that $I\cap R$ is zero-dimensional. Suppose $x_1^m\in\mathrm{Lt}(I)$, $x_1^{m-1}\notin\mathrm{Lt}(I)$. Then every $f\in I$ with $\deg(f)<m$ is a zero-divisor modulo $I\cap R$.*

PROOF. Let $L \subset R$ be the ideal generated by the leading coefficients of elements of I of degree less than m. We claim that if $f \in I$ has degree less than m then $f \equiv 0 \bmod L$. Let $f = c_1 x_1^{m-1} + \cdots + c_m$. Then c_1 is either 0 or it is the leading coefficient of f, so $c_1 \in L$. By assumption, there exists a $g \in I$ with $\mathrm{lt}(g) = x_1^m$. Let $f' = x_1 f - c_1 g$. Then $f' \in I$ and $f' = c'_1 x_1^{m-1} + \cdots + c'_m$ with $c'_i \equiv c_{i+1} \bmod L$. It follows by induction that $c_i \in L$ for all i, proving the claim. Now if $L = (1)$ then I would contain a monic polynomial of degree less than m, contrary to assumption. Thus L is a proper ideal. Since $I \cap R \subset L$ with $I \cap R$ zero-dimensional, L is contained in some associated prime of $I \cap R$. Thus there exists an $a \notin I \cap R$ such that $aL \subset I \cap R$. Then $af \equiv 0 \bmod I \cap R$ whenever $\deg(f) < m$. □

LEMMA 5.7. *Let $I \subset R[x_1]$ be a zero-dimensional ideal such that $I \cap R$ is zero-dimensional primary. Let G be a minimal Gröbner basis for I and let $g_1 \in G$ be as in Proposition* 5.5. *Then* $\sqrt{I} = \sqrt{(g_1, I \cap R)}$.

PROOF. Let $\mathrm{lt}(g_1) = c_1 x_1^{m_1}$. By assumption, c_1 is a unit modulo $I \cap R$, so $x_1^{m_1} \in \mathrm{Lt}(g_1, I \cap R) \subset \mathrm{Lt}(I)$. $\mathrm{Lt}(I)$ cannot contain any smaller powers of x_1 since otherwise g_1 would be reducible, contradicting the minimality of G. Thus by Lemma 5.6, every $f \in I$ of degree less than m_1 is a zero-divisor modulo $I \cap R$. But since $I \cap R$ is primary, the set of zero-divisors modulo $I \cap R$ is exactly $\sqrt{I \cap R}$. Thus $f \in I$, $\deg(f) < m_1$ implies $f \equiv 0 \bmod \sqrt{I \cap R}$. Let $f \in I$. By Proposition 2.5, there exists $f' \equiv f \bmod (g_1, I \cap R)$ such that f' is reduced modulo $(g_1, I \cap R)$. Since $x_1^{m_1} \in \mathrm{Lt}(g_1, I \cap R)$, f' has degree less than m_1 so $f' \equiv 0 \bmod \sqrt{I \cap R}$. Thus $f \in (g_1, I \cap R) + \sqrt{I \cap R} = (g_1, \sqrt{I \cap R})$. In other words we have

$$I \subset (g_1, \sqrt{I \cap R}) \subset \sqrt{I}.$$

Taking radicals proves the lemma. □

We are now able to completely characterize zero-dimensional primary ideals in terms of verifiable conditions on their lexicographical Gröbner bases.

PROPOSITION 5.8. *Let $I \subset R[x]$ be a zero-dimensional ideal such that $I \cap R$ is zero-dimensional primary. Let G be a minimal Gröbner basis for I with respect to the lexicographical order with $x_1 > \cdots > x_n$, and let $g_1, \ldots, g_n \in G$ be as in Proposition* 5.5. *Then I is primary if and only if for all i, g_i is a power of an irreducible polynomial modulo $\sqrt{I \cap R[x_{i+1}, \ldots, x_n]}$. If this is the case then for every $h \in G \cap R[x_i, \ldots, x_n] - \{g_i\}$, $h \equiv 0$ mod $\sqrt{I \cap R[x_{i+1}, \ldots, x_n]}$.*

PROOF. Let $R' = R[x_2, \ldots, x_n]$, $I' = I \cap R'$. In view of Proposition 3.1, we may proceed by induction to conclude that the proposition holds for I' and $g_2, \ldots, g_n \in G \cap R'$. Thus we only need to show that I is primary if and only if I' is primary and g_1 is the power of an irreducible polynomial modulo $\sqrt{I'}$, in which case $h \equiv 0 \bmod \sqrt{I'}$ for $h \in G - \{g_1\}$.

Clearly if I is primary then so is I', so assume I' is primary. Let $\mathrm{lt}(g_1) = c_1 x_1^{m_1}$. If h is an element of G other than g_1, then it must have degree less than m_1 in x_1, since otherwise it would be reducible by (g_1, I'). Thus by Lemma 5.6 (and the assumption that I' is primary) $h \equiv 0 \bmod \sqrt{I'}$, proving the second part of the proposition. Since I is zero-dimensional, it is primary if and only if its radical is prime. By Lemma 5.7, $\sqrt{I} = \sqrt{(g_1, I')} = \sqrt{(g_1, \sqrt{I'})}$. Thus I is primary if and only if $(g_1, \sqrt{I'})$ is primary, or equivalently, if and only if the ideal generated by g_1 in $(R'/\sqrt{I'})[x_1]$ is primary. □

PROPOSITION 5.9. *Let $I \subset R[x]$ be a zero-dimensional ideal such that $I \cap R$ is zero-dimensional prime. Let G be a minimal Gröbner basis for I with respect to the lexicographical order with $x_1 > \cdots > x_n$, and let $g_1, \ldots, g_n \in G$ be as in Proposition 5.5. Then I is prime if and only if for all i, g_i is irreducible modulo $I \cap R[x_{i+1}, \ldots, x_n]$. If this is the case then $G = \{g_1, \ldots, g_n\} \cup (G \cap R)$.*

PROOF. Suppose I is prime. By Proposition 5.8, $g_i \equiv h_i^{k_i}$ for some h_i irreducible modulo $I \cap R[x_{i+1}, \ldots, x_n]$. Since I is prime, we must have $h_i \in I$. If $k_i > 1$ then g_i would be reducible by h_i, an element of smaller degree, contradicting the minimality of G. Thus $k_i = 1$ and so g_i is irreducible mod $I \cap R[x_{i+1}, \ldots, x_n]$.

Conversely, suppose $I \cap R[x_{i+1}, \ldots, x_n]$ is prime and g_i is irreducible modulo $I \cap R[x_{i+1}, \ldots, x_n]$. Then $(g_i, I \cap R[x_{i+1}, \ldots, x_n]) \subset R[x_i, \ldots, x_n]$ is prime. Furthermore, if h is an element of $G \cap R[x_i, \ldots, x_n]$ other than g_i, then by the previous proposition $h \equiv 0 \bmod I \cap R[x_{i+1}, \ldots, x_n]$. In particular h is reducible modulo $G \cap R[x_{i+1}, \ldots, x_n]$, so from the minimality of G it follows that $h \in G \cap R[x_{i+1}, \ldots, x_n]$. Thus $G \cap R[x_i, \ldots, x_n] = \{g_i\} \cup (G \cap R[x_{i+1}, \ldots, x_n])$ and consequently $I \cap R[x_i, \ldots, x_n] = (g_i, I \cap R[x_{i+1}, \ldots, x_n])$ is prime. The proposition now follows by induction. □

6. Zero-dimensional Primary Decomposition

In this section we assume that for any given maximal ideal $M \subset R$, it is possible to factor univariate polynomials over finitely generated algebraic extensions of R/M. This will be the case for instance if R is a finitely generated algebra over a prime field or **Z** (see Davenport & Trager, 1981).

We now present an algorithm for computing the irredundant primary decomposition of zero-dimensional ideals in $R[x]$. The algorithm works by computing the primary decomposition of $I \cap R[x_n]$, extending it to a (not necessarily primary) decomposition of all of I, and then proceeding by induction to construct a complete primary decomposition of each component. The following proposition describes the induction step.

PROPOSITION 6.1. *Let $I \subset R[x]$ be a zero-dimensional ideal such that $I \cap R$ is M-primary, where $M \subset R$ is a maximal ideal. Then one can construct zero-dimensional $I_1, \ldots, I_m \subset R[x]$ and distinct maximal ideals $M_1, \ldots, M_m \subset R[x_n]$ such that $I = \bigcap_i I_i$ and $I_i \cap R[x_n]$ is M_i-primary.*

PROOF. Let $I^c = I \cap R[x_n]$. By Lemma 5.7, we can find $g \in I^c$ such that $\sqrt{I^c} = \sqrt{(g, M)}$. Let $g(x_n) \equiv \Pi\, p_i(x_n)^{s_i} \bmod M$ be the complete factorization of g modulo M, that is the images of $p_i(x_n)$ in $(R/M)[x_n]$ are pairwise comaximal irreducible non-units. Since $\Pi\, p_i^{s_i} \in (g, M) \subset \sqrt{I^c}$, $(\Pi\, p_i^{s_i})^s \in I^c$ for some s. Now since p_i, p_j are comaximal modulo M, and I contains a power of M, p_i, p_j are comaximal mod I. Thus $\bigcap_i (p_i^{s_i s}, I) = (\Pi\, p_i^{s_i s}, I) = I$. Let $I_i = (p_i^{s_i s}, I)$, $M_i = (p_i, M)R[x_n]$. M_i is clearly maximal, and since $I_i \cap R[x_n]$ contains a power of M_i, it is either M_i-primary or the unit ideal. We have $\prod_{j \neq i} p_j^{s_j s} I_i \subset I$, so if $I_i = (1)$ then $\prod_{j \neq i} p_j \in \sqrt{I^c} = \sqrt{(g, M)}$. This contradicts the assumption that p_i is not a unit modulo M. Thus I_i is M_i-primary. □

By recursively applying the proposition to M_i, I_i over the base ring $R[x_n]$, we can compute the complete primary decomposition of I along with the associated primes.

ALGORITHM ZPD (R; x; M); *Zero-dimensional primary decomposition*

Input: Ring R; variables $\mathrm{x} = x_1, \ldots, x_n$; ideal $I \subset R[\mathrm{x}]$; ideal $M \subset R$
Assumptions: M is maximal, I is zero-dimensional, $I \cap R$ is M-primary.
Output: $\{(Q_1, M_1), \ldots, (Q_m, M_m)\}$, Q_i, M_i ideals in $R[\mathrm{x}]$ such that M_i is maximal, $M_i \neq M_j$, Q_i is M_i-primary, and $I = \bigcap_i Q_i$.

Step 1: *If* $n = 0$ then return $\{(I, M)\}$
Step 2: Compute a minimal Gröbner basis G for $I \cap R[x_n]$. [Proposition 3.1(ii)]
Step 3: Select the $g \in G$ of largest degree.
Step 4: Compute the complete factorization of g mod $M, g = \Pi\, p_i^{s_i}$ in $(R/M)[x_n]$, $p_i \in R[x_n]$.
Step 5: Find s such that $(\Pi\, p_i^{s_i})^s \in I \cap R[x_n]$.
Step 6: Let $I_i = (p_i^{s_i s}, I)$, $M_i = (p_i, M)R[x_n]$.
Step 7: Return $\bigcup_i$ ZPD $(R[x_n]; x_1, \ldots, x_{n-1}; I_i; M_i)$.

REMARK. The union in Step 7 is disjoint, that is, it is not necessary to check for and remove duplicates.

7. Zero-dimensional Ideals Over Fields of Characteristic 0

In this section we assume that K is a field of characteristic zero and that all Gröbner bases G are normalised so that $\mathrm{lc}(g) = 1$ for all $g \in G$.

If I is an ideal in $K[\mathrm{x}] = K[x_1, \ldots, x_n]$, let us denote $I_i = I \cap K[x_i, \ldots, x_n]$. If I is a zero-dimensional prime then by Proposition 5.9 every minimal lexicographical Gröbner basis for I has the form $\{g_1(x_1, \ldots, x_n), g_2(x_2, \ldots, x_n), \ldots, g_n(x_n)\}$, with g_i monic as a polynomial in x_i and irreducible modulo I_{i+1}. We can in fact obtain the following stronger result:

PROPOSITION 7.1 *Let I be a prime zero-dimensional ideal in $K[\mathrm{x}]$, $G = \{g_1(x_1, \ldots, x_n), \ldots, g_n(x_n)\}$ a minimal Gröbner basis for I with respect to the lexicographical order. Then "almost all" linear transformations of coordinates, $g_i = x_i - p_i(x_{i+1}, \ldots, x_n)$ for $i < n$.*

PROOF. By (the proof of) the primitive element theorem (Zariski & Samuel, 1975), for almost all $a_1, \ldots, a_n \in K$,

$$K[\mathrm{x}]/I \simeq K\left(\sum a_i x_i\right).$$

If we choose new coordinates $z_1, \ldots, z_n$ such that $z_n = \Sigma\, a_i x_i$, then we have:

$$K[z_1, \ldots, z_n]/I \simeq K(z_n).$$

Since $z_i \in K(z_n)$ for every i, we have that $z_i = f_i(z_n)$ holds in $K[z_1, \ldots, z_n]/I$ and hence I contains polynomials of the form $z_i - f_i(z_n)$ for all $i < n$. If G is a Gröbner basis relative to coordinates $z_1, \ldots, z_n$ then $z_i - f_i(z_n)$ is reducible mod G. Since the only element of G which could reduce z_i is g_i, we have $\mathrm{lt}(g_i) = z_i$ as required. □

We can now introduce the notion of "general position".

DEFINITION 7.2. If I is a prime zero-dimensional ideal in $K[\mathrm{x}]$ such that its lexicographical minimal Gröbner basis satisfies Proposition 7.1, we say that I is in general position.

We say that I, an arbitrary zero-dimensional ideal, is in general position if all of its associated primes are in general position and their contractions to $K[x_n]$ are pairwise comaximal.

COROLLARY 7.3. *If I is a primary zero-dimensional ideal in general position, then the g_i in Proposition 5.8 are powers of linear equations modulo $\sqrt{I_{i+1}}$ for $i<n$.*

As an example, consider the ideal $I=(x_1^2+1, x_2) \subset Q[x_1, x_2]$. x_2 is irreducible over Q and x_1^2+1 is irreducible over $Q[x_2]/(x_2)$, so by Proposition 5.9 I is a zero-dimensional prime ideal. It is not in general position since x_1^2+1 is not linear in x_1. If we make the substitution $x_2=ax_1+x_2$ and consider the ideal $I_a=(ax_1+x_2, x_1^2+1)$, we find that $G_a=\{x_2^2+a^2, ax_1+x_2\}$ is the Gröbner basis for I_a whenever $a \neq 0$. In that case G_a is as required by Definition 7.2, so we see that any non-zero value of a is sufficient to bring I into general position.

REMARK. From the proof of Proposition 7.1 it follows that in order to put a zero-dimensional prime ideal in general position it is sufficient to replace x_n by $x_n+\Sigma c_i x_i$ for random $c_i \in K$. We remark also that it is always possible to put any zero-dimensional ideal in general position. The intent is to separate all the zeros in an algebraic closure by the last coordinate. To do so, one simply chooses c_i such that the values $x_n+\Sigma c_i x_i$ are distinct as $(x_1, \ldots, x_n)$ ranges over the set of zeros of the ideal in the algebraic closure of K. The set of "bad" choices form a proper algebraic subset of K^{n-1} and thus "almost all" choices of c_i are good.

PROPOSITION 7.4. *Let $I \subset K[\mathrm{x}]$ be a zero-dimensional ideal in general position, G a lexicographical Gröbner basis for I, and let $g_1, \ldots, g_n \in G$ be in Proposition 5.5. If $g_n=\Pi p_i^{s_i}$ is the irreducible decomposition of g_n then $I=\bigcap_i(p_i^{s_i}, I)$ is the primary decomposition of I.*

PROOF. $(p_i^{s_i}, I)$ is a zero-dimensional ideal and by definition of general position it is contained in exactly one prime ideal. Thus it must be a primary ideal. □

If we are given a zero-dimensional ideal I, not necessarily in general position, then the above construction will yield a decomposition but not necessarily into primary components. If the minimal Gröbner basis for $(p_i^{s_i}, I)$ is not of the form predicted by Corollary 7.3, then I is not in general position. We can then proceed by choosing a different set of coordinates (or by reverting to the non-probabilistic algorithm ZPD). We remark however that a random substitution "almost always" works.

ALGORITHM ZPDF $(K; \mathrm{x}; I)$. *Zero-dimensional primary decomposition over a field*

Input: Field K; variables $\mathrm{x}=x_1, \ldots, x_n$; ideal $I \subset K[\mathrm{x}]$
Assumptions: K is a field of characteristic zero, I is zero-dimensional.
Output: $\{Q_1, \ldots, Q_m\}$ such that $Q_i \subset K[\mathrm{x}]$ is a primary ideal, $I=\bigcap_i Q_i$ and $\sqrt{Q_i} \neq \sqrt{Q_j}$.
Step 1: Select random $c_1, \ldots, c_{n-1} \in K$ and replace x_n by $x_n+\Sigma c_i x_i$.
Step 2: Compute $I \cap K[x_n]=(g)$. [Proposition 3.1(ii)]
Step 3: Compute the complete factorization of g, $g=\Pi p_i^{s_i}$
Step 4: If $(p_i^{s_i}, I)$ is not a primary ideal in general position [Corollary 7.3] then go to Step 1.

Step 5: Replace x_n by $x_n - \Sigma c_i x_i$.
Step 6: Return $\{(p_i^{s_i}, I)\}$.

REMARK. In Step 4, it would be sufficient to test $(p_i^{s_i}, I)$ for being primary (using Proposition 5.8), but since the simpler test of Corollary 7.3 will be satisfied in almost all cases, it is preferable.

8. Primary Decomposition in Principal Ideal Domains

In this section we show how to reduce the general primary decomposition problem to the zero-dimensional case when the coefficient ring is a PID.

LEMMA 8.1. *Let S be a multiplicatively closed subset of R,* $s \in S$. *If* $S^{-1}I \cap R \subset (I : s)$ *then*

$$I = (I : s) \cap (I, s).$$

PROOF. $\subset$ is obvious. To prove $\supset$, suppose $f \in (I : s) \cap (I, s)$, so that $f = i + as$ with $i \in I$. Then $i + as \in (I : s) \Rightarrow is + as^2 \in I \Rightarrow as^2 \in I \Rightarrow a \in S^{-1}I \cap R \Rightarrow a \in (I : s) \Rightarrow as \in I \Rightarrow f \in I$. □

Combining the lemma with the construction of Proposition 3.7 we obtain the following fundamental decomposition mechanism.

PROPOSITION 8.2. *Let R be an integral domain,* $(p) \subset R$ *a principal prime ideal. For any given ideal* $I \subset R[\mathrm{x}]$ *it is possible to find* $r \in R - (p)$ *such that*

$$I = (I, r) \cap I^{ec}$$

where $I^{ec} = IR_{(p)}[\mathrm{x}] \cap R[\mathrm{x}]$.

PROOF. By Proposition 3.7 we can find $s \in R - (p)$ such that $I^{ec} = IR_s[\mathrm{x}] \cap R[\mathrm{x}]$. Thus we can compute I^{ec} by the method of Corollary 3.2(v). Since R is Noetherian, there exists an m such that $s^m I^{ec} \subset I$. Given a basis G for I^{ec}, we can compute m by testing whether $s^m G \subset I$ for successive values of m. By the lemma, $r = s^m$ is as required. □

PROPOSITION 8.3. *Let R be a PID, I an ideal in* $R[\mathrm{x}]$, $(p) \subset R$ *a maximal ideal. If* $I \cap R$ *is* (p)*-primary then it is possible to compute a primary decomposition for I.*

PROOF. If I is zero-dimensional then we can compute its decomposition using one of the algorithms of previous sections. Otherwise, by Proposition 5.5 we can find an i such that $I \cap R[x_i]$ is not zero-dimensional. Let $R' = R[x_i]$ and $\mathrm{x}' = x_1, \ldots, x_{i-1}, x_{i+1}, \ldots, x_n$, so that $R'[\mathrm{x}'] = R[\mathrm{x}]$ and $I \cap R'$ is not zero-dimensional. Applying Proposition 8.2, we can find $r' \in R' - (p)R'$ such that $I = (I, r') \cap I^{ec} = IR'_{(p)}[\mathrm{x}'] \cap R'[\mathrm{x}']$. Thus to decompose I it is sufficient to separately decompose (I, r') and I^{ec}.

Since $(I, r') \cap R'$ contains both the (p)-primary ideal $I \cap R$ and the element $r' \notin (p)R'$, either $(I, r') \cap R'$ is zero-dimensional or it is the unit ideal. In the former case, we can compute the primary decomposition of (I, r') by induction on the number of x_k such that the contraction of the ideal to $R[x_k]$ is not zero-dimensional. In the latter case $I = I^{ec}$ and so we only need to compute the decomposition I^{ec}.

In order to decompose I^{ec} we only need to decompose $I^e = IR'_{(p)}[\mathrm{x}']$ and then contract the decomposition back to $R'[\mathrm{x}']$ using Proposition 3.7. Note that $R'_{(p)}$ is again a PID, and $(p)R'_{(p)}$ is a maximal ideal. We claim that $I^e \cap R'_{(p)}$ is $(p)R'_{(p)}$-primary. Since $I \cap R$ is (p)-primary, I(and hence $IR'_{(p)}$) contains a power of p. Thus it is sufficient to show that $IR'[\mathrm{x}'] \cap R' \subset (p)R'$. Let P be a non-zero-dimensional associated prime of $I \cap R'$. Then $P \supset (p)R'$. But $(p)R'$ is one-dimensional, so $P = (p)R'$, which proves the claim. Thus $I^e \subset R'_{(p)}[\mathrm{x}']$ satisfies the hypotheses of the proposition and so we may decompose it by induction on the number of variables. □

COROLLARY 8.4. *If K is a field then it is possible to compute the primary decomposition of any ideal in $K[\mathrm{x}]$.*

PROOF. Take $p = 0$ in the proposition. □

We remark that in the field case, the ring $R'_{(p)} = R[x_i]_{(p)}$ appearing in the algorithm described above is simply the field $K(x_i)$. Thus if the initial problem is presented over a coefficient field then all computations, including all the recursive invocations of the decomposition algorithm, take place with a field as the coefficient ring.

PROPOSITION 8.5. *Let R be a PID, I an ideal in $R[\mathrm{x}]$. Then it is possible to compute a primary decomposition for I.*

PROOF. If $I \cap R$ is not zero-dimensional (i.e. $I \cap R = (0)$ and R is not a field) then apply Proposition 8.2 to $(0) \subset R$ to find $r \neq 0$ such that $I = (I, r) \cap (IR_{(0)}[\mathrm{x}] \cap R[\mathrm{x}])$. Since $R_{(0)}$ is a field (the quotient field of R), $IR_{(0)}[\mathrm{x}]$ can be decomposed using the field algorithm above, and the results contracted to $R[\mathrm{x}]$ using Proposition 3.7. We are then left with (I, r), which contracts to a zero-dimensional ideal in R.

Thus we may assume that $I \cap R$ is zero-dimensional, say $I \cap R = (\Pi\, p_i^{m_i})$ where $(p_i)R$ is maximal. Then $(p_i^{m_i}, I) \cap R$ is (p_i)-primary, so $(p_i^{m_i}, I)$ can be decomposed using the algorithm of Proposition 8.3. Since $I = \bigcap_i (p_i^{m_i}, I)$ we get a decomposition for I. □

REMARK. The decomposition obtained above is not irredundant.

ALGORITHM PPD $(R; \mathrm{x}; I)$: *Primary decomposition over a PID*

> *Input*: Ring R; variables $\mathrm{x} = x_1, \ldots, x_n$; ideal $I \subset R[\mathrm{x}]$.
> *Assumptions*: R is a PID.
> *Output*: $\{Q_1, \ldots, Q_m\}$ such that $Q_i \subset R[\mathrm{x}]$ is primary and $I = \bigcap_i Q_i$.
> Step 1: Find $r \neq 0$ such that $I = (I, r) \cap (IR_{(0)}[\mathrm{x}] \cap R[\mathrm{x}])$. [Proposition 8.2]
> Step 2: Let $\{Q_1, \ldots, Q_k\} = \text{PPD-0}\ (R_{(0)}; x; IR_{(0)}[\mathrm{x}]; 0)$.
> Step 3: Let $Q_i^c = Q_i \cap R[\mathrm{x}]$. [Proposition 3.7]
> Step 4: Compute $(I, r) \cap R = (r')$.
> Step 5: If r' is a unit, return $\{Q_1^c, \ldots, Q_m^c\}$.
> Step 6: Factor $r' = \Pi\, p_i^{m_i}$, p_i irreducible.
> Step 7: For each i let $\{Q_1^i, \ldots, Q_{k_i}^i\} = \text{PPD-0}\ (R; \mathrm{x}; (I, p_i^{m_i}); p_i)$.
> Step 8: Return $\{Q_1^c, \ldots, Q_k^c\} \cup \bigcup_i \{Q_1^i, \ldots, Q_{k_i}^i\}$.

ALGORITHM PPD-0 $(R; \mathrm{x}; I; p)$: *Primary decomposition over a PID, primary contraction case*

> *Input*: Ring R; variables $\mathrm{x} = x_1, \ldots, x_n$; ideal $I \subset R[\mathrm{x}]$; $p \in R$

Assumptions: R is a PID, $(p)R$ is maximal, $I \cap R$ is (p)-primary.
Output: $\{Q_1, \ldots, Q_m\}$ such that $Q_i \subset R[\mathrm{x}]$ is primary and $I = \cap Q_i$.
Step 1: If I is zero-dimensional [Proposition 5.5] then return its decomposition using ZPD or ZPDF.
Step 2: Find i such that $I \cap R[x_i]$ is not zero-dimensional [still Proposition 5.5]
Step 3: Let $R' = R[x_i]$, $\mathrm{x}' = x_1, \ldots, x_{i-1}, x_{i+1}, \ldots, x_n$, $I^e = IR'_{(p)}[\mathrm{x}']$.
Step 4: Find $r' \in R' - (p)R'$ such that $I = (I, r') \cap (I^e \cap R'[\mathrm{x}'])$. [Proposition 8.2]
Step 5: Let $\{Q_1, \ldots, Q_m\} =$ PPD-0 $(R'_{(p)}; \mathrm{x}'; I^e; p)$.
Step 6: Let $Q_i^c = Q_i \cap R'[\mathrm{x}']$. [Proposition 3.7]
Step 7: If $(I, r') = (1)$ then return $\{Q_1^c, \ldots, Q_m^c\}$.
Step 8: Let $\{Q'_1, \ldots, Q'_k\} =$ PPD-0 $(R; \mathrm{x}; (I, r'); p)$.
Step 9: Return $\{Q_1^c, \ldots, Q_m^c, Q'_1, \ldots, Q'_k\}$.

9. Applications to Computing Radicals and Associated Primes

The algorithm of the preceding section depends on repeated applications of the following formula

$$I = (I, s) \cap (I : s) \qquad (*)$$

to reduce the dimension of I. s is chosen so that the dimension of (I, s) is strictly less than that of I, and $(I : s) = I^{ec}$ is the contraction of the extension of I to a polynomial ring of lower dimension.

This reduction strategy can be applied to other constructions provided they are well-behaved under the basic operations employed by the reduction process. As an example, we consider the computation of the radical and of ideal. We first observe that (*) implies that

$$\sqrt{I} = \sqrt{(I, s)} \cap \sqrt{I : s}).$$

But $\sqrt{(I : s)} = \sqrt{I^{ec}} = (\sqrt{I^e})^c$. Thus computing radicals commutes with our reduction strategy. At the point where algorithm PPD-0 is ready to call ZPD or ZPDF, we have reduced the problem to a zero-dimensional ideal whose contraction to the underlying PID R is (p)-primary. Since algorithm ZPD can also compute the associated primes in the situation, we can simply compute the radical as the intersection of the associated primes.

Using ZPD, however, makes radical computation no easier than primary decomposition. Since square-free factorization of polynomials over perfect fields reduces to greatest common divisor computations, which are in general easier than polynomial factorization, we could hope for an easier way to compute radicals of ideals. Once we have arrived at the situation where we have an ideal I such that $I \cap R$ is (p)-primary, we can adjoin p to I and assume $I \cap R$ is maximal. We can now reduce I modulo p, which brings us to the case of zero-dimensional ideals in a polynomial ring over a field. When this field is perfect, there is a much simpler radical construction based on Lemma 92 of Seidenberg (1974). Since I is zero-dimensional, it contains non-constant univariate polynomials $f_i(x_i)$ in each variable x_i. We define

$$g_i = f_i / gcd(f_i, f'_i)$$

where f'_i is the derivative of f_i taken with respect to x_i. Since our coefficient field is perfect, g_i will have all distinct roots in any splitting field. Seidenberg shows that $\sqrt{I} = (I, g_1, \ldots, g_n)$. Note that the f_i can be found using a single Gröbner basis computation along with the solution of linear equations, as observed by Buchberger (1985).

References

Ayoub, C. (1982). The decomposition theorem for ideals in polynomial rings over a domain. *J. Algebra*, **76,** 99–110.

Ayoub, C. (1983). On constructing bases for ideals in polynomial rings over the integers. *J. Number Theory*, **17,** 204–225.

Buchberger, B. (1965). *Ein algorithmus zum Auffinden der Basiselemente des Restklassenringes nach einem nulldimensionalen Polynomideal.* Ph.D Thesis, Universitat Innsbruck.

Buchberger, B. (1970). Ein algorithmisches Kriterium für die Lösbarkeit eines algebraischen Gleichungssystems. *Aequationes Math.* **4,** 374–383.

Buchberger, B. (1976). A theoretical basis for the reduction of polynomials to canonical forms. *ACM SIGSAM Bulletin*, **39,** 19–29.

Buchberger, B. (1979). A criterion for detecting unnecessary reductions in the construction of Gröbner bases. In: *Symbolic and Algebraic Computation*, Lecture notes in computer science, Springer-Verlag, Heidelberg. Vol. 72, pp. 3–21.

Buchberger, B. (1985). Gröbner bases: An algorithmic method in polynomial ideal theory. In: (Bose, N. K., ed.) *Multidimensional Systems Theory*, D. Reidel Publishing Co., pp. 184–232.

Davenport, J., Trager, B. (1981). Factorization over finitely generated fields. *Proceedings of the 1981 Symposium on Symbolic and Algebraic Computation — Snowbird, Utah*, pp. 200–205.

Kaplansky, I. (1968). *Commutative Rings.* Queen Mary College Math Notices, London.

Lazard, D. (1985). Ideal bases and primary decomposition: case of two variables, *J. of Symb. Comp.* **1,** 261–270.

Richman, F. (1974). Constructive aspects of Noetherian rings. *Proc. Am. Math. Soc.* **44,** 436–441.

Seidenberg, A. (1974). Constructions in algebra. *Trans. Am. Math. Soc.* **197,** 273–313.

Seidenberg, A. (1978). Constructions in a polynomial ring over the ring of integers. *Am. J. Math.* **100,** 685–703.

Seidenberg, A. (1984). On the Lasker-Noether Decomposition Theorem. *Am. J. Math.* **106,** 611–638.

Spear, D. (1977). A constructive approach to commutative ring theory. *Proc. 1977 MACSYMA Users' Conference*, 369–376.

Trinks, W. (1978). Über B. Buchbergers Verfahren, Systeme algebraischer Gleichungen zu lösen. *J. Number Theory*, **10,** 475–488.

Zacharias, G. (1978). *Generalized Gröbner bases in Commutative Polynomial Rings.* Bachelor's Thesis, MIT.

Zariski, O., Samuel, P. (1975). *Commutative Algebra, Volume I.* Graduate Texts in Mathematics Volume 28, Springer-Verlag, Neidelberg.

Lifting Canonical Algorithms from a Ring R to the Ring $R[x]$

R. SHTOKHAMER†

Department of Computer and Information Sciences, University of Delaware, Newark, DE 19716, U.S.A.

Given ample sets in a ring R we define related ample sets in the polynomial ring $R[x]$. For Noetherian rings we state natural and sufficient conditions for canonical form algorithms to be lifted to canonical form algorithms in $R[x]$. It is possible then to construct syzygies as well as to perform general division with respect to any finite set in $R[x]$. The construction is based on the notion of a Szekeres basis.

Introduction

In computer algebra the problem of canonical simplification and zero recognition are of sufficient importance to justify investigations of several different approaches. In this paper we address ourselves to the problem of lifting canonical forms from a ring R to the polynomial ring $R[x]$.

According to van der Waerden (1970), in 1903 König found that it is possible to decide in a finite number of steps if a polynomial over a field belongs to a given ideal (i.e. ideals are *detachable*), but only in the past ten years have effective algorithms been implemented. Most of the constructed algorithms are based on the original solution of Buchberger (1970) for polynomials (in several variables) over a field. In general terms Buchberger's algorithm is based on a completion procedure similar to the one used by Knuth & Bendix (1970) for first order terms and found to be in a general class of critical-pair/completion algorithms (Loos, 1981; Buchberger & Loos, 1982; Winkler, 1984).

Richman (1974) has suggested a different approach for deciding if an ideal in $R[x]$ is detachable. Independently, the author (Shtokhamer, 1975) has given an algorithm to determine canonical representations in $R[x]$ for simplification rings. The solution proposed in Shtokhamer (1975) was based on the construction of a Szekeres (1959) basis. Richman's lemma 5 and the related theorem (Richman, 1974) define such a basis. However, from an algorithmic point of view we must be able to construct a finite basis for the R-module N_j defined in the proof of Richman's lemma 5. It is not clear from his paper how such a construction can be performed. It was pointed out (Shtokhamer, 1975) that such a construction is possible once a certain finite basis for syzygies is constructable in R. The algorithm suggested, in addition to solving the detachability problem, constructs a basis for syzygies for any finite set of polynomials (see also Buchberger, 1985, problem 6.18) as well as enables one to perform a generalised division with respect to a given set of polynomials.

† On leave from A.D.A. and Technion, Haifa, Israel.

Ayoub (1983) based her results on Richman's theorem and constructed an algorithm for a detaching basis and for a canonical form in the polynomial ring $Z[x_1, \ldots, x_n]$ over the integers. Her approach in working with power products resembles the method used by Buchberger (1984) for reduction rings. Other related works are those of Hurd (1970) and Sims (1978). The algorithm we suggest is of a different nature.

This paper is an elaboration and refinement of the results in the previous report (Shtokhamer, 1976). We try to show how such a construction arises naturally when the lifting problem is addressed. In the lifting problem one starts with a canonical algorithm viewed as a black box. Denote by $\mathcal{I}_R$ the set of all ideals in R. The subscript R will be omitted when it is clear from the context. The black box is a general-canonical algorithm $G: \mathcal{I} \times R \to R$ that for a given pair (I, r) maps r to a unique representative of the residue class $r+I = \{r+s : s \in I\}$ to which r belongs. This algorithm trivially solves the problem of detachability of ideals in R, i.e. every ideal in R is detachable. For let $r \in R$, then $r \in I$ if and only if $G(I, r) = G(I, 0)$. One tries then to impose some additional conditions on the ring R to enable the canonical algorithms to be lifted to $R[x]$. To achieve a finite lifting process we require some properties on the bases of ideals. The first required properties are that the ring is commutative and Noetherian implying that every ideal has a finite basis.

The paper is organised as follows: In the first section we introduce the usual notions of ample sets in R and of the associated canonical maps. In the second section we lift the ample sets and maps previously introduced to the ring $R[x]$. The third section introduces the notion of a simplification ring. Section four discusses the notion of a Szekeres basis and its relevance to the problem of lifting canonical algorithms from R to $R[x]$. The outline of the algorithm for constructing a Szekeres basis is given in the appendix. The fifth section introduces the notion of a reduced basis. We have collected some examples in the last section.

1. Ample Sets and Canonical Maps in R

Two elements $r_1, r_2 \in R$ are said to be *equivalent* modulo I (or simply equivalent) denoted by $r_1 \equiv r_2(I)$ if and only if $r_1 - r_2 \in I$. Clearly $r+I = \{s : s \equiv r(I)\}$.

A subset A of R will be called an *ample set* or a *system of representatives* for the equivalence classes R/I if and only if its intersection with every equivalence class consists of exactly one element. A map $F: R \to A$, A an ample set for the ideal I, such that $F(r) \equiv r(I)$ will be called a *canonical map* associated with I and A.

A canonical map has the following properties:

(1) F is idempotent, i.e. $F^2(r) = F(r)$.
(2) $r_1 \equiv r_2(I) \leftrightarrow F(r_1) = F(r_2)$.

Any map having property (1) and (2) defines uniquely an ample set by: $A = F(R)$ (the image of R under F). Hence, we can interchange the roles between ample sets and their associated canonical maps. If $0 \in A$ then the associated canonical map is called *zero recognising*.

From now on we assume that all canonical maps are zero recognising and consequently all ample sets contain zero. Let $I(F) = \{a : F(a) = 0\}$ denote the ideal of F, and $A(F)$ be the ample set of F. Let $\mathcal{F}$ be a set of canonical maps $\{F\}$; let $\mathcal{I}(\mathcal{F}) = \{J : J = I(F), F \in \mathcal{F}\}$.

A set $\mathscr{F}$ of canonical maps will be called *compatible* if and only if:

$$F_1 \in \mathscr{F},\ F_2 \in \mathscr{F} \quad \text{and} \quad I(F_1) \supset I(F_2) \to A(F_2) \supset A(F_1).$$

Notice that compatible maps have the property $I(F_1) \supset I(F_2) \to F_1 F_2 = F_2 F_1 = F_1$ for all $r \in R$.

LEMMA. *Given a ring* R *there exists a set* $\mathscr{F}$ *of compatible canonical maps such that* $\mathscr{I}(\mathscr{F}) = \mathscr{I}_R$.

PROOF. Every set can be well ordered. Assume $>$ is a well ordering for R with the property that $a > 0$ if $a \neq 0$ (such an order always exists). Let I be an ideal in R. Choose from each equivalence class the smallest element. The resulting set forms an ample set and the associated map is zero recognising. Obviously the resulting set of canonical maps for all ideals in R is compatible.

We would like to rephrase now the definitions given above in terms of algorithms instead of maps. We assume throughout the paper that R is a commutative Noetherian ring, hence every ideal has a finite basis, and we can use such bases to represent (not uniquely) ideals. To say that we have canonical algorithms in R means that we have an effective *general-canonical* algorithm whose input is a pair of: (ideal specification, element in R) and whose output (in a finite number of steps) is a canonical representative of the equivalence class to which this element belongs. As in the introduction we denote this algorithm by G, $G: \mathscr{I} \times R \to R$. If I is some ideal (represented by a basis), then by our notation $G(I,*)$ is the canonical map associated with I, and $G(I, R)$ is its ample set. We will say that G is compatible if the set of canonical maps $G(I,*)$, (for all ideals I in R) is compatible.

2. Ample Sets and Canonical Maps in $R[x]$

Suppose we are given ample sets for ideals in R. We shall show that these sets introduce natural definitions of ample sets in $R[x]$. Let J be an ideal in $R[x]$. Denote by $I_m(J)$ the ideal spanned by all the leading coefficients of polynomials of degree m in J, and by $A_m(J)$ the corresponding ample set in R. Let k be the minimal m such that $I_m(J)$ is non-zero ($k \geqslant 0$). Let $p = ax^n + b$, $\deg(b) < n$ and $a \neq 0$.

We now define recursively the set $A(J)$.

$$p \in A(J) \leftrightarrow \begin{cases} p = 0 & \text{or} \\ n < k & \text{or} \\ n \geqslant k,\ a \in A_n(J) & \text{and} \quad b \in A(J). \end{cases}$$

LEMMA. *The set* $A(J)$ *is an ample set for the ideal* J.

PROOF. We need to show two things:

(a) for every p in $R[x]$ there exists q in $A(J)$ such that $p - q$ is in J, and
(b) if p and q are both in $A(J)$ then $p \equiv q(J) \to p = q$.

(a) The proof is by induction on the degree of $p = ax^n + b$, $a \neq 0$.

For $n=0$, consider 2 cases:

(1) $k=0$, $b=0$, and $a \in R$, take q in $A_0(J)$ such that $a-q=p-q$ in $I_0(J) \subset J$.
(2) $k>0$, p is in the ample set by definition, and $p-p \in J$.

For $n>0$,

Assume the lemma is true for all $m<n$. There exists a' from $A_n(J)$ *s.t.* $a-a'$ is in $I_n(J)$, let $(a-a')=c$ and the corresponding polynomial in J be $f=c\,.\,x^n+d$. We shall refer to f as an *adjusting polynomial* of p.

$$p = a\,.\,x^n+b = (a'+c)\,.\,x^n+b = a'\,.\,x^n+(b-d)+(c\,.\,x^n+d) = a'\,.\,x^n+b'+f,$$
$$a' \in A_n(J), \quad \deg(b') < n, \quad \text{and} \quad f \in J.$$

By the induction hypothesis there exists d' in $A(J)$ such that $f' = (b'-d')$ is in J. Summing up: $p=(a'\,.\,x^n+d')+(f+f') = q+h$, where q is in $A(J)$ and h is in J.

(b) This is easy to prove following the same arguments.

Having defined the ample sets in $R[x]$, we can now define the general-canonical map in $R[x]$. Let G be the corresponding map in R.

Let $p=ax^n+b$ be a polynomial in $R[x]$ and $f=cx^n+d$ be the adjusting polynomial of p.

$$G_x(J, p) = \begin{cases} G(I_n(J), a)x^n + G_x(J, b-d); \; n \geqslant k \\ p; \; n<k. \end{cases}$$

We must show that G_x is well defined.

A problem arises because the adjusting polynomial is not unique. It is obvious that G_x is well defined for polynomials of degree k and projects them into $A(J)$. Suppose by induction that G_x is well defined for all polynomials up to degree $n-1 \geqslant 0$. Consider now the polynomial $p=ax^n+b$. Let $f_1=a_1x^n+d_1$ and $f_2=a_2x^n+d_2$ be two different adjusting polynomials. By definition there exist c_1 and c_2 both in $I_n(J)$ such that $a-a_1=c_1$ and $a-a_2=c_2$, hence $(a_1-a_2)=-(c_1-c_2)$ is in $I_n(J)$ as well. Therefore $a_1 \equiv a_2(I_n(J))$ but a_1 and a_2 are in the same ample set therefore $a_1=a_2$. The polynomials $b-d_1$ and $b-d_2$ are of degree $n-1$ at most and $(b-d_1)-(b-d_2)=(d_2-d_1)=f_2-f_1 \in J$. Hence by induction $G_x(b-d_2)=G_x(b-d_1)$ and therefore $G_x(p)$ is well defined.

Since we have assumed that R is Noetherian then so is $R[x]$. Hence the above map defines for every polynomial a reduction scheme to its canonical form in at most n steps. Because R is Noetherian, we actually need only a finite set of representatives for $I_n(J)$'s (for fixed J), i.e. there exists n such that if $m>n$ we have $I_m(J)=I_n(J)$. Notice that if G is compatible so is G_x. However, to convert the map to an algorithm we must construct the adjusting polynomial f. This is the main reason we introduce the notion of a *simplification basis* in the next section.

3. Simplification Basis and Simplification Ring

Let $(a_1, a_2, \ldots, a_n)$ be a finite basis for an ideal I in R. We shall use the vector notation $\mathbf{a}$ to denote the ordered set (n-tuple) given above and shall say that $\mathbf{a}$ spans (or generates) the ideal I if the components of the n-tuple generate the ideal I. If y is in I then there exists a vector $\mathbf{b}$ such that $y=\mathbf{ba}$, $(y=\Sigma b_i \times a_i)$. We call the vector $\mathbf{b}$ a (non-unique) *B-quotient* of y (with respect to the n-tuple $\mathbf{a}$).

If $\mathbf{d} = (d_1, \ldots, d_t)$ is another vector spanning the ideal I then there exists a set of vectors $\boldsymbol{\alpha}_1, \ldots, \boldsymbol{\alpha}_n$ (each of dimension t) such that: $a_i = \mathbf{d}\boldsymbol{\alpha}_i$ for $1 \leqslant i \leqslant n$. We denote by α the $t \times n$ matrix whose columns are the vectors $\boldsymbol{\alpha}_1, \ldots, \boldsymbol{\alpha}_n$, $\mathbf{a} = \mathbf{d}\boldsymbol{\alpha}$, and shall say that the matrix α transforms the basis $\mathbf{d}$ to the basis $\mathbf{a}$.

The set of *syzygies* of $\mathbf{a}$, i.e. the set $\{\mathbf{z} : \mathbf{az} = 0\}$, form an orthogonal R-module.

R is Noetherian therefore there exists a finite basis for this R-module. So there exists a finite set of vectors $\mathbf{v}_1, \ldots, \mathbf{v}_m$, and $\mathbf{av}_j = 0$, for $j = 1, \ldots, m$ such that if some vector $\mathbf{w}$ is annihilating (orthogonal to) $\mathbf{a}$ (i.e. $\mathbf{aw} = 0$) then there exist elements $b_1, \ldots, b_m$ in R *s.t.* $\mathbf{w} = \Sigma b_j \mathbf{v}_j$. We shall denote the set of vectors $\mathbf{v}_j, \ldots, \mathbf{v}_m$ by the $n \times m$ matrix V. ($\mathbf{a}V = \mathbf{0}$, where $\mathbf{0}$ is the m dimensional zero vector). The existence of the above vectors and matrices is assured because R is Noetherian, we must, however, ensure constructibility!

A basis $(a_1, \ldots, a_l) = \mathbf{a}$ for an ideal I in R will be called a *simplification basis* if and only if B-quotients (for elements in I) and a basis for syzygies of $\mathbf{a}$ are computable.

Recall that R is assumed to have a canonical algorithm for every ideal I in R. Hence, it is always possible to decide if a given element d belongs to a specific ideal I, however, the canonical algorithm does not assure constructibility of a B-quotient. For a simplification basis we can define a general division. Let d be an element in R and let $d' = G(I, d)$ be its canonical form, then $d - d' = c \in I$. We can compute a B-quotient $\mathbf{b}$, $c = \mathbf{ab}$, and get $d = d' + \mathbf{ab}$.

In general it is easy to compute B-quotients. For example in any reduction ring (Buchberger, 1984), we can follow the reduction path of any given element to get the result. Constructing a basis for the syzygies is more problematic. For reduction rings one shows that any S-polynomial gives an orthogonal vector—follow the two independent paths of reductions leading to the same result and form the difference (Zacharias, 1978). It is easy to construct such a basis in a unique factorisation ring for which an extended GCD algorithm is available. Independent of these considerations the important fact is that those properties can be lifted to $R[x]$.

A Noetherian ring R will be called a *simplification ring* if there exists a general-canonical algorithm for ideals of R and for any finite set of generators it is possible to compute a transformation matrix to a simplification basis (for the same ideal).

Observations.

(1) Any field is a simplification ring. The proof is trivial because the only ideals are $\{0\}$ and the field itself generated by $\langle 1 \rangle$.

(2) Any Euclidean domain is a simplification ring. To prove it recall that every ideal in an Euclidean domain is a principal ideal and the generating element (and the transformation matrix) can be computed by computing the *gcd* of the basis elements using the extended Euclidean algorithm.

Lemma 1. *A ring R (with a general-canonical algorithm) is a simplification ring if and only if every finite set of generators is a simplification basis.*

Proof. The if part is trivial; the non-trivial part of the only-if part is to construct a basis for the syzygies of a given vector $\mathbf{g}$. By assumption we can compute a transformation matrix α to a basis $\mathbf{a}$ such that $\mathbf{a} = \mathbf{g}\alpha$ is a simplification basis. We can now compute a basis for the syzygies of $\mathbf{a}$, and a transformation matrix G such that $\mathbf{g} = \mathbf{a}G$, i.e. $\mathbf{g} = \mathbf{g}\alpha G$. Now use the construction suggested by Zacharias (1978) and by Buchberger (1985) (problem 6.18) to construct a basis for the syzygies of **g**. Let H be a matrix columns of which form a basis for the syzygies of **a**. Construct the matrix D by augmenting the columns of $I-\alpha G$ to the columns of αH, i.e.

$$D = |I-\alpha G|\alpha H|.$$

It follows that the columns of the matrix D form a basis for the syzygies of **g**.

We can now state the main result for simplification rings (Shtokhamer, 1975).

PROPOSITION. *Let R be a simplification ring then the ring $R[x]$ is also a simplification ring.*

The proof of this proposition relies on the construction of a Szekeres basis, which is defined in the next section. It turns out that a Szekeres basis is a simplification basis, and that there is an algorithm constructing such a basis for any finite set of generators. In the appendix the outline of an algorithm constructing a Szekeres basis is repeated from Shtokhamer (1976) and a proof is given that the algorithm constructs a basis for the syzygies.

COROLLARY. *$F[x_1, \ldots, x_n]$ where F is a field or an Euclidean domain is a simplification ring.*

To prove the corollary notice that $F[x_1, \ldots, x_n]$ is isomorphic to $F[x_1][x_2, \ldots, x_n]$.

4. Szekeres Basis

Let S be a set of polynomials generating an ideal J in $R[x]$. We partition this set into vectors $\mathbf{M}_k, \ldots, \mathbf{M}_n$ such that the polynomials in each vector are of the same degree, i.e. if $\mathbf{M}_j = (q_{j1}, \ldots, q_{jm})$, then all q_{jk}'s, $k = 1, \ldots, m$, are of the same degree j.

We shall also use the vector notation $\mathbf{M}_j = x^j . \mathbf{LC}_j + \mathbf{R}_j$, where $\mathbf{LC}_j$ denotes the vector of leading coefficients and $\mathbf{R}_j$ denotes that of the reducta.

Let M be the R-module spanned by the set S. According to lemma 5 (in Richman, 1974) M consists of all polynomials of degree $\leqslant n$ in J if and only if for $f \in M$ and $\deg(f) < n$, $xf \in M$. It then follows that $\mathbf{LC}_n$ spans the ideal of all the leading coefficients of polynomials in J of degree n. We can now define R-modules for polynomials of maximal degree $j < n$. Let k be the minimal j such that the corresponding R-module is non-trivial (i.e. not equal to the $\{0\}$ ideal). Let also n be the minimal j such that $\langle \mathbf{M}_k, \ldots, \mathbf{M}_n \rangle$ spans the ideal J in $R[x]$. We have thus arrived at the following useful definition of a Szekeres sequence and of a Szekeres basis.

A sequence $\mathbf{M}_k, \ldots, \mathbf{M}_n$ $(0 \leqslant k \leqslant n)$ of vectors will be called a *Szekeres sequence* for an ideal J in $R[x]$ if and only if:

(1) $\mathbf{LC}_i$ spans the ideal of all the leading coefficients of polynomials of degree i in J, for $k \leqslant i \leqslant n$.

(2) k is the minimal degree of non-zero polynomials in J.

A Szekeres sequence is a *Szekeres basis* if and only if the ideal spanned by $\mathbf{LC}_{n-1}$ is properly included in the ideal spanned by $\mathbf{LC}_n$.

Properties of a Szekeres basis:

(1) If p is one of the polynomials in the set (vector) $\mathbf{M}_i$, $p = a . x^i + b$, then there exists a

constant vector (i.e. with entries from R) $\mathbf{C}$ *s.t.* $a = \mathbf{CLC}_{i+1}$, and there exist constant vectors $\mathbf{C}_k, \ldots, \mathbf{C}_i$ such that

$$\mathbf{CM}_{i+1} - xp = \mathbf{C}_k\mathbf{M}_k + \ldots + \mathbf{C}_i\mathbf{M}_i.$$

(2) If $\mathbf{V}$ is a constant vector *s.t.* $\mathbf{LC}_i\mathbf{V} = 0$ then there exists a sequence of constant vectors $\mathbf{C}_k, \ldots, \mathbf{C}_{i-1}$ such that

$$\mathbf{VM}_i = \mathbf{C}_k\mathbf{M}_k + \ldots + \mathbf{C}_{i-1}\mathbf{M}_{i-1}.$$

The above properties follow directly from the definition.

If a basis for an ideal J in $R[x]$ satisfies the properties (1) and (2) then for any non-zero p from J, $m = \deg(p)$, we have:

(a) $k \leqslant m \leqslant n \rightarrow p = \mathbf{C}_m\mathbf{M}_m + \ldots + \mathbf{C}_k\mathbf{M}_k$ (for some constant vectors $\mathbf{C}_i$).
(b) $m > n \rightarrow p = \mathbf{q}(x)\mathbf{M}_n + \mathbf{C}_{n-1}\mathbf{M}_{n-1} + \ldots + \mathbf{C}_k\mathbf{M}_k$, for some constant $\mathbf{C}_i$'s and $\deg(\mathbf{q}(x)) = m - n$.
(c) $I_i(J) = I_n(J)$ for $i > n$, i.e. the chain of ideals of the leading coefficients stops at I_n.

One notices that properties (1) and (2) can be checked by relations in the ring R only. Using (a) and (b) one can easily prove the following variant of lemma 5.

LEMMA 2. *If a sequence of polynomial vectors* $\mathbf{M}_k, \ldots, \mathbf{M}_n$ *satisfy properties* (1) *and* (2) *then the sequence is a Szekeres sequence for the ideal spanned by those polynomials.*

Assume now that R is a simplification ring, hence there exists a general-canonical algorithm $G(I,*)$ for any ideal I in R, and (see the appendix) a Szekeres basis can be constructed in which the vectors $\mathbf{LC}_k, \ldots, \mathbf{LC}_n$ are simplification bases for ideals in R.

Let $p = ax^m + b$, $a \neq 0$, $\deg(b) < m$, be a polynomial in $R[x]$.

With the help of the Szekeres basis we construct the canonical representative of the polynomial p with respect to the ample set defined in section 2.

(1) If $m > n$ then $I_m(J) = I_n(J)$ and therefore $a' = G(I_n(J), a)$ is the canonical form of a in R and $a - a'$ is in the ideal $I_n(J)$.
Compute a vector $\boldsymbol{\alpha}_n$ such that $a' = a - \mathbf{LC}_n\boldsymbol{\alpha}_n$.
We also compute an adjusting polynomial f of p by letting $f = \mathbf{M}_n\boldsymbol{\alpha}_n$.
(2) If $k \leqslant m \leqslant n$, the construction is as before but use $f = \mathbf{M}_n\boldsymbol{\alpha}_m$ as the adjusting polynomial.

By computing the adjusting polynomials the canonical form of the polynomial p follows in a finite number of steps. Notice that because of compatibility, the reduction of a' with respect to $\mathbf{LC}_{n-1}$ will not change its canonical form. A more formalistic definition of a general-canonical algorithm for so-called reduced bases will be given in the next section.

The above procedure enables us to compute B-quotients with respect to a Szekeres basis. Let the polynomial p be as before and assume that p is in the ideal J in $R[x]$. The B-quotient of p (denote it by $B(p)$) with respect to the Szekeres basis can be computed (using obvious notation) as follows:

$$B(p) = \begin{cases} (0, \ldots, 0);\ p = 0 \\ (\boldsymbol{\alpha}x^{m-n}, 0, \ldots, 0) + B(bx^{m-n} - \mathbf{R}_n\boldsymbol{\alpha}_n x^{m-n});\ m \geqslant n \\ (0, \ldots, \boldsymbol{\alpha}_m 0, \ldots, 0) + B(b - \mathbf{R}_m\boldsymbol{\alpha}_m);\ k \leqslant m < n. \end{cases}$$

Use the above, the algorithm and the proof outlined in the appendix to conclude that the Szekeres basis is a simplification basis. In addition the algorithm in the appendix constructs for any basis for an ideal a Szekeres basis spanning the same ideal, and a transformation matrix mentioned in definition 8, hence for a simplification ring R, $R[x]$ is also a simplification ring.

5. Reduced Bases

In the previous chapter we have shown that the Szekeres basis enables us to lift canonical algorithms from R to $R[x]$. The constructed Szekeres basis is quite uneconomical. It might happen, for example, that $\mathbf{M}_j = xC\mathbf{M}_{j-1}$, hence the vector $\mathbf{M}_j$ is superfluous. Moreover, an ideal in $R[x]$ might have different Szekeres bases. To have an economical basis and to achieve uniqueness (i.e. an algorithm constructing a unique basis for an ideal J given any finite set of generators of J) we shall introduce a *reduced* basis.

Let $\mathbf{M}_n, \ldots, \mathbf{M}_k$ be a Szekeres basis for an ideal in $R[x]$. First we assume that we can compute reduced unique bases for ideals in R and if I is an ideal in R we denote such a basis by $S(I)$. If for some I, $a \in S(I)$ then we require that a is in the ample set (i.e. reduced) corresponding to the ideal spanned by $\langle S(I) - \{a\} \rangle$. The algorithm for constructing a Szekeres basis chooses some bases for ideal is R, certainly we can choose them to be the reduced bases as defined above. We can assume therefore that the vectors $\mathbf{LC}_j$'s are reduced bases in R. Notice that the sequence $\mathbf{M}_j, \ldots, \mathbf{M}_k$ is also a Szekeres sequence.

Let J_j denote the ideal spanned by $\langle \mathbf{M}_k, \ldots, \mathbf{M}_j \rangle$. We reduce the vectors $\mathbf{M}_j$ as follows:

$$\mathbf{M}'_j = \begin{cases} G_x(J_{j-1}, \mathbf{M}_j); \; k < j \leqslant n \\ \mathbf{M}_k; \; j = k. \end{cases}$$

First it is clear that the new reduced sequence is basis for the ideal J spanned by the original sequence. The number of non-zero entries in $\mathbf{M}'_j$ (i.e. the vector dimension) might be less than in $\mathbf{M}_j$, and some of the vectors might be zero. The vectors $\mathbf{M}'_n$ and $\mathbf{M}'_k$ are, however, non-zero vectors.

Let $\mathbf{LC}'_j$ and $\mathbf{R}'_j$ denote the vectors corresponding to the non-zero leading coefficients and to the reducta respectively. The vector $\mathbf{LC}'_n$ does not in general form a basis for the ideal spanned by the leading coefficients of polynomials of degree n, however, the union $LC'_n \cup \ldots \cup LC'_k$ is such a basis.

Let $S(m) = \langle M_{j_1}, \ldots, M_{j_m} \rangle$ be a reduced basis where $j_m = n$ and $j_1 = k$. The reduced basis for J has the following properties:

(1) $S(m)$ spans the ideal J.
(2) k is the minimal degree of polynomials in J.
(3) If $q \in M_{ji}$ then q is in the ample set of the ideal spanned by $S(i-1)$.
(4) If p is in J and the degree t of p is such that $j_{i+1} < t \leqslant j_i$ then p is in the ideal J_{j_i}.

We claimed that a basis constructed by the above procedure is uniquely defined by the ideal. It means that starting from two different Szekeres bases for the same ideal and performing the above construction we end up with the same result. Recall that the vectors $\mathbf{LC}_j$ were chosen to be unique (reduced) bases in R, hence the polynomials in the Szekeres basis might be different only in their reducta. The reduction process produces the same non-zero leading coefficients $\mathbf{LC}_j$'s, hence the two reduced (in $R[x]$) Szekeres basis can differ in their reducta only. Let $S(i)$ and $S_1(i)$ be two corresponding reduced bases and let $p \in S(i)$ and $p_1 \in S1(i)$ be two corresponding polynomials of degree j_i each and having the same leading coefficient a and reducta r and r_1 respectively. Then $r_1 - r$ is a polynomial in J of lower degree than j_i, hence it is in the ideal spanned by $S(i-1)$ (or by $S_1(i-1)$ which spans the same ideal!), therefore r_1 and r have the same canonical form modulo this ideal, but both of them are in the ample set corresponding to those ideal (by construction), hence they are equal. Therefore the two reduced basis must be equal. Let $S(i) = \mathbf{LC}_k \cup \ldots \cup \mathbf{LC}_{j_i}$ be the unique reduced basis for the ideal in R of the leading coefficients of polynomials of degree j_i in $R[x]$. Using the same arguments as above one can show that a basis with minimal m and having the sets $S(i)$ as above is uniquely defined by properties (1)–(4).

In the previous section we defined a reduction process of polynomials to their canonical form using a Szekeres basis. As $R[x]$ is a simplification ring the reduced basis is also a simplification basis and hence an algorithm for reduction with respect to this basis can be constructed. This algorithm will in general be not efficient because it will proceed through the full Szekeres basis.

We construct now a more efficient canonical form algorithm. Let $I \supseteq J$ be two ideals in R and S be the unique simplification basis for the ideal I. Let $\mathbf{S}_J$ be the reduced set with respect to the ideal J, i.e. $\mathbf{S}_J = G(J, \mathbf{S})$.

We assume that for every $d \in R$ we can compute a vector $\mathbf{D}$ such that

$$G(I, d) = G(J, d) - \mathbf{S}_J\mathbf{D}.$$

Let the reduced Szekeres basis for an ideal Id in $R[x]$ be $\mathbf{M}'_k, \ldots, \mathbf{M}'_n$. We consider only those vectors $\mathbf{M}'_j$ for which the vectors $\mathbf{LC}'_j$ are not zero. Let J_j denote the ideal spanned by $\langle \mathbf{M}'_k, \ldots, \mathbf{M}'_j \rangle$, and let I_j denote the ideal in R spanned by $\langle \mathbf{LC}'_k, \ldots, \mathbf{LC}'_j \rangle$. Let $p = ax^m + b$ be a polynomial in $R[x]$. We define first the simplification associated with a non-zero vector $\mathbf{M}'_j$.

$$G'_x(\mathbf{M}'_j, p) = \begin{cases} p : j > m \\ G(I_j, a) \,.\, x^m + G'_x(\mathbf{M}'_j, b - b') : j \leq m. \end{cases}$$

In the above we used $G(I_j, a) = G(I_i, a) - \mathbf{LC}'_j\mathbf{D}$ and $b' = x^{(m-j)}\mathbf{R}'_j\mathbf{D}$, where $\mathbf{M}'_i$ is the first non-vanishing vector for $i < j$. In case of $j = k$, $G(I_k, a)$ is the canonical form of a (modulo I_k).

We define now the canonical algorithms $F(J_j, p)$ associated with the ideals J_j in $R[x]$.

$$F(J_j, p) = \begin{cases} G'_x(\mathbf{M}_k, p) : j = k \\ G'_x(\mathbf{M}'_j, F(J_i, p)) : j > k. \end{cases}$$

Here, again, i is the first index for which the vector $\mathbf{M}'_i$ is not zero. The final algorithm for the ideal Id is obtained by:

$$G_x(Id, p) = F(J_n, p).$$

We notice that the canonical simplification can be obtained by systematic simplifications starting at $\mathbf{M}'_k$ and progressing up to $\mathbf{M}'_n$ (in that order). In other words the sequence of simplifications is dictated by the reduced basis. We do not have to guess or to try which reduction to apply, the sequence of reductions is predetermined. We have been able to construct the reduced basis, first constructing the full Szekeres basis, and then applying some reduction algorithm to that basis. We lack at this time a direct algorithm for constructing the reduced basis, which would make the construction more efficient than the one suggested here.

6. Examples

We will consider three sets of generators (with the order $z > y > x$):

(1) $\langle 3x^2y-2, 4xy^2+1\rangle$
(2) $\langle 3x^2y+2xy+y+9x^2+5x-3, 2x^3y-xy-y+6x^3-2x^2-3x+3, x^3y+3x^3+x^2y+2x^2\rangle$
(3) $\langle -z+xy^2+4x^2+1/4, y^2z+2x+1/2, x^2z-y^2-x/2\rangle$

Examples (2) and (3) are taken from Buchberger (1985).

Assume first that the basic domain is the field of rationals. Then the Szekeres basis for (1) becomes $\langle y+8/3x, x^3+16/9\rangle$. Notice that this basis is already reduced. The space of syzygies is clearly one dimensional and spanned by the two-tuple: $(x^3+16/9, -y-3/8x)$.

For the set (2) the Szekeres basis is $\langle y-14/3x^3+38/3x^2+61/2x-3, x^3-5/2x^2-5/2x\rangle$. This basis is not reduced!

Here, again, the space of syzygies is one dimensional.

For the set (3) we get a basis of the form $\langle z+A, y^2+B, C\rangle$, where C is a seventh degree polynomial in x, while A and B are sixth degree polynomials in x.

The syzygies are spanned by three linear independent (over $Q[x, y]$) three-tuples.

In all of the above examples notice how easy it is to extract the zeros of the corresponding ideals from the (reduced) corresponding basis. (Detailed discussion on this subject can be found in Buchberger (1985).

Now assume that the basic domain is the ring of integers. As no divisions are allowed the Szekeres basis will change. We will consider the example (1) only.

The full Szekeres basis becomes:

$$\langle x^2y^2+2y+x, 4xy^2+1, 8y^2+3xy, x^2-3x^3-6, 8y+3x, 9x^3+16\rangle.$$

The coefficients of y^2, i.e. $\langle x^2, 4x, 8\rangle$ and of y, i.e. $\langle x^2, 8\rangle$ are themselves Szekeres bases. Check that the requirements of Szekeres basis are really satisfied.

The syzygies are spanned now by six vectors:

$$(4, -x, 0, 0, -1, 0), (0, 2, -x, 3, 0, 1), (0, 0, 0, 8, -x^2, 3),$$
$$(0, 0, 0, 9x, 2, 3x-y), (0, 0, 1, 0, -y, 0), (1, 0, 0, -y-3x, -1, -x).$$

The reduced basis becomes:

$$\langle 4xy^2+1, x^2y+6x^3+10, 8y+3x, 9x^3+16\rangle.$$

Notice that it is different from the full Szekeres basis.

The author wishes to thank Professor B. Caviness for his kind hospitality, support and appreciated help in preparing this manuscript.

References

Ayoub, C. W. (1983). On constructing bases for ideals in polynomial rings over the integers. *J. Numbers Th.* **17,** 204–225.

Buchberger, B. (1970). Ein Algorithmus zum Auffinden der Basiselemente des Restklassenringes nach einem nulldimensionalen Polynomideal. Dissertation, Univ. Innsbruck, 1965, see also *Aequationes Mathematicae* **4/3,** 374–383.

Buchberger, B., Loos, R. (1982). Algebraic simplification. In: (Buchberger, B., Collins, G. E., Loos, R, eds) *Computer Algebra—Symbolic and Algebraic Computations,* Computing Suppl. 4, pp. 11–43. New York: Springer-Verlag.

Buchberger, B. (1984). A critical-pair/completion algorithm for finitely generated ideals in rings, logics and machines: decision problems and complexity (Börger, E., Hasenjaeger, G., Rödding, D., eds), *Springer Lect. Notes Comp. Sci.* **171**.

Buchberger, B. (1985). Gröbner bases: an algorithmic method in polynomial ideal theory. In: (Bose, N. K., ed.) *Recent Trends in Multidimensional Systems Theory*, Chap. 6. New York: D. Reidel.

Hurd, G. B. (1970). Concerning ideals in $Z[x]$ and $Z_{pn}[x]$. PhD dissertation, Pennsylvania State University, Park, PA.

Knuth, D. E., Bendix, P. B. (1970). Simple word problems in universal algebra. In: (Leech, J., ed.) *Proc. Conf. on Computational Problems in Abstract Algebra, Oxford 1967*. Oxford: Pergamon Press.

König, J. (1903). Einleitung in die allgemeine Theorie der algebraischen Grössen (Leipzig, B. G. Teubner).

Lauer, M. (1976). Kanonische Repräsentanten für die Restklassen nach einem Polynomideal. Diplomarbeit, Univ. Kaiserslautern.

Loos, R. (1981). Term reduction systems and algebraic algorithms. *Proc. 5th Workshop Artif. Intell., Bad Honef 1981*, Informatik, Fachberichte 47, pp. 214–234. Berlin: Springer-Verlag.

Richman, F. (1974). Constructive aspects of Noetherian rings. *Proc. Am. Math. Soc.* **44,** 436–441.

Sims, C. (1978). The role of algorithms in teaching of algebra. In (Newman, M. F., ed.) *Topics in Algebra, Springer Lect. Notes Math.*, No. 697, pp. 95–107. New York: Springer-Verlag.

Szekeres, G. A. (1952). A canonical basis for the ideals of polynomial domain. *Am. Math. Monthly* **59,** 319–386; and Redei, L. (1959). *Algebra* 1, Leipzig.

Shtokhamer, R. (1975). Simple ideal theory. Some applications to algebraic simplification. Univ. of Utah, Salt Lake City (Computer algebra group) *Tech. Rep. UCP-36* (1975); and A canonical form of polynomials in the presence of side relations. Technion, Haifa, Israel, *Tech. Rep. PH-76-25* (1976).

van der Waerden, B. L. (1970). *Algebra II*, p. 159. Frederick Ungar.

Winkler, F. (1984). The Church–Rosser property in computer algebra and special theorem proving: An investigation of critical-pair/completion algorithms. Dissertation Johannes Kepler Universität, Linz. VWGO, Wien.

Zacharias, G. (1978). Generalized Gröbner bases in commutative polynomial rings. Bachelor thesis, M.I.T. Dept. of Computer Science.

Appendix

Outline of an algorithm constructing a Szekeres basis, a transformation matrix to this basis and a basis for its syzygies.

We assume that the underlying ring R is a simplification ring and therefore the following algorithms are available:

$(r1, \mathbf{D}) \leftarrow$ REDUCE0($\mathbf{S}$, r)

input: $\mathbf{S}$, a finite basis for an ideal in R.
r, an element in R.

output: $r1 =$ a canonical form of r with respect to the ideal generated by $\mathbf{S}$.
$\mathbf{D} =$ a B-quotient of r *w.r.t.* $\mathbf{S}$, $r = r1 + \mathbf{SD}$.

$\mathbf{V} \leftarrow$ ORTOG0($\mathbf{S}$)

input: $\mathbf{S}$, a finite basis for an ideal in R.

output: $\mathbf{V}$, a matrix whose columns form a basis for the syzygies of $\mathbf{S}$.

For notational convenience we will use a vector version of the algorithm REDUCE0, i.e. an algorithm whose input is a vector $\mathbf{r}$ instead of a scalar r, and whose output is therefore a pair of (vector, matrix), and we shall denote this algorithm by VREDUCE0.

$(\mathbf{r1}, \mathbf{D}) \leftarrow$ VREDUCE0($\mathbf{S}$, $\mathbf{r}$), and $\mathbf{r} = \mathbf{r1} + \mathbf{SD}$

We shall construct a Szekeres sequence.

It suffices to notice that after the Szekeres sequence (together with the transformation matrix and the basis for syzygies) are obtained, the final requested result (for a Szekeres basis) can be obtained by simple elimination of some of the entries in the output.

Let $\mathbf{S}_n$ be a vector of polynomials whose highest degree is the parameter n. As in section 4 we partition this set into the vectors: $\mathbf{M}_k, \ldots, \mathbf{M}_n$, in which $\mathbf{M}_j \neq \mathbf{0}$, for $k \leqslant j \leqslant n$. If necessary, this is achieved by multiplying some of the $\mathbf{M}_i$'s $\neq \mathbf{0}$ by x. Let $\mathbf{M}_j = x^j \mathbf{LC}_j + \mathbf{R}_j$. I will denote a unit matrix whose dimension can be deduced from the context it appears. By recursion (and section 4) we have also available the algorithm VREDUCE whose first argument is a finite basis of an ideal in $R[x]$ and whose second argument is a vector of elements in $R[x]$.

$(\mathbf{S}_n, \mathbf{O}, \mathbf{T}) \leftarrow$ SZEKERES($\mathbf{S1}_n$)

input: $\mathbf{S1}_n$, vector of polynomials (of maximal degree n) generating an ideal Id in $R[x]$.

output: $\mathbf{S}_n$, a Szekeres basis generating the ideal Id,
$\mathbf{O}$, a matrix whose columns form a basis for the syzygies of $\mathbf{S}_n$,
$\mathbf{T}$, a transformation matrix, $\mathbf{S}_n = \mathbf{S1}_n \mathbf{T}$.

Begin:

(1) if $n = 0$ then return ($\mathbf{S1}_n$, ORTOG0($\mathbf{S1}_n$), $\mathbf{1}$)
$\mathbf{T} \leftarrow \mathbf{1}$

(2) $(\mathbf{S}_{n-1}, \mathbf{O1}, \mathbf{A1}) \leftarrow$ SZEKERES($\mathbf{S}_{n-1}$)

(3) $\mathbf{T} \leftarrow \mathbf{T}\,\mathbf{T1}$

(4) $(\mathbf{C1}, \mathbf{D1}) \leftarrow$ VREDUCE0($\mathbf{LC}_n$, $\mathbf{LC}_{n-1}$)
$\mathbf{B} \leftarrow x\mathbf{M}_{n-1} - \mathbf{M}_n \mathbf{D1}$

(5) if $\mathbf{C1} \neq \mathbf{0}$ then ($\mathbf{M}_n \leftarrow \mathbf{B} \oplus \mathbf{M}_n$; $\mathbf{T} \leftarrow \mathbf{T2}$; goto (4))

(6) $(\mathbf{B}, \mathbf{D2}) \leftarrow$ VREDUCE($\mathbf{S}_{n-1}$, $\mathbf{B}$)

(7) if $\mathbf{B} \neq \mathbf{0}$ then ($\mathbf{S}_{n-1} \leftarrow \mathbf{B} \oplus \mathbf{S}_{n-1}$; $\mathbf{T} \leftarrow \mathbf{T}\,\mathbf{T3}$; goto (2))

(8) $\mathbf{Q} \leftarrow$ ORTOG0($\mathbf{LC}_n$)
$(\mathbf{C2}, \mathbf{D3}) \leftarrow$ VREDUCE($\mathbf{S}_{n-1}$, $\mathbf{M}_n \mathbf{Q}$)

(9) if $\mathbf{C2} \neq \mathbf{0}$ then ($\mathbf{S}_{n-1} \leftarrow \mathbf{C2} \oplus \mathbf{S}_{n-1}$; $\mathbf{T} \leftarrow \mathbf{T4}$; goto (2))

(10) /* compose the output: $\mathbf{S}_n$, $\mathbf{O}$, $\mathbf{T}$ */
return ($\mathbf{S}_n$, $\mathbf{O}$, $\mathbf{T}$)

end;

Remarks:

(1) The highest degree is checked and if $n = 0$ the input is from R, hence the output is trivial.

(2) The algorithm is applied recursively on a vector whose entries are polynomials of lower degree $n-1$.

(3) The new transformation matrix is obtained by matrix multiplication.

$\mathbf{T1}$ is a matrix such that: $(\mathbf{M}_n, \mathbf{S}_{n-1})\,\mathbf{T1} = (\mathbf{M}_n, \mathbf{S}_{n-1}\mathbf{A1})$

therefore with obvious notations and dimensions $\mathbf{T1} = \begin{bmatrix} \mathbf{I} & 0 \\ 0 & \mathbf{A1} \end{bmatrix}$

In the above $\mathbf{S}_{n-1}$ is the argument to the procedure SZEKERES in step (2).

(4) We begin to check if the properties (1) and (2) of a Szekeres sequence are satisfied.

(5) In this case the ideal generated by $\mathbf{LC}_{n-1}$ is not included by the ideal generated by $\mathbf{LC}_n$, so we have to update the vector $\mathbf{M}_n$. The $\oplus$ sign indicates that only non-zero entities of $\mathbf{B}$ should be considered, when adding $\mathbf{B}$ to the entries of the vector $\mathbf{S}_{n-1}$ and forming the new vector $\mathbf{S}_{n-1}$.

The matrix **T2** (in the case all entries of **B** are non-zero) is a matrix such that:

$(\mathbf{B}, \mathbf{M}_n, \mathbf{M}_{n-1}, \mathbf{S}_{n-2}) = (x.\,\mathbf{M}_{n-1} - \mathbf{M}_n\,\mathbf{D1}, \mathbf{M}_{n-1}, \mathbf{S}_{n-2}) = (\mathbf{M}_n, \mathbf{M}_{n-1}, \mathbf{S}_{n-2})\,\mathbf{T2}.$

(6) Now we check if the entries of the vector **B** are in the ideal generated by $\mathbf{S}_{n-1}$.

(7) If the new vector **B** is non-zero we must update the polynomials in $\mathbf{S}_{n-1}$, and repeat the construction.
The matrix **T3** is such that:

$(\mathbf{M}_n, \mathbf{B}, \mathbf{S}_{n-1}) = (\mathbf{M}_n, -\mathbf{M}_n\,\mathbf{D1} + \mathbf{x}.\,\mathbf{M}_{n-1} - \mathbf{S}_{n-1}\,\mathbf{D2}, \mathbf{S}_{n-1}) = (\mathbf{M}_n, \mathbf{M}_{n-1})\,\mathbf{T3}.$

(8) Construct a basis for the syzygies of $\mathbf{LC}_n$.

(9) And check if property (2) of a Szekeres sequence is satisfied.
If property (2) is not satisfied then update the vector $\mathbf{S}_{n-1}$ and repeat the construction. The matrix **T4** is such that: (in case the entries of **C2** are non-zero, and permutation of entries is not needed)

$(\mathbf{M}_n, \mathbf{M}_n\,\mathbf{Q} - \mathbf{S}_{n-1}\,\mathbf{D3}, \mathbf{S}_{n-1}) = (\mathbf{M}_n, \mathbf{S}_{n-1})\,\mathbf{T4}.$

(10) Now we have verified that properties (1) and (2) are satisfied and hence we can compose the final result:

$\mathbf{S}_n \leftarrow (\mathbf{M}_n, \mathbf{S}_{n-1}),$

$$\mathbf{O} = \begin{bmatrix} \mathbf{0} & \mathbf{D1} & \mathbf{Q} \\ \mathbf{O1} & \mathbf{D2} - \mathbf{x1} & \mathbf{D3} \end{bmatrix}$$

The dimensions of the blocks are such that: $(\mathbf{M}_n, \mathbf{S}_{n-1})\,\mathbf{O} = \mathbf{0}$. In the above the matrix $\mathbf{1}'$ is a zero matrix except at the leading minor corresponding in length to the length of the vector $\mathbf{M}_{n-1}$, at which it is a unit matrix.

The termination of the algorithm follows essentially from the fact that R and $R[x]$ are Noetherian domains. If the algorithm does not terminate then one of the steps (2) or (4) are executed indefinitely. In either case it would mean that we get some infinite ascending (by inclusion) chain of ideals, which is impossible in Noetherians domains.

ASSERTION. *The basis* $\mathbf{S}_n$ *is a simplification basis.*

We must show that the returned basis is a simplification basis. We have shown in section 4 how a canonical forms and B-quotients can be constructed using the Szekeres basis. Here we shall prove that the returned set **O** is actually a basis for the syzygies of the Szekeres basis $\mathbf{S}_n$.

Let **a** be such that $\mathbf{a}.\,\mathbf{S}_n = 0$. This can be written as $\Sigma\, \mathbf{a}_i(x^i\mathbf{LC}_i + \mathbf{R}_i)$.

First we show that by subtracting vectors in the span of the vectors in **O** we can get a vector **b**, such that $\mathbf{b} = (\mathbf{d}_n, \mathbf{c}_{n-1}, \ldots, \mathbf{c}_k)$, where all the vectors $\mathbf{c}_i$ $i = n-1, \ldots, k$ are constant vectors (i.e. with entries from R). Notice that after step 4 and step 7 (see the remark to step 11) the following vectors have been added to **O**: $x.(\mathbf{0}, 0, \ldots, 1, 0, 0, \ldots, 0) - \mathbf{C}$ where **C** is a constant vector. Hence by subtracting combinations of such a vector we may reduce the degree of all the components to zero, except at those entries corresponding to the components of $\mathbf{M}_n$.

Let t be the highest degree of x in **b**. $\mathbf{b} = x^t(\mathbf{b}_t, 0, \ldots, 0) + \ldots + (\mathbf{b}_0, \mathbf{c}_{n-1}, \ldots, \mathbf{c}_k)$, where $\mathbf{b}_t$ is a constant vector. If follows that $\mathbf{b}_t\,\mathbf{LC}_n = 0$. Therefore $\mathbf{b}_t$ is a syzygy of $\mathbf{LC}_n$. By subtracting now the vectors constructing at step 8 and at step 9 (see again the explanation to step 10) we arrive to the new vector **b**, such that $\mathbf{b} = (\mathbf{0}, \mathbf{d}_{n-1}, \ldots, \mathbf{d}_k)$. The new vector **b** is again a syzygy but now of $\mathbf{S}_{n-1}$.

And the rest follows from step 2 of the algorithm by induction. (The $n = 0$ is trivial.)

The Gröbner Fan of an Ideal

TEO MORA AND LORENZO ROBBIANO

Dipartimento di Matematica dell'Università di Genova, Via L. B. Alberti 4, 16132 Genova, Italy

To every ideal I in the polynomial ring $\mathbf{A} := k[X_1, \ldots, X_n]$ new invariants are attached, such as the Gröbner Fan $\mathbf{F}(I)$, the Gröbner region $\mathbf{G}(I)$ and the set $\mathbf{ATO}(I)$ of the almost term-orderings of I, i.e. orderings which "behave like" term-orderings with respect to I. These invariants arise by considering the reduced Gröbner bases of I with respect to all the term-orderings. Moreover $\mathbf{F}(I)$, $\mathbf{G}(I)$, $\mathbf{ATO}(I)$ can be got in a constructive way, as we show by producing a suitable algorithm which computes them.

Introduction

The starting point of this paper was the hope of understanding the effect of changing orderings to the computation of Gröbner bases of ideals in the polynomial ring. So far this question was not very much considered in the literature (see for instance Kollreider, 1978) and also current implemented algorithms which produce Gröbner bases allow the user to dispose of a very limited choice of orderings.

In this context we are referring to total orderings on T_A, the monoid of terms in the polynomial ring $A := k[X_1, \ldots, X_n]$, which are compatible with the product; since T_A is isomorphic to $\mathbb{N}^n$ via the canonical map *log*, which associates to every term $X_1^{a_1} \ldots X_n^{a_n}$ the n-tuple of exponents $(a_1, \ldots, a_n)$, the study of these orderings can be done in $\mathbb{N}^n$.

Every compatible ordering on $\mathbb{N}^n$ uniquely extends to a compatible ordering on $\mathbb{Z}^n$ and $\mathbb{Q}^n$ and the solution to the problem of describing them was given in Robbiano (1985) (see also Robbiano, 1986) by means of suitable orthogonal real matrices.

Although this result was not essentially new (see for instance Trevisan (1953) for an abstract solution and Kolchin (1973) for a suggested vectorial approach) and very elementary, it became the initial step of new research, including the investigations performed in this paper.

As a natural extension of total orderings, we introduce in the first section the notion of *ordering of linear type*. These (partial) orderings are described by arrays of real vectors, while the class of partial orderings is much wider (see for instance Dress & Schiffels, 1987).

The first section deals with the problem of checking whether two arrays of real vectors give rise to the same ordering: this goal is achieved by Theorem 1.6.

The use of vectors turns out to be suitable to express concepts as refinement of orderings; moreover, since to every ordering a half-line in $\mathbb{R}^n$ is canonically associated, namely the half-line of vectors which can be taken as *first vectors* of the ordering itself, to every set of orderings we may associate a *cone* in $\mathbb{R}^n$.

Section 2 begins with the description of a suitable polyhedral cone associated to all the orderings for which a fixed finite set of vectors is positive (see Theorem 2.1).

Polyhedral cones become then the main tool and the first main result is Theorem 2.5, which essentially asserts that if we are given an ideal I in $A := k[X_1, \ldots, X_n]$, a term-ordering σ and the reduced Gröbner basis G of I with respect to σ, then the set of orderings τ with $M_\tau(I) = M_\sigma(I)$ is the set of orderings such that the behaviour of τ and σ with respect to G is in some sense the same. Moreover all the term-orderings in this set yield the same reduced Gröbner basis, and all the orderings in this set have this reduced Gröbner basis as a standard set for which the Buchberger reduction procedure is noetherian and confluent.

Since the set $\mathrm{Mon}^+(I)$ of monomial ideals which are maximal monomial ideals of I with respect to term-orderings is finite (we present here a simple proof of this fact due to A. Logar) (see 2.6), we can use Theorem 2.5 to describe a set $\mathbf{F}(I)$ of polyhedral cones, which are in correspondence with $\mathrm{Mon}^+(I)$. It turns out that $\mathbf{F}(I)$ is a fan (see the description given in Theorems 2.7 and 4.3) and it is termed the *Gröbner Fan* of I. The union $\mathbf{G}(I)$ of these cones is termed the *Gröbner region* of I.

The polyhedral cone decomposition is related to the polytopes independently studied in Bayer & Morrison (1986).

Section 3 is a technical interlude, which treats the homogenisation of vectors and extension of orderings from $\mathbb{Z}^n$ to $\mathbb{Z}^{n+1}$; these tools are used to study the behaviour of Gröbner bases and of reduced Gröbner bases with respect to the homogenisation process of ideals.

The results of this section (in particular 3.10) allow to see what happens also to those orderings whose associated half-line is outside $\mathbf{G}(I)$ (see Theorem 4.3) and it turns out that suitable polyhedral cones can be defined, which unfortunately depend on the choice of the homogenisation process.

This is the content of section 4, while in section 5 we focus our attention on the description of $\mathbf{G}(I)$ and we show that it describes the set of orderings, which behave with respect to I as term-orderings; we also show, by exhibiting suitable examples, that $\mathbf{G}(I)$ can be bigger than $(\mathbb{R}^n)^+$ and in the case of ideals which are homogeneous with respect to positive weights of the indeterminates it coincides with $\mathbb{R}^n$.

The final section describes a prototype of a "Buchberger parallel algorithm", which yields the Gröbner Fan and all the reduced Gröbner bases of a given ideal I.

Therefore our results are strongly related with the concept of "universal Gröbner bases" independently studied in Schwartz (1986), Schemmel (1987) and Weispfenning (1987).

Our research shows that given a problem about the polynomial ring, which can be solved by means of a Gröbner basis computation, there exists a "shortest" reduced Gröbner basis which solves the problem and which can be got constructively. At present we do not have criteria for determining *a priori* such shortest path in the computational tree, but we hope that our work can serve as a good basis for such investigation.

Finally, we want to remark that this paper can be viewed both as a theoretic study in Commutative Algebra, since it presents new invariants of ideals, and as a background for further development in algorithmic theory of the polynomial ring.

1. Orderings

Let us start by recalling that by a *partial* or *non strict ordering* on an abelian group G we always mean a partial ordering which is compatible with the group structure and such that $\{v \in G / v > 0\}$ spans G.

Now we borrow some terminology and some results from Robbiano (1985). If $v \in \mathbb{R}^n$, we denote by $d(v)$ the dimension of the $\mathbb{Q}$-subvectorspace of $\mathbb{R}$ spanned by the coordinates of v and we call it the *rational dimension* of v. Let now V be a $\mathbb{Q}$-subvectorspace of $\mathbb{Q}^n$, such that $\dim(V) = d$, and denote by $V_{\mathbb{R}} := V \otimes \mathbb{R}$. Let $\mathbf{u} = (u_1, \ldots, u_s)$ be an array of vectors in $V_{\mathbb{R}}$ and denote by $d_i := d(u_i)$; let $\{\lambda_{i1}, \ldots, \lambda_{id_i}\}$ be a basis of the $\mathbb{Q}$-vectorspace generated by the coordinates of u_i, $i = 1 \ldots s$; then there exists a uniquely determined set of vectors $u_{ij} \in V$ so that

$$u_i = \Sigma\, \lambda_{ij} u_{ij} \quad \text{where} \quad u_{ij} \in V,\ \lambda_{ij} \in \mathbb{R}.$$

DEFINITION 1.

(a) The array $\mathbf{u} = (u_1, \ldots, u_s)$ is said to be an array of *rationally independent* vectors if the $\Sigma\, d_i$-tuple $(u_{11}, \ldots, u_{1d_1}, u_{21}, \ldots, u_{2d_2}, \ldots, u_{sd_s})$ is an array of linearly independent vectors over $\mathbb{Q}$.

(b) The set $\{u_1, \ldots, u_s\}$ is said to be a set of *rational generators* of V if $\{u_{ij}\}$ is a set of generators of V over $\mathbb{Q}$.

(c) The array $\mathbf{u} = (u_1, \ldots, u_s)$ is said to be a *rational basis* of V if the $\Sigma\, d_i$-tuple $(u_{11}, \ldots, u_{1d_1}, u_{21}, \ldots, u_{2d_2}, \ldots, u_{sd_s})$ is a basis of V over $\mathbb{Q}$.

(d) As a generalisation of the case $s = 1$, the dimension of the $\mathbb{Q}$-vectorspace spanned by $(u_{11}, \ldots, u_{1d_1}, u_{21}, \ldots, u_{2d_2}, \ldots, u_{sd_s})$ is termed the *rational dimension* of $\mathbf{u}$.

(e) In general we denote by $W_{\mathbb{Q}}(\mathbf{u})$ or simply by $W(\mathbf{u})$ the $\mathbb{Q}$-subvectorspace of V generated by the u_{ij}'s, hence $W_{\mathbb{Q}}(\mathbf{u})$ is rationally generated by $\{\mathbf{u}\}$.

REMARK. It is clear that the notions given in Definition 1 are independent of the choice of $\{\lambda_{ij}\}$.

PROPOSITION 1.1. *Let* $\mathbf{u} = (u_1, \ldots, u_s)$ *be as before; then*

(1) $\{u_1, \ldots, u_s\}$ *is a set of rational generators of* V *iff*

$$(*) \quad u \in V \quad \text{and} \quad u \,.\, u_1 = \cdots = u \,.\, u_s = 0 \Rightarrow u = 0$$

(2) $(u_1, \ldots, u_s)$ *is a rational basis of* V *iff* (*) *holds and* $\Sigma\, d(u_i) = d$.

PROOF. Easy exercise. □

REMARK 1. We must be careful about Definition 1. Namely we cannot substitute "the $\Sigma\, d_i$-tuple $(u_{11}, \ldots, u_{1d_1}, u_{21}, \ldots, u_{2d_2}, \ldots, u_{sd_s})$ is an *array* of linearly independent vectors over $\mathbb{Q}$" with "the set $\{u_{ij}\}$ is a *set* of linearly independent vectors over $\mathbb{Q}$" as the following example shows.

EXAMPLE 1. Let $u_1 := (1, \sqrt{2}, \sqrt{2})$; $u_2 := (0, \sqrt{2}, \sqrt{2} + 1)$. Then $u_1 = (1, 0, 0) + \sqrt{2}(0, 1, 1)$; $u_2 = (0, 0, 1) + \sqrt{2}(0, 1, 1)$, hence $d_1 = d_2 = 2$, $d_1 + d_2 > 3$ and $\mathbf{u} := (u_1, u_2)$ has not to be considered as a rational basis of $\mathbb{Q}^3$.

REMARK 2. Given an array $\mathbf{u} = (u_1, \ldots, u_s)$ of vectors of $V_{\mathbb{R}}$, after selecting s bases $\{\lambda_{i1}, \ldots, \lambda_{id_i}\}$ $i = 1 \ldots s$ of the $\mathbb{Q}$-vectorspaces spanned by the coordinates of the u_i's, we may get as before an array $(u_{11}, \ldots, u_{1d_1}, u_{21}, \ldots, u_{2d_2}, \ldots, u_{sd_s})$ of vectors of V.

From that, we may start a procedure which eliminates a vector if it is linearly dependent with the preceding ones. When this operation is performed, we drop u_i if every vector u_{ij} was eliminated and we substitute $u_i = \Sigma_j \lambda_{ij} u_{ij}$, $j = 1 \ldots d_s$ with $v_i := \Sigma \lambda_{ij} u_{ij}$ where the sum is taken over the indexes j's such that u_{ij} survived. *The new array* $(v_1, \ldots, v_t)$, *where* $t \leqslant s$, *is an array of rationally independent vectors. Moreover we remark that* $u_1 = v_1$.

DEFINITION 2. We say that the described procedure, which takes an array $\mathbf{u}$ of vectors of $V_{\mathbb{R}}$ and produces an array $\mathbf{v}$ of rationally independent vectors, *simplifies* $\mathbf{u}$ to $\mathbf{v}$ or that $\mathbf{v}$ is a *simplification* of $\mathbf{u}$.

We say that an array $\mathbf{u}$ is *essential* if the procedure does not kill any vector of $\mathbf{u}$.

An essential array is not necessarily an array of rationally independent vectors as Example 1 shows.

COROLLARY 1.2. *If* $\mathbf{u} = (u_1, \ldots, u_s)$ *is an array of rationally independent vectors of* $V_{\mathbb{R}}$, then $u_1, \ldots, u_s$ *are linearly independent over* $\mathbb{R}$.

PROOF. We choose λ_{ij} and u_{ij} as we did before and we denote by U the $\mathbb{Q}$-vectorspace spanned by the u_{ij}'s. Then the set $\{u_{ij}\}$ is a basis of U, hence a basis of $U_{\mathbb{R}}$; let $\Sigma \lambda_i u_i = 0$; then $\Sigma \lambda_i \lambda_{ij} u_{ij} = 0$, whence $\lambda_i \lambda_{ij} = 0$; in conclusion $\lambda_i = 0$, $i = 1 \ldots s$. □

Let now $\mathbf{u} = (u_1, \ldots, u_s)$ be an essential array of vectors in $V_{\mathbb{R}}$ and consider the map

$$\deg_{\mathbf{u}}: V \longrightarrow \mathbb{R}^s$$

$$v \dashrightarrow (v \,.\, u_1, \ldots, v \,.\, u_s)$$

which is called the "**u**-(multi)-degree map".

If we endow $\mathbb{R}^s$ with the ordering **lex** (i.e. $(a_1, \ldots, a_s) > 0$ iff the first nonzero a_i from the left is positive), then we get an ordering on V, which we denote by **ord**($\mathbf{u}$) and which is in general partial; however

COROLLARY 1.3. *Given an essential array* $\mathbf{u} = (u_1, \ldots, u_s)$ *of vectors in* $V_{\mathbb{R}}$, *the following conditions are equivalent*

(*a*) **ord**($\mathbf{u}$) *is a total ordering*
(*b*) $\{u_1, \ldots, u_s\}$ *is a rational set of generators of* V.

PROOF. It follows immediately from Proposition 1.1. □

DEFINITION 3. We denote by $\mathbf{A}(V)$ the set of arrays of essential vectors of $V_{\mathbb{R}}$, by $\mathbf{A}_{\text{gen}}(V)$ the subset of arrays whose underlying set is a set of rational generators.

We denote by $\mathbf{A}_{\text{ind}}(V)$ the set of arrays of rationally independent vectors of $V_{\mathbb{R}}$, by $\mathbf{A}_{\text{bas}}(V)$ the set of arrays of rational bases of V. We denote by **simp** the maps

$$\textbf{simp}: \mathbf{A}(V) \longrightarrow \mathbf{A}_{\text{ind}}(V) \quad \text{and} \quad \textbf{simp}: \mathbf{A}_{\text{gen}}(V) \longrightarrow \mathbf{A}_{\text{bas}}(V)$$

which were described in Definition 2. We denote by $\mathbf{O}(V)$ the set of partial orderings σ on V such that there exists $u \in \mathbf{A}(V)$ and $\sigma = \mathbf{ord}(\mathbf{u})$. Such orderings are called *orderings of linear type*. We denote by **ord** the maps

$$\mathbf{A}(V) \longrightarrow \mathbf{O}(V); \quad \mathbf{A}_{\text{ind}}(V) \longrightarrow \mathbf{O}(V); \quad \mathbf{A}_{\text{gen}}(V) \longrightarrow \mathbf{O}(V); \quad \mathbf{A}_{\text{bas}}(V) \longrightarrow \mathbf{O}(V),$$

which were described by Corollary 1.3 and the preceding discussion, and by $\mathbf{O}_{\text{tot}}(V)$ the set of total orderings on V.

LEMMA 1.4.

(*a*) *The maps* **simp:** $\mathbf{A}(V) \longrightarrow \mathbf{A}_{\text{ind}}(V)$ *and* **simp:** $\mathbf{A}_{\text{gen}}(V) \longrightarrow \mathbf{A}_{\text{bas}}(V)$ *are surjective.*

(*b*) $\mathbf{O}_{\text{tot}}(V) = \mathbf{ord}(\mathbf{A}_{\text{gen}}(V)) = \mathbf{ord}(\mathbf{A}_{\text{bas}}(V))$. *In particular for every total ordering σ on V there exists* **u** *such that* $\sigma = \mathbf{ord}(\mathbf{u})$ *i.e.* $v_1 < v_2$ *in* σ *iff* $\deg_{\mathbf{u}}(v_1) < \deg_{\mathbf{u}}(v_2)$ *in the lex ordering.*

(*c*) *If* $\mathbf{u} = (u_1, \ldots, u_s)$ *and* $\mathbf{v} = (v_1, \ldots, v_t)$ *are in* $\mathbf{A}(V)$ *and* $\mathbf{ord}(\mathbf{u}) = \mathbf{ord}(\mathbf{v})$, *then there exists* $\lambda \in \mathbb{R}^+$ *such that* $u_1 = \lambda v_1$.

PROOF.

(a) is clear.

(b) This is a consequence of Robbiano (1985).

(c) Assume the contrary: then there exists a rational vector u such that $u \,.\, u_1 \geqslant 0$ and $u \,.\, v_1 < 0$, hence $\mathbf{ord}(\mathbf{u}) \neq \mathbf{ord}(\mathbf{v})$, a contradiction. □

DEFINITION 4. If $\mathbf{u} = (u_1, \ldots, u_s)$ is an array of (essential) vectors of $V_{\mathbb{R}}$ and $\mathbf{ord}(\mathbf{u}) = \sigma$ we say that **u** is a *representation* of σ.

COROLLARY 1.5. *Associated with every ordering of linear type σ, there is a half-line $L(\sigma)$, which is characterized by:*

for every representation $\mathbf{u} = (u_1, \ldots, u_s)$ *of* σ, $L(\sigma)$ *is spanned by* u_1.

PROOF. It follows from Lemma 1.4. □

THEOREM 1.6. *Let* $\mathbf{u} = (u_1, \ldots, u_s)$ *and* $\mathbf{v} = (v_1, \ldots, v_t)$ *be elements of* $\mathbf{A}(V)$ *i.e. arrays of essential vectors of* $V_{\mathbb{R}}$; *let* W_{i-1} *be the* $\mathbb{Q}$-*subspace of* V *of vectors orthogonal to* $(u_1, \ldots, u_{i-1})$ (*in particular* $W_0 = V$), *let* $(W_{i-1})_{\mathbb{R}}$ *denote* $W_{i-1} \otimes \mathbb{R}$, *and let* pr_i *be the orthogonal projection from* $V_{\mathbb{R}}$ *to* $(W_{i-1})_{\mathbb{R}}$, $i = 1 \ldots s$. *Then the following conditions are equivalent*

(a) $\mathbf{ord}(\mathbf{u}) = \mathbf{ord}(\mathbf{v})$

(b) $s = t$ *and there exist positive real numbers* λ_i, *such that* $pr_i(u_i) = \lambda_i pr_i(v_i)$ *for* $i = 1 \ldots s$.

PROOF. (b) ⇒ (a) This follows from the definition of **ord**.

(a) ⇒ (b) $i = 1$ is the case treated by Lemma 1.4(c). So we may assume that $pr_j(u_j) = \lambda_j pr_j(v_j)$ for $j = 1 \ldots i-1$. By assumption $\mathbf{ord}(\mathbf{u})$ and $\mathbf{ord}(\mathbf{v})$ induce the same ordering on W_i which we denote by $\mathbf{ord}_i$.

u and **v** being essential, if we think of $\mathbf{ord}_i$ as induced by $\mathbf{ord}(\mathbf{u})$ it can be represented by an array with first vector $pr_i(u_i)$ while if we think of $\mathbf{ord}_i$ as induced by $\mathbf{ord}(\mathbf{v})$ it can be represented by an array with first vector $pr_i(v_i)$. By Lemma 1.4 we get $pr_i(u_i) = \lambda_i pr_i(v_i)$ and the conclusion follows. □

COROLLARY-DEFINITION 1.7.

(a) *Let σ be an ordering of linear type on V and let* $\sigma = \mathbf{ord}(\mathbf{u})$. *Then* $W_{\mathbb{Q}}(\mathbf{u})$ (*see Definition* 1(*e*)) *is independent of the representation of* σ; *it will be denoted by* $W_{\mathbb{Q}}(\sigma)$ *or simply* $W(\sigma)$.

(b) *There is a canonical map* $\mathbf{O}(V) \longrightarrow \{$*Subspaces of* $V\}$, *whose constant value on* $\mathbf{O}_{\mathrm{tot}}(V)$ *is* V.

PROOF.
(a) Let $W = \{w \in V / \deg_{\mathbf{u}} w = \mathbf{0}\}$ and $W' = \{w \in V / \deg_{\mathbf{v}} w = \mathbf{0}\}$, where $\sigma = \mathbf{ord}(\mathbf{u}) = \mathbf{ord}(\mathbf{v})$. It is then clear that $W = W'$, while it is also clear that $W_{\mathbb{Q}}(\mathbf{u})$ is the orthogonal complement of W and $W_{\mathbb{Q}}(\mathbf{v})$ the one of W'. The conclusion follows.
(b) Of course the map is defined by sending σ to $W_{\mathbb{Q}}(\sigma)$ and if σ is total, $\sigma = \mathbf{ord}(\mathbf{u})$, then $W_{\mathbb{Q}}(\sigma) = V$ by Corollary 1.3. □

We now denote by $A(d)$ the quotient set of vectors of rational dimension d modulo the equivalence relation given by: $v \sim v'$ iff there exists $\lambda \in \mathbb{R}^+$ with $v' = \lambda v$.

THEOREM 1.8. *If* V *is a subgroup of* $\mathbb{Z}^n$ *of rank* d *(or a subvectorspace of* $\mathbb{Q}^n$ *of dimension* d*), a partial ordering* $<$ *of linear type on* V *is given by the following data*:

the type, i.e. an integer s *with* $1 \leqslant s \leqslant d$,
the partition type of $<$, *i.e. an array* $(d_1, \ldots, d_s)$ *of natural numbers such that* $\Sigma\, d_i \leqslant d$,
an element $(\bar{u}_1, \ldots, \bar{u}_s) \in A(d_1) \times \cdots \times A(d_s)$ *such that for every* $i = 1 \ldots s$ *if* W_{i-1} *denotes the subgroup of* $\mathbb{Z}^n$ *(or the subvectorspace of* $\mathbb{Q}^n$*) of the vectors orthogonal to* $(u_1, \ldots, u_{i-1})$, *then* $u_i \in (W_{i-1})_{\mathbb{R}}$

Moreover every total ordering is given as before, with $\Sigma\, d_i = d$.

PROOF. It follows from 1.6 and Robbiano (1985). □

COROLLARY 1.9. *If* σ *is an ordering on* V *of type* s *and partition type* $(d_1, \ldots, d_s)$ *and* $\mathbf{u} = (u_1, \ldots, u_r)$ *is an essential array of vectors of* $V_{\mathbb{R}}$ *representing* σ, *then*

(*a*) $r = s$
(*b*) $d(u_i) \geqslant d_i$ *for* $i = 1 \ldots s$

If moreover $\mathbf{u}$ *is an array of rationally independent vectors representing* σ, *then* $d(u_i) = d_i$ *for* $i = 1 \ldots s$.

PROOF.
(a) follows from Theorem 1.6 and Theorem 1.8.
(b) Using the same notations as before, by Theorem 1.8 it is sufficient to show that $d(u_i) \geqslant d(pr_i(u_i))$. But $(W_{i-1})_{\mathbb{R}}$ is generated by W_{i-1} hence it has an orthonormal basis of rational vectors; the elementary formula for projections gives the desired inequality. Given $\mathbf{u} = (u_1, \ldots, u_s)$ essential, we get $\bar{\mathbf{u}} = (\bar{u}_1, \ldots, \bar{u}_s)$ where $\bar{u}_i = pr_i(u_i)$, and clearly $\sigma = \mathbf{ord}(\mathbf{u}) = \mathbf{ord}(\bar{\mathbf{u}})$; moreover the partition type of σ is $(d(\bar{u}_1), \ldots, d(\bar{u}_s))$ by Theorem 1.8. Therefore if $\mathbf{u}$ is an array of rationally independent vectors representing σ, then $\Sigma\, d(u_i) = \Sigma\, d(\bar{u}_i) = \Sigma\, d_i$. Combining with (b) we get the conclusion. □

DEFINITION 5. If $\mathbf{u} = (u_1, \ldots, u_s)$ and $\mathbf{v} = (v_1, \ldots, v_t)$ are two arrays of vectors of V, we get the new ordering $\mathbf{ord}(\mathbf{u}, \mathbf{v}) = \mathbf{ord}(\mathbf{simp}(\mathbf{u}, \mathbf{v}))$, where of course $(\mathbf{u}, \mathbf{v})$ denotes the array

$(u_1, \ldots, u_s, v_1, \ldots, v_t)$. Sometimes $\mathbf{ord}(\mathbf{u}, \mathbf{v})$ is referred to as "the refinement of $\mathbf{ord}(\mathbf{u})$ obtained with $\mathbf{ord}(\mathbf{v})$" (compare with Bayer & Stillman, 1985).

REMARK. A partial ordering defined by a single vector in $V_{\mathbb{R}}$ is what is usually termed a *degree*. For instance if *revlex* denotes the reverse lexicographic total ordering on $\mathbb{Q}^n$, then

revlex $= \mathbf{ord}(-e_n, -e_{n-1}, \ldots, -e_1)$ where $(e_1, e_2, \ldots, e_n)$ is the canonical basis. If we refine the degree $w = (1, 1, \ldots, 1)$ with *revlex*, we get $\mathbf{ord}(w, -e_n, -e_{n-1}, \ldots, -e_2, -e_1) = \mathbf{ord}(w, -e_n, -e_{n-1}, \ldots, -e_2)$.

COROLLARY 1.10. *Every partial ordering of linear type on V can be refined to a total ordering. Moreover this can be done by refining with an array of vectors in V.*

PROOF. Let $\sigma = \mathbf{ord}(\mathbf{u})$ be a partial ordering of linear type on V. Then let $\mathbf{f} = (f_1, f_2, \ldots, f_d)$ be a basis of V as a $\mathbb{Q}$-vectorspace and simply consider $\mathbf{ord}(\mathbf{u}, \mathbf{f})$. □

2. Cones of Orderings and the Restricted Gröbner Fan

Henceforth we are going to use freely some notions and results from the theory of polyhedral cones (see for instance Kuhn & Tucker, 1956). With $\langle w_1, \ldots, w_r \rangle$ we denote the *polyhedral cone* $\{\Sigma\, r_i w_i / r_i \in \mathbb{R}^+\}$ spanned by $w_1, \ldots, w_r$ and with $\langle w_1, \ldots, w_r \rangle^* := \{v \in \mathbb{R}^n / v \,.\, w_i \geqslant 0$ for all $i\}$ the *polar* of $\langle w_1, \ldots, w_r \rangle$; we recall that, if V is a polyhedral cone, its *vertex* is the maximal subspace of $\mathbb{R}^n$ contained in V. Let now V be as before a subvectorspace of $\mathbb{Q}^n$ of dimension d. If σ is an ordering of linear type of V then the half-line $L(\sigma)$ is well defined by Corollary 1.5. This leads to the following:

DEFINITION 1. If $S \subseteq \mathbf{O}(V)$, then the *associated cone to S* is the subset $C(S)$ of $V_{\mathbb{R}}$ of all the half-lines $L(\sigma)$, $\sigma \in S$.

DEFINITION 2. If $E \subseteq V$, the set of partial orderings of linear type (resp. total orderings) positive on E is denoted by $\mathbf{O}(+, E)$ (resp. $\mathbf{O}_{\text{tot}}(+, E)$). If $V = \mathbb{Q}^n$, then $\mathbf{O}_{\text{tot}}(+, \mathbb{N}^n)$ is termed "the set of term-orderings".

THEOREM 2.1. *Let $E := \{w_1, \ldots, w_r\}$ be a finite subset of V; let $E^* := \langle E \rangle^*$ (i.e. $E^* = \{v \in V / v \,.\, w_i \geqslant 0, i = 1 \ldots r\}$); let $(E^*)^\circ$ be the interior of E^*. Then*

(*a*) $(E^*)^\circ \subseteq C(\mathbf{O}(+, E)) \subseteq E^*$
(*b*) *If* $\dim(E^*) < d$ *then* $\mathbf{O}(+, E) = \emptyset$
(*c*) *If* $\dim(E^*) = d$ *then* $C(\mathbf{O}(+, E)) = E^*$
(*d*) *Let* $\sigma = \mathbf{ord}(\mathbf{u})$ $\mathbf{u} = (u_1, \ldots, u_s)$. *Then* $\sigma \in \mathbf{O}(+, E) \Leftrightarrow \Sigma\, \varepsilon^{i-1} u_i \in (E^*)^\circ$ *for every small* $\varepsilon > 0$.

PROOF.
(a) The first inclusion follows from the fact that $(E^*)^\circ = \{v \in V / v \,.\, w_i > 0, i = 1 \ldots r\}$, while the second one is obvious.
(b) It is known that

$$\dim(E^*) < d \Leftrightarrow \dim(\text{Vertex of } \langle E \rangle) > 0$$

hence there exists a non zero r-tuple $(a_1, \ldots, a_r)$ of non negative rational numbers such that $\Sigma\, a_i w_i = 0$. If $\sigma \in \mathbf{O}(+, E)$ then $w_i > 0$ in σ for every i, hence $0 = \Sigma\, a_i w_i > 0$ in σ, a contradiction.

(c) If $\dim(E^*) = d$, then $(E^*)^\circ \neq \emptyset$ hence $C(\mathbf{O}(+, E)) \neq \emptyset$ by a). Therefore it is sufficient to show that

$$\mathbf{O}(+, E) \neq \emptyset \Rightarrow C(\mathbf{O}(+, E)) = E^*.$$

After a) it is sufficient to show that if $v \in E^* \backslash (E^*)^\circ$, then $v \in C(\mathbf{O}(+, E))$. Let $0 \neq v \in E^* \backslash (E^*)^\circ$, then v belongs to a proper face of E^*, hence there exists a non empty subset $F = \{w_1, \ldots, w_s\}$ of E such that $v \,.\, w_i = 0\; i = 1 \ldots s$, and $v \,.\, w_i > 0\; i = s+1 \ldots r$. Being $\mathbf{O}(+, E)$ non empty, it follows that also $\mathbf{O}(+, F)$ is non empty. Let us denote by W the orthogonal to the subvectorspace generated by v; then $F \subseteq W$ and every ordering τ in $\mathbf{O}(+, F)$ induces an ordering $\tau|_W$ on W, which is positive on F. Let $\mathbf{w}' = (w'_2, \ldots, w'_n)$ represent one of these orderings $\tau|_W$ on W. Then $\sigma = \mathbf{ord}(v, \mathbf{w}') \in \mathbf{O}(+, E)$ hence $v \in C(\mathbf{O}(+, E))$.

(d) $\sigma \in \mathbf{O}(+, E) \Leftrightarrow \deg_{\mathbf{u}}(w_i) > 0$ in the lex order, $i = 1 \ldots r \Leftrightarrow w_i \,.\, \Sigma\, \varepsilon^{j-1} u_j$ is positive for ε small and positive, $i = 1 \ldots r \Leftrightarrow \Sigma\, \varepsilon^{j-1} u_j \in (E^*)^\circ$. □

REMARK. Let $\sigma \in \mathbf{O}(V)$ and let $\mathbf{u} = (u_1, \ldots, u_s)$ represent σ. If $u_1 \in (E^*)^\circ$ then we have seen that $\sigma \in \mathbf{O}(+, E)$, while of course not all of the orderings τ such that $L(\tau) \subseteq E^* \backslash (E^*)^\circ$ are in $\mathbf{O}(+, E)$, as the following example shows: Let $V = \mathbb{R}^2$, $E = \{e_1\}$; then $E^* = \{(x, y)/x \geqslant 0\}$; hence $C(\mathbf{O}(+, E)) = E^*$. Consider the orderings $\alpha = \mathbf{ord}(e_2, e_1)$, $\beta = \mathbf{ord}(e_2, -e_1)$. It is then clear that $\alpha \in \mathbf{O}(+, E)$, while $\beta \notin \mathbf{O}(+, E)$.

COROLLARY 2.2. *Let $E = \{w_1, \ldots, w_s\} \subseteq \mathbb{Z}^n$ and suppose that there exists a term-ordering which is positive on E. Then there is an infinite set of n-tuples $q = (q_1, \ldots, q_n)$ in $(\mathbb{N}^+)^n$ such that $q \,.\, w_i > 0$, $i = 1 \ldots s$.*

PROOF. The assumption means that $\mathbf{O}(+, E \cup \mathbb{N}^n) \neq \emptyset$; but $\mathbf{O}(+, E \cup \mathbb{N}^n) = \mathbf{O}(+, E')$ where $E' := E \cup \{e_1, \ldots, e_n\}$ ($e_1, \ldots, e_n$ is the canonical basis of $\mathbb{Z}^n$). By Theorem 2.1, $\dim(E')^* = n$ and of course $(E')^* \cap (\mathbb{N}^+)^n$ is infinite. □

Now we are ready to apply the above developed machinery to the polynomial rings. So let $A := k[X_1, \ldots, X_n]$, $T_A :=$ the set of terms of A (i.e. commutative words in $X_1, \ldots, X_n$). The map

$$\log\colon T_A \longrightarrow \mathbb{N}^n \subseteq \mathbb{Z}^n \quad \text{defined by} \quad \log(X_1^{a_1} \ldots X_n^{a_n}) := (a_1, \ldots, a_n)$$

is an isomorphism of monoids and allows us to transfer from $\mathbb{Z}^n$ to T_A (hence to A) all the terminology and the results obtained so far. For instance, we draw from 2.2 the following consequences: If σ is a term-ordering positive on a finite set E, then there are infinitely many n-tuples of weights $(q_1, \ldots, q_n) \in (\mathbb{N}^+)^n$ such that the *degrees of the elements of E, computed with respect to the weights $q_1, \ldots, q_n$ of the variables $X_1, \ldots, X_n$* are *positive*. Moreover we can talk about cones of orderings, multidegrees, term-orderings on T_A or on A. Given a polynomial $f = \Sigma\, c(f, m) m \in A$ $(m \in T_A)$, we denote by $\mathrm{Supp}(f) := \{m/c(f, m) \neq 0\}$. If σ is an ordering of linear type on T_A and $0 \neq f \in A$ then $M_\sigma(f) = \Sigma\, c(f, m) m$ where the sum is taken over those m's such that $c(f, m) \neq 0$ and $m <_\sigma m'$ does not hold for every m' with $c(f, m') \neq 0$. If σ is a total ordering, of course

$M_\sigma(f)$ is a single monomial. If I is an ideal, $M_\sigma(I)$ denotes the ideal generated by $\{M_\sigma(f)/f \in I\}$.

DEFINITION 3. Let $I \subseteq A$ be an ideal and let $\sigma \in \mathbf{O}(\mathbb{Z}^n)$. A finite set $\{f_1, \ldots, f_t\} \subseteq I - \{0\}$ is termed a σ-standard set of I if $\{M_\sigma(f_1), \ldots, M_\sigma(f_t)\}$ generates $M_\sigma(I)$. If σ is a term-ordering then a σ-standard set is usually termed a σ-Gröbner basis and it is easy to prove that it is also a basis of I (see for instance Möller & Mora, 1986).

PROPOSITION–DEFINITION 2.3. *If σ is a term-ordering, then there exists a unique set $\{f_1, \ldots, f_r\} \subseteq I - \{0\}$ such that*

(*a*) $f_i = m_i + R_i$, $m_i \in T_A$ *for every* i
(*b*) $\{m_1, \ldots, m_r\}$ *minimally generates* $M_\sigma(I)$
(*c*) $m_i = M_\sigma(f_i)$
(*d*) $\text{Supp}(R_i) \cap M_\sigma(I) = \emptyset$.

Such a Gröbner basis is termed the reduced Gröbner basis of w.r.t. σ.

PROOF. See Buchberger (1976b). □

REMARK. If σ is *not* a term-ordering, bad things can happen. For instance if $A := k[X]$, $\sigma := \mathbf{ord}(-1)$, $f := X - X^2$, $I := (f)$ then for every $g(X)$ with $g(0) \neq 0$ the set $\{gf\}$ is a standard set of I and it is also a basis iff $g \in k^*$. Moreover for every such g, if $R(X) := gf - M_\sigma(gf)$, then there exists $a > 1$ such that $X^a \in \text{Supp}(R(X))$; obviously $X^a \in M_\sigma(I)$, too.

In the remaining part of this section we deal only with total orderings (unless the contrary is specified), hence we use the following notations

$$\mathbf{O} := \mathbf{O}_{\text{tot}}(\mathbb{Z}^n) \qquad \mathbf{TO} := \mathbf{O}_{\text{tot}}(+, \mathbb{N}^n).$$

Let us start by discussing the following notion of reduction: Let $f_1, \ldots, f_r \in A - \{0\}$, $I := (f_1, \ldots, f_r)$. Let $m_1, \ldots, m_r \in T_A$ be such that $m_i \in \text{Supp}(f_i)$ $i = 1 \ldots r$ and let $c_i := c(f_i, m_i)$. Assume that $\mathbf{\Sigma} := \{\sigma \in \mathbf{O}/M_\sigma(f_i) = m_i\ i = 1 \ldots r\} \neq \emptyset$ and let $g \in A$. We say that g *reduces* to h with respect to $\wedge := (f_1, \ldots, f_r; m_1, \ldots, m_r)$, and we denote it by $g \mathbin{-\wedge\rightarrow} h$, iff there exist $t \in T_A$ and $i \in \{1, \ldots, r\}$ with $t \,.\, m_i \in \text{Supp}(g)$ such that $h = g - (1/c_i) \,.\, c(g, t \,.\, m_i) \,.\, t \,.\, f_i$. Let $\mathbin{-\wedge\rightarrow^*}$ denote the transitive-reflexive closure of $\mathbin{-\wedge\rightarrow}$; we say that $h \in A$ is $\wedge$*-irreducible* iff $h \mathbin{-\wedge\rightarrow^*} g$ implies $g = h$; we say that $\mathbin{-\wedge\rightarrow^*}$ is *noetherian* iff for each infinite sequence $(g_0, \ldots, g_i, \ldots)$ with $g_{i-1} \mathbin{-\wedge\rightarrow^*} g_i$ for every i, there exists N such that $g_i = g_N$ if $i > N$. We write

$$\text{RED}(f_1, \ldots, f_r; m_1, \ldots, m_r; g) := \{h \in A : h \text{ is } \wedge\text{-irreducible and } g \mathbin{-\wedge\rightarrow^*} h\}.$$

We explicitly remark that if σ is a term-ordering and $\wedge := (f_1, \ldots, f_r; M_\sigma(f_1), \ldots, M_\sigma(f_r))$, then $\mathbin{-\wedge\rightarrow}$ is the usual noetherian reduction relation introduced by Buchberger (1965) (see also Buchberger, 1970; 1976a) in connection with his Gröbner basis algorithm; essentially the only difference is that we don't require that a total ordering is imposed on T_A.

LEMMA 2.4.
(*a*) *If* $0 \in \text{RED}(f_1, \ldots, f_r; m_1, \ldots, m_r; g)$, *then* $g \in I$.
(*b*) *If* $f \in \text{RED}(f_1, \ldots, f_r; m_1, \ldots, m_r; g)$, *then* $\text{Supp}(f) \cap (m_1, \ldots, m_r) = \emptyset$.

(c) *If* Σ *contains a term-ordering* σ, *and we write* $\wedge := (f_1, \ldots, f_r; m_1, \ldots, m_r)$, *then* $-\wedge\to^*$ *is noetherian.*

(d) *If* Σ *contains a term-ordering* σ *and* $\{f_1, \ldots, f_r\}$ *is a Gröbner basis for I with respect to* σ, *then for every* $g \in A - \{0\}$ *there is a unique* $h \in \mathrm{RED}(f_1, \ldots, f_r; m_1, \ldots, m_r; g)$, *which we will denote by* $\mathrm{red}(f_1, \ldots, f_r; m_1, \ldots, m_r; g)$. *Moreover if* $g \in I$ *then* $\mathrm{red}(f_1, \ldots, f_r; m_1, \ldots, m_r; g) = 0$.

(e) *For each ordering* σ *which is not a term-ordering, there are* $f_1, \ldots, f_r$ *such that, denoting* $\wedge := (f_1, \ldots, f_r; M_\sigma(f_1), \ldots, M_\sigma(f_r))$ *then* $-\wedge\to^*$ *is not noetherian.*

PROOF. (a) and (b) are clear. If Σ contains a term-ordering σ, then $\wedge = (f_1, \ldots, f_r; M_\sigma(f_1), \ldots, M_\sigma(f_r))$ and $-\wedge\to^*$ is Buchberger's reduction relation. Hence (c) and (d) are well-known properties (cf. Buchberger, 1976b). (e) Namely there exists a term N, such that $N < 1$ in σ; then let $r := 1, f_1 := N - N^2, g_i := N^i$ for $i \geqslant 1$; then $g_i -\wedge\to g_{i+1}$ for all i. □

We can extend the usual Buchberger reduction algorithm to our notion; we remark that the resulting procedure can be guaranteed to halt returning an element in $\mathrm{RED}(f_1, \ldots, f_r; m_1, \ldots, m_r; g)$ if and only if $-\wedge\to^*$ is noetherian.

DEFINITION 4. Let I be an ideal of A. We put

$$\mathrm{Mon}(I) := \{\mathsf{m}/\mathsf{m} = M_\sigma(I), \sigma \in \mathbf{O}\},$$

$$\mathrm{Mon}^+(I) := \{\mathsf{m}/\mathsf{m} = M_\sigma(I), \sigma \in \mathbf{TO}\}.$$

Given a monomial ideal m, we put

$$\mathbf{O}(I, \mathsf{m}) := \{\sigma \in \mathbf{O}/M_\sigma(I) = \mathsf{m}\}$$

$$\mathbf{TO}(I, \mathsf{m}) := \{\sigma \in \mathbf{TO}/M_\sigma(I) = \mathsf{m}\}.$$

Let now $\mathsf{m} \in \mathrm{Mon}^+(I)$ and $\sigma \in \mathbf{TO}(I, \mathsf{m})$. Let $G = G(\sigma) = \{g_1, \ldots, g_n\}$ be the reduced Gröbner basis of I with respect to σ and let us write $g_i = m_i + R_i$, with $m_i = M_\sigma(g_i)$. Then let $\mathrm{DIFF}(G) := \bigcup_{i=1\ldots n} \{\log(m_i) - \log(t)/t \in \mathrm{Supp}(R_i)\}$.

THEOREM 2.5.

(a) $\mathbf{O}(I, \mathsf{m}) = \mathbf{O}(+, \mathrm{DIFF}(G))$.

(b) *For every* $\tau \in \mathbf{TO}(I, m)$, *the reduced Gröbner basis of I with respect to* τ *is G.*

(c) *For every* $\tau \in \mathbf{O}(I, m)$, *G is a standard set of I with respect to* τ, *such that if we write* $\wedge := (g_1, \ldots, g_n; m_1, \ldots, m_n)$, *then* $-\wedge\to^*$ *is noetherian and for each* $g \in A$, $\mathrm{RED}(g_1, \ldots, g_n; m_1, \ldots, m_n; g)$ *contains a unique element.*

PROOF.

(a) Let $\tau \in \mathbf{O}(I, \mathsf{m})$; then $M_\tau(I) = \mathsf{m}$; now G is a Gröbner basis of I, hence a basis; in particular $g_i \in I, i = 1 \ldots n$, hence $M_\tau(g_i) \in M_\tau(I) = \mathsf{m}$ and of course $M_\tau(g_i) \in \mathrm{Supp}(g_i)$. Moreover G is a reduced Gröbner basis of I, hence $\mathsf{m} \cap \mathrm{Supp}(g_i) = \{m_i\}$. Therefore $M_\tau(g_i) = m_i$, whence $\tau \in \mathbf{O}(+, \mathrm{DIFF}(G))$. Conversely let $\tau \in \mathbf{O}(+, \mathrm{DIFF}(G))$; then of course $M_\tau(g_i) = m_i$; hence $M_\tau(I) \supseteq (m_1, \ldots, m_n) = M_\sigma(I) = \mathsf{m}$; on the other hand let $g \in I$ be such that $M_\tau(g) \notin \mathsf{m}$. By Lemma 2.4(d), $\mathrm{RED}(g_1, \ldots, g_n; m_1, \ldots, m_n; g)$ consists of a unique element $\mathrm{red}(g)$, and $M_\tau(\mathrm{red}(g)) = M_\tau(g) \notin \mathsf{m}$. But again by Lemma 2.4(d), $\mathrm{red}(g) = 0$, a contradiction. Therefore $M_\tau(I) = \mathsf{m}$.

(b) If $\tau \in \mathbf{TO}(I, \mathsf{m})$, then $\tau \in \mathbf{O}(+, \mathrm{DIFF}(G))$ by (a), hence G is also a Gröbner basis of I with respect to τ and then it is clearly the reduced Gröbner basis.
(c) That G be a standard set is obvious. The rest follows from (b) and 2.5(d). □

LEMMA 2.6. *For every ideal I of A, the set* $\mathrm{Mon}^+(I)$ *is finite.*

PROOF. The result could be achieved as a consequence of Möller & Mora (1984), but we present here an elementary proof which has been proposed to us by Alessandro Logar (Logar, 1986). Let $(f_1, \ldots, f_r)$ be a basis of I and assume that $\mathrm{Mon}^+(I)$ is infinite; for each element $\mathsf{m} \in \mathrm{Mon}^+(I)$ we can choose a term-ordering $\tau(\mathsf{m}) \in \mathbf{TO}(I, \mathsf{m})$; let $\Sigma := \{\tau(\mathsf{m}) / \mathsf{m} \in \mathrm{Mon}^+(I)\}$.

Since $\mathrm{Supp}(f_i)$ is finite, there is an infinite subset Σ_1 of Σ, and terms $m_i \in \mathrm{Supp}(f_i)$, such that for each $\tau \in \Sigma_1$, $M_\tau(f_i) = m_i$. Let $\mathsf{m}_1 := (m_1, \ldots, m_r)$ and let us choose $\tau \in \Sigma_1$; either $(f_1, \ldots, f_r)$ is a Gröbner basis of I with respect to τ, or we find a non-zero polynomial $f_{r+1} \in I$, such that $\mathrm{Supp}(f_{r+1}) \cap \mathsf{m}_1 = \emptyset$. In the latter case, reasoning as before, we see that there is an infinite subset Σ_2 of Σ_1, and a term $m_{r+1} \in \mathrm{Supp}(f_{r+1})$, such that for each $\tau \in \Sigma_2$, $M_\tau(f_i) = m_i$, for $i = 1 \ldots r+1$. Also, $\mathsf{m}_2 := (m_1, \ldots, m_{r+1}) \supset \mathsf{m}_1$. Let us choose $\tau_2 \in \Sigma_2$ and apply the same argument as before.

By noetherianity, after a finite number of steps, we get a set $G := (f_1, \ldots, f_s)$, terms $m_i \in \mathrm{Supp}(f_i)$ for $i = 1 \ldots s$, and an infinite subset Σ' of Σ such that $m_i = M_\sigma(f_i)$ for all $\sigma \in \Sigma'$, and G is a Gröbner basis for at least one $\sigma \in \Sigma'$. Since, for all $\sigma \in \Sigma'$, $\sigma \in \mathbf{O}(+, \mathrm{DIFF}(G))$, by Theorem 2.5, G is a Gröbner basis for all $\sigma \in \Sigma'$, a contradiction. □

This result enables us to denote $\mathrm{Mon}^+(I)$ by $\{\mathsf{m}_1, \ldots, \mathsf{m}_r\}$. For every $i = 1 \ldots r$ we can choose as before a term-ordering $\sigma_i \in \mathbf{TO}(I, \mathsf{m}_i)$, its reduced Gröbner basis $G_i = \{g_{i1}, \ldots, g_{in_i}\}$ where $g_{ij} = m_{ij} + R_{ij}$, $m_{ij} = M_\sigma(g_{ij})$ and we denote by $D_i := \mathrm{DIFF}(G_i) = \bigcup_j \{\log(m_{ij}) - \log(t) / t \in \mathrm{Supp}(R_{ij})\}$. We recall once more that D_i^* means $\langle D_i \rangle^*$, the polar cone of D_i and that $(\mathbb{R}^n)^+$ denotes $\{(a_1, \ldots, a_n) \in \mathbb{R}^n / a_i \geqslant 0 \; i = 1 \ldots n\}$.

THEOREM 2.7. *Let I be an ideal of $A = k[X_1, \ldots, X_n]$. Then*

(*a*) $\dim(D_i^* \cap (\mathbb{R}^n)^+) = n$ *for* $i = 1 \ldots r$
(*b*) $(\mathbb{R}^n)^+ = \bigcup_i (D_i^* \cap (\mathbb{R}^n)^+)$
(*c*) *For every i, j $i \neq j$, $D_i^* \cap D_j^*$ is part of a proper face of D_i^* and D_j^*.*
(*d*) *Given a term-ordering $\sigma = \mathbf{ord}(\mathbf{u})$, where $\mathbf{u} = (u_1, \ldots, u_s)$, then the following conditions are equivalent*

  (*i*) $\sigma \in \mathbf{O}(+, D_i)$
  (*ii*) $\sum_k \varepsilon^{k-1} u_k \in (D_i^* \cap (\mathbb{R}^n)^+)^\circ$ *for every small* $\varepsilon > 0$.
  (*iii*) $M_\sigma(I) = \mathsf{m}_i$ *and G_i is the reduced Gröbner basis of I with respect to σ.*

PROOF.

(a) Follows from 2.1(b).
(b) Every non zero vector $u \in (\mathbb{R}^n)^+$ can be considered as the first vector of a representation of a term-ordering σ. If $M_\sigma(I) = \mathsf{m}_i$, then by 2.5(a), $\sigma \in \mathbf{O}(+, D_i)$, hence $u \in D_i^*$.
(c) It is sufficient to show that $\dim(D_i^* \cap D_j^*) < n$. And indeed if $\dim(D_i^* \cap D_j^*) = n$,

then by 2.1(c) $D_i^* \cap D_j^* = C(\mathbf{O}(+, D_i \cup D_j))$, while the latter has to be empty, since by 2.5 an ordering σ in $\mathbf{O}(+, D_i \cup D_j)$ would give rise to $M_\sigma(I) = \mathsf{m}_i = \mathsf{m}_j$.

(d) The equivalence of (i) and (ii) follows from 2.1(d). The equivalence of (i) and (iii) follows from 2.5. □

DEFINITION 5. The set of polyhedral cones

$$\mathbf{F}^+(I) := \{D_1^* \cap (\mathbb{R}^n)^+, \ldots, D_r^* \cap (\mathbb{R}^n)^+\}$$

will be called the *restricted Gröbner Fan* of I.

EXAMPLE 1. Let $A := k[X_1, X_2]$, $f_1 := X_1^2 - X_1$, $f_2 := X_2 - X_1$, $I := (f_1, f_2)$, $f_3 := X_2^2 - X_2$. Let $\mathbf{C}_1 := \{(x, y) \in (\mathbb{R}^2)^+ : x \geqslant y\}$, $\mathbf{C}_2 := \{(x, y) \in (\mathbb{R}^2)^+ : x \leqslant y\}$, $\mathsf{m}_1 := (X_1, X_2^2)$, $\mathsf{m}_2 := (X_1^2, X_2)$, $G_1 := \{-f_2, f_3\}$, $G_2 := \{f_1, f_2\}$, $D_i := \mathrm{DIFF}(G_i)$. It is easy to verify that $F^+(I) = \{C_1, C_2\}$, $C_i = (D_i)^* \cap (\mathbb{R}^n)^+$, and that D_i, G_i, m_i satisfy Theorem 2.7.

3. Homogenization

DEFINITION 1. Given $E \subseteq V$ we put $D(E) := \{u - v / u, v \in E\}$; let $\mathbf{u} = (u_1, \ldots, u_s)$, $u_i \in V_{\mathbb{R}}$, let $\sigma := \mathbf{ord}(\mathbf{u})$ and let $W := W_{\mathbb{Q}}(\sigma)$ (see 1.7). A subset E of V is said to be $\mathbf{u}$-homogeneous (or σ-homogeneous or W-homogeneous) if $\deg_{\mathbf{u}}(D(E)) = 0$. We remark that E is W-homogeneous iff $D(E)$ is orthogonal to W.

Of course, while the concept of $\mathbf{u}$-degree depends on $\mathbf{u}$, the concept of $\mathbf{u}$-homogeneity depends only on W. In particular, different orderings may give rise to the same notion of homogeneity as the following easy examples show:

EXAMPLE 1. Let $u \neq 0$; then $\mathbf{ord}(u) \neq \mathbf{ord}(-u)$, while $W(u) = W(-u)$.

EXAMPLE 2. Let $u := (\pi, 1, 1)$ $v_1 := (1, 0, 0)$ $v_2 := (0, 1, 1)$ $\mathbf{v} := (v_1, v_2)$; then $\mathbf{ord}(u) \neq \mathbf{ord}(\mathbf{v})$, but $W(u) = W(\mathbf{v})$.

EXAMPLE 3. By 1.7(b) all the total orderings give rise to the same notion of homogeneity, being V the associated vectorspace; besides, if σ is a total ordering, E is σ-homogeneous iff $\#(E) \leqslant 1$.

REMARK. If $E \subseteq V$, then the set of vectors $v \in V_{\mathbb{R}}$ such that E is v-homogeneous, is a real subvectorspace of $V_{\mathbb{R}}$ of type $W \otimes_{\mathbb{Q}} \mathbb{R}$. Namely, if $v := \Sigma \lambda_i u_i$, $u_i \in \mathbb{Q}^n$, $\lambda_1, \ldots, \lambda_n$ linearly independent over $\mathbb{Q}$, $\mathbf{u} := (u_1, \ldots, u_s)$, then v-homogeneous $\Leftrightarrow$ $\mathbf{u}$-homogeneous.

PROPOSITION 3.1. *Let $E \subseteq V$, $u_1, \ldots, u_s, v_1, \ldots, v_t \in V_{\mathbb{R}}$ and put $\mathbf{u} := (u_1, \ldots, u_s)$, $\mathbf{v} := (v_1, \ldots, v_t)$, $\mathbf{u} \wedge \mathbf{v}$ a permutation of $(u_1, \ldots, u_s, v_1, \ldots, v_t)$, whose restriction to $(v_1, \ldots, v_t)$ is the identity. If E is $\mathbf{u}$-homogeneous, then*

$$\mathbf{ord}(\mathbf{u} \wedge \mathbf{v})|_E = \mathbf{ord}(\mathbf{v})|_E.$$

PROOF. Let $\sigma := \mathbf{ord}(\mathbf{u} \wedge \mathbf{v})|_E$, $\tau := \mathbf{ord}(\mathbf{v})|_E$; then if $w_1, w_2 \in E$ we have $w_1 <_\sigma w_2 \Leftrightarrow 0 <_\sigma w_2 - w_1 \Leftrightarrow 0 <_{\mathrm{lex}} \deg_{\mathbf{u} \wedge \mathbf{v}}(w_2 - w_1) \Leftrightarrow 0 <_{\mathrm{lex}} \deg_{\mathbf{v}}(w_2 - w_1) \Leftrightarrow w_1 <_\tau w_2$ since $u_i . (w_2 - w_1) = 0$ for every i. □

Let now $i\colon \mathbb{R}^n \longrightarrow \mathbb{R}^{n+1}$ be the map defined by $i(a_1, \ldots, a_n) := (0, a_1, \ldots, a_n)$. If $\mathbf{u} = (u_1, \ldots, u_t)$ is an array of vectors of $\mathbb{R}^n$, we denote by $i(\mathbf{u})$ the array $(i(u_i), \ldots, i(u_t))$. Similarly let ${}^h\colon \mathbb{R}^n \longrightarrow \mathbb{R}^{n+1}$ be the map defined by ${}^h(a_1, \ldots, a_n) := (1, a_1, \ldots, a_n)$; as before, ${}^h(\mathbf{u})$ denotes $({}^h u_1, \ldots, {}^h u_n)$. Finally let ${}^a\colon \mathbb{R}^{n+1} \longrightarrow \mathbb{R}^n$ be the map defined by ${}^a(a_0, a_1, \ldots, a_n) := (a_1, \ldots, a_n)$. If $\mathbf{u} = (u_1, \ldots, u_s)$ is an array of vectors of $\mathbb{R}^{n+1}$, we denote by ${}^a(\mathbf{u}) := ({}^a u_1, \ldots, {}^a u_s)$.

LEMMA-DEFINITION 3.2.

(*a*) *Let* $\sigma \in \mathbf{O}(\mathbb{Z}^n)$, $\sigma = \mathbf{ord}(\mathbf{u})$. *Then* $\mathbf{ord}(i(\mathbf{u}))$ *is independent of the representation of* σ *and it will be denoted by* $i(\sigma)$.

(*b*) *Let* $\tau \in \mathbf{O}(\mathbb{Z}^{n+1})$, $\tau := \mathbf{ord}(\mathbf{u})$. *Then* $\mathbf{ord}({}^a(\mathbf{u}))$ *is independent of the representation of* τ *and it will be denoted by* ${}^a(\tau)$.

PROOF.

(a) Let $\sigma = \mathbf{ord}(\mathbf{u}) = \mathbf{ord}(\mathbf{v})$ and let $u \in \mathbb{R}^{n+1}$; then $\deg_{i(\mathbf{u})} u = \deg_{\mathbf{u}} {}^a(u) = \deg_{\mathbf{v}} {}^a(u) = \deg_{i(\mathbf{v})} u$ and the conclusion follows.

(b) Similar to (a). □

REMARK. We observe that ${}^h(\sigma)$ cannot be defined in the same way, since if $\sigma = \mathbf{ord}(\mathbf{u}) = \mathbf{ord}(\mathbf{v})$, then $\mathbf{ord}({}^h(\mathbf{u}))$ and $\mathbf{ord}({}^h(\mathbf{v}))$ may be different, as the following example shows. Let $u_1 := (1, 0)$, $u_2 := (0, 1)$, $v_1 := (1, 0)$, $v_2 := (-1, 1)$, $\mathbf{u} := (u_1, u_2)$, $\mathbf{v} := (v_1, v_2)$. Then $\mathbf{ord}(\mathbf{u}) = \mathbf{ord}(\mathbf{v})$, but $\mathbf{ord}({}^h(\mathbf{u})) \neq \mathbf{ord}({}^h(\mathbf{v}))$; for instance $(-2, 2, 3)$ is positive in the first and negative in the second one.

Let now $d \in \mathbb{R}^n$, $\Delta := {}^h d = (1, d) \in \mathbb{R}^{n+1}$, $\sigma \in \mathbf{O}(\mathbb{Z}^n)$, $W := \{u \in \mathbb{R}^{n+1} / \Delta \,.\, u = 0\}$.

DEFINITION 2. We denote by $\mathrm{EXT}(\sigma, \Delta)$ the set $\{\tau \in \mathbf{O}(\mathbb{Z}^{n+1}) / \tau|_W = i(\sigma)|_W\}$ and we term it the set of the Δ-*extensions of* σ. In other words $\tau \in \mathrm{EXT}(\sigma, \Delta)$ iff: for every $u, v \in \mathbb{R}^{n+1}$, with $\deg_\Delta u = \deg_\Delta v$, $u <_\tau v$ iff ${}^a u <_\sigma {}^a v$.

DEFINITION 3. Let $u \in \mathbb{R}^{n+1}$; then $p(u)$ is the vector of $\mathbb{R}^n$ defined by

$$u = i(p(u)) + \lambda\Delta, \quad \lambda \in \mathbb{R}$$

If $\mathbf{u}$ is an array of vectors of $\mathbb{R}^{n+1}$ then we have the notion of $p(\mathbf{u})$.

PROPOSITION-DEFINITION 3.3. *If* $\tau = \mathbf{ord}(\mathbf{u})$, *then* $\mathbf{ord}(p(\mathbf{u}))$ *is independent of the representation of* τ *and it will be denoted by* $p(\tau)$.

PROOF. Let us first prove the following

CLAIM: If $u \in W := \{u \in \mathbb{R}^{n+1} / \Delta \,.\, u = 0\}$ and $v \in \mathbb{R}^{n+1}$, then

$$u \,.\, v = i({}^a u) \,.\, p(v)$$

Namely $u = i({}^a u) + \lambda e_0$, $v = p(v) + \mu\Delta$, $e_0 := (1, 0, \ldots, 0)$. Therefore $u \,.\, v = u \,.\, p(v) = (i({}^a u) + \lambda e_0) \,.\, p(v) = i({}^a u) \,.\, p(v)$.

Now let $\tau = \mathbf{ord}(\mathbf{u}) = \mathbf{ord}(\mathbf{v})$ and let $w \in \mathbb{R}^n$. Then there exists $z \in W$ such that $w = i({}^a z)$ whence $\deg_{p(\mathbf{u})} w = p(\mathbf{u}) \,.\, w = p(\mathbf{u}) \,.\, i({}^a z) = \mathbf{u} \,.\, z$ by the claim, and $\deg_{p(\mathbf{v})} w = p(\mathbf{v}) \,.\, w = p(\mathbf{v}) \,.\, i({}^a z) = \mathbf{v} \,.\, z$ by the claim. The conclusion follows. □

LEMMA 3.4. *Let V be a subvectorspace of $\mathbb{Q}^n$, $U \subseteq V$ a subvectorspace of V. Let $\sigma \in \mathbf{O}(V)$ $\sigma = \mathbf{ord}(\mathbf{u})$. Then*

(*a*) *$\sigma|_U = \mathbf{ord}(pr_U(\mathbf{u}))$ where $pr_U(\mathbf{u})$ denotes the array of orthogonal projections of the vectors of $\mathbf{u}$ over U.*
(*b*) *$\mathbf{ord}(pr_U(\mathbf{u}))$ does not depend on the representation of σ, hence it can be denoted by $pr_U(\sigma)$, and (a) can be phrased by*

$$\sigma|_U = pr_U(\sigma).$$

PROOF. Easy from the definition, because if $w \in U$ then $\mathbf{u} \,.\, w = (pr_U(\mathbf{u})) \,.\, w$. □

PROPOSITION 3.5. *Let $\tau \in \mathbf{O}(\mathbb{Z}^{n+1})$, $\sigma \in \mathbf{O}(\mathbb{Z}^n)$. Then the following conditions are equivalent*

(1) "$\tau \in \mathrm{EXT}(\sigma, \Delta)$
(2) $p(\tau) = \sigma$

PROOF. (1) ⇒ (2) After observing that for a vector u, $p(u) = p(pr_W(u))$, by 3.3 we have $p(\tau) = p(pr_W(\tau))$ and $\sigma = p(i(\sigma)) = p(pr_W(i(\sigma)))$; by 3.4 we conclude that $p(\tau) = p(\tau|_W)$ and $\sigma = p\ (i(\sigma)|_W)$. Since $\tau \in \mathrm{EXT}(\sigma, \Delta)$ iff $\tau|_W = i(\sigma)|_W$, it follows that $p(\tau) = \sigma$.

(2) ⇒ (1) $p(\tau) = \sigma \Rightarrow i(p(\tau)) = i(\sigma) \Rightarrow (i(p(\tau))|_W = i(\sigma)|_W$. But it is clear that $pr_W(i(p(\tau)) = pr_W(\tau)$, so by 3.4 $(i(p(\tau))|_W = \tau|_W$ and we are done. □

EXAMPLE. Let $e_1 := (1, 0)$, $e_2 := (0, 1)$, $\mathbf{u} := (e_1, e_2)$, $\sigma := \mathbf{ord}(\mathbf{u})$, $d := (1, 1)$, $\Delta := (1, 1, 1)$. Then $\tau_1 := \mathbf{ord}(\Delta, i(e_1), i(e_2))$, $\tau_2 := \mathbf{ord}(-\Delta, i(e_1), i(e_2))$, $\tau_3 := \mathbf{ord}(i(e_1), -\Delta, i(e_2))$, $\tau_4 := \mathbf{ord}(i(e_1), i(e_2), \Delta + i(e_1))$ are all examples of elements of $\mathrm{EXT}(\sigma, \Delta)$.

Inside $\mathrm{EXT}(\sigma, \Delta)$ there is a distinguished element, namely $\mathbf{ord}(\Delta, i(\sigma))$, which we denote by $H_\Delta(\sigma)$. Hence we get a map, which is denoted by H_Δ.

COROLLARY 3.6.

(*a*) *If $d \in (\mathbb{N}^+)$, then $\mathrm{Im}(H_\Delta) \subseteq \mathbf{O}(+, \mathbb{N}^{n+1})$, the set of the term-orderings on $\mathbb{Z}^{n+1}$.*
(*b*) *The maps H_Δ and $p\colon \mathbf{O}(\mathbb{Z}^{n+1}) \longrightarrow \mathbf{O}(\mathbb{Z}^n)$ are such that $p \cdot H_\Delta = id$.*
(*c*) *If we denote by $T_\Delta := \{\tau \in \mathbf{O}_{\mathrm{tot}}(\mathbb{Z}^{n+1}) / u_1(\tau) = \Delta\}$, then the restriction of $H_\Delta\colon \mathbf{O}_{\mathrm{tot}}(\mathbb{Z}^n) \longrightarrow T_\Delta$ is a bijection, whose inverse is the restriction of p.*

PROOF. (a) is clear; (b) and (c) are then consequences of Proposition 3.5. □

Let us now apply the concepts introduced in this paragraph to the theory of polynomial ideals. As usual, we denote by A the polynomial ring $k[X_1, \ldots, X_n]$; when we say that $f \in A$ is $\mathbf{u}$-homogeneous (or σ-homogeneous or W-homogeneous) we mean that $\mathrm{Supp}(f)$ is $\mathbf{u}$-homogeneous. Also, an ideal I of A is said to be $\mathbf{u}$-homogeneous if for every $f \in I, f = \Sigma f_i$ with f_i $\mathbf{u}$-homogeneous and $\deg_{\mathbf{u}} f_i \neq \deg_{\mathbf{u}} f_j$ for $i \neq j$, then $f_i \in I$ for every i. For the sake of completeness we record the following easy facts

PROPOSITION 3.7.

(*a*) *I is $\mathbf{u}$-homogeneous ⇔ I has a basis of $\mathbf{u}$-homogeneous elements.*
(*b*) *I is monomial ⇔ I is $\mathbb{Q}^n$-homogeneous.*
(*c*) *If I is $\mathbf{u}$-homogeneous then there exists $v \in \mathbb{Z}^n - \{0\}$ such that I is v-homogeneous.*

(*d*) *if I is* **u**-*homogeneous, σ is a term-ordering and $\{f_1, \ldots, f_r\}$ is the reduced G-basis of I with respect to σ, then f_i is* **u**-*homogeneous for every i.*

PROOF. The proofs are standard. Let us only remark that (b) $\Leftarrow$ is a consequence of the fact that if $f \in I$ and $\#(\mathrm{Supp}(f)) > 1$ then a condition of u-homogeneity of f yields a non empty set of linear constraints on u, whence the conclusion follows immediately.

(c) is a consequence of the remark before Proposition 3.1. □

PROPOSITION 3.8. *Let* **u**, **v** *be two arrays of vectors in $\mathbb{R}^n$, $\sigma := \mathbf{ord}(\mathbf{v})$, $\tau := \mathbf{ord}(\mathbf{u}, \mathbf{v})$. Then*

(*a*) *If $f \in A$ is* **u**-*homogeneous, then $M_\tau(f) = M_\sigma(f)$.*

(*b*) *If I is a* **u**-*homogeneous ideal, then $M_\tau(I) = M_\sigma(I)$.*

PROOF.

(a) It is an easy consequence of 3.1.

(b) Let $f \in I, f = f_1 + \cdots + f_s$ with f_i **u**-homogeneous and $\deg_{\mathbf{u}} f_1 < \cdots < \deg_{\mathbf{u}} f_s$. Then $M_\tau(f) = M_\tau(f_s) = M_\sigma(f_s)$ by (a), whence $M_\tau(I) \subseteq M_\sigma(I)$. Conversely of course $M_\sigma(f) = \Sigma\, \varepsilon_i M_\sigma(f_i)$ where $\varepsilon_i \in \{0, 1\}$ for $i = 1, \ldots, s$ and $f_i \in I$. But this means that $M_\sigma(I)$ is generated by $\{M_\sigma(f)/f$ **u**-homogeneous$\}$, hence $M_\sigma(I) \subseteq M_\tau(I)$ by (a). □

Let now $A := k[X_1, \ldots, X_n]$, $B := k[X_0, X_1, \ldots, X_n]$, $d := (d_1, \ldots, d_n) \in (\mathbb{N}^+)^n$, $\Delta := (1, d) = {}^h d$. To every non zero $f \in A$ we associate ${}^h f \in B$, where

$${}^h f = X_0^{\deg_d(f)} f(X_1/X_0^{d_1}, \ldots, X_n/X_0^{d_n})$$

and we put ${}^h 0 = 0$. To every $F \in B$, F Δ-homogeneous, we associate ${}^a F := f(1, X_1, \ldots, X_n)$. To every ideal $I \subseteq A$ we associate ${}^h I$, the ideal generated by $\{{}^h f / f \in I\}$. To every Δ-homogeneous ideal $J \subseteq B$ we associate ${}^a J := \{{}^a f / f \in J, f\ \Delta\text{-homogeneous}\}$. Moreover if $\mathbf{f} = (f_1, \ldots, f_t)$ $f_i \in A$ then we put ${}^h\mathbf{f} := ({}^h f_1, \ldots, {}^h f_t)$. If $I = (\mathbf{f})$ then ${}^h I \supseteq ({}^h\mathbf{f})$ and $I = {}^{ah} I = {}^{ah}(\mathbf{f})$. In general ${}^h I = \bigcup_n (({}^h\mathbf{f}) : X_0^n) = ({}^h\mathbf{f}) : X_0^N$ for $N \gg 0$.

PROPOSITION 3.9. *Let $\sigma \in \mathbf{O}_{\mathrm{tot}}(\mathbb{Z}^n)$, $d \in (\mathbb{N}^+)^n$, $\Delta := {}^h d$, $\tau \in \mathrm{EXT}(\sigma, \Delta)$, I an ideal of A, and let $\mathbf{f} := (f_1, \ldots, f_r)$ be such that $I = (\mathbf{f})$. Then*

(*a*) *$M_\tau({}^h g) = X_0^s M_\sigma(g)$ for every $g \in A$.*

(*b*) *$M_\sigma({}^a F) = {}^a(M_\tau(F))$ for every Δ-homogeneous $F \in B$.*

(*c*) *${}^a(M_\tau({}^h\mathbf{f})) = {}^a(M_\tau({}^h I)) = M_\sigma(I)$.*

(*d*) *If $G := \{G_1, \ldots, G_t\}$ is a τ-standard set of ${}^h I$ (resp. of ${}^h(\mathbf{f})$) and G_i is Δ-homogeneous for every i, then $\{{}^a G_1, \ldots, {}^a G_t\}$ is a σ-standard set of I.*

PROOF. (see Lazard 1983; Möller & Mora, 1984).

(a) and (b) follow from the very definition of $\mathrm{Ext}(\sigma, \Delta)$.

(c) Clearly ${}^a(M_\tau({}^h\mathbf{f})) \subseteq {}^a(M_\tau({}^h I))$. Moreover $M_\tau({}^h g) = X_0^r M_\sigma(g)$ for every g by (a), whence ${}^a(M_\tau({}^h I)) \subseteq M_\sigma(I)$. On the other hand, if $f \in I$, then $X_0^r\, {}^h f \in {}^h(\mathbf{f})$ for some r and, by (b), $M_\tau({}^h f) = X_0^u M_\sigma(f)$, hence $M_\sigma(f) = {}^a(M_\tau(X_0^{r+s}\, {}^h f)) \in {}^a(M_\tau({}^h\mathbf{f}))$.

(d) We have

$$
\begin{aligned}
M_\sigma(I) &= {}^a(M_\tau({}^h I)) && \text{by (c)}\\
&= {}^a(M_\tau(G_1), \ldots, M_\tau(G_t)) && \text{with } G_i\ \Delta\text{-homogeneous by assumption}\\
&= ({}^a M_\tau(G_1), \ldots, {}^a M_\tau(G_t)) && \text{by standard properties of affinisation}\\
&= (M_\sigma({}^a G_1), \ldots, M_\sigma({}^a G_t)) && \text{by (b).} \ \square
\end{aligned}
$$

REMARK. If $\tau \notin \mathrm{EXT}(\sigma, \Delta)$ then there exists an ideal I such that ${}^a(M_\tau({}^hI)) \neq M_\sigma(I)$. Namely there exist $M, N \in T_A$ with $\deg_\Delta(X_0^s M) = \deg_\Delta(N)$, $X_0^s M >_\tau N$, $M <_\sigma N$. Then let $I := (f)$ where $f := M - N$; we get ${}^hI = ({}^hf) = (X_0^s M - N)$; $M_\tau({}^hI) = (X_0^s M)$ hence ${}^a(M_\tau({}^hI)) = (M)$, while $M_\sigma(I) = (N)$.

PROPOSITION 3.10. *Let* $d \in (\mathbb{N}^+)^n$, $\Delta := {}^hd$, $\sigma \in \mathbf{O}_{\mathrm{tot}}(\mathbb{Z}^n)$ *with* $u_1(\sigma) = d$, $\tau \in \mathrm{EXT}(\sigma, \Delta)$, *I an ideal of A. Then*

(*a*) *If* $G := \{g_1, \ldots, g_t\}$ *is a Gröbner basis of I with respect to* σ, *then* ${}^hG = \{{}^hg_1, \ldots, {}^hg_t\}$ *is a* τ*-standard set of* ${}^hI$.
If, moreover, τ *is a term-ordering (e.g.* $\tau = H_\Delta(\sigma)$*), then*
(*b*) If $\mathbf{G} := \{G_1, \ldots, G_t\}$ *is the* Δ*-homogeneous reduced Gröbner basis of* ${}^hI$ *with respect to* τ, *then* ${}^a\mathbf{G}$ *is the reduced Gröbner basis of I with respect to* σ.
(*c*) *If* $G := \{g_1, \ldots, g_t\}$ *is the reduced Gröbner basis of I with respect to* σ, *then* ${}^hG = \{{}^hg_1, \ldots, {}^hg_t\}$ *is the* Δ*-homogeneous reduced Gröbner basis of* ${}^hI$ *with respect to* τ.

PROOF. The assumption that $u_1(\sigma) = d$ implies that $M_\tau({}^hg) = M_\sigma(g)$ for every $g \in A$.

(a) It is clearly sufficient to show that, if $F \in {}^hI, F = {}^hf$, then $M_\tau(F) \in (M_\tau({}^hg_1), \ldots, M_\tau({}^hg_t))$, and indeed $M_\tau({}^hf) = M_\sigma(f) = M \,.\, M_\sigma(g_i) = M \,.\, M_\tau({}^hf_i)$ for some i.
(b) By 3.9 (d) we get that ${}^a\mathbf{G}$ is a standard set, hence, in this case, a Gröbner basis of I with respect to σ. If we assume that $\{M_\sigma({}^aG_1), \ldots, M_\sigma({}^aG_t)\}$ is not a minimal basis of $M_\sigma(I)$, then there exist $i, j, i \neq j$, such that $M_\sigma({}^aG_i) = NM_\sigma({}^aG_j)$; hence $M_\tau(G_i) = NM_\tau(G_j)$, a contradiction. If $M \in \mathrm{Supp}({}^aG_i) \cap M_\sigma(I)$, then $X_0^r M \in \mathrm{Supp}(G_i) \cap M_\tau({}^hI)$; therefore $X_0^r M = M_\tau(G_i)$, hence $r = 0$ and $M = M_\sigma({}^aG_i)$.
(c) ${}^hG$ is a standard set of ${}^hI$ with respect to τ by (a), but τ is a term-ordering, hence it is a Gröbner basis. If $M_\tau({}^hg_i) = NM_\tau({}^hg_j)$, $i \neq j$, then $M_\sigma(g_i) = NM_\sigma(g_j)$, a contradiction. If $M \in \mathrm{Supp}({}^hg_i) \cap M_\tau({}^hI)$, then $M = X_0^r N$, with $N \in \mathrm{Supp}(g_i)$ and $M = X_0^s M_\tau({}^hf) = X_0^s M_\sigma(f)$, with $f \in I$. Therefore $r = s$ and $N = M_\sigma(f) \in \mathrm{Supp}(g_i) \cap M_\sigma(I)$. Therefore $N = M_\sigma(g_i) = M_\tau({}^hg_i)$ and so $M = N$. □

EXAMPLE 1. $I := (g_1, g_2)$, $g_1 := X_1^2 - X_2^3$, $g_2 := X_3^2 - X_4^3$, $\sigma = \mathbf{ord}(e_1, e_2, e_3, e_4)$, $d := (1, 1, 1, 1)$, $\tau := H_\Delta(\sigma)$. We have that $\{g_1, g_2\}$ is a Gröbner basis of I with respect to σ, while $\{{}^hg_1, {}^hg_2\}$ is *not* a Gröbner basis of ${}^hI$ with respect to τ. This happens since $u_1(\sigma) \neq d$.

EXAMPLE 2. $I := (g_1, g_2)$, $g_1 := X_2 - X_1^2$, $g_2 := X_2^2$, $\sigma := \mathbf{ord}(e_2, e_1)$, $d := (1, 1)$, $\tau := H_\Delta(\sigma)$. We have that ${}^hI = ({}^hg_1, {}^hg_2)$, the reduced Gröbner basis of I with respect to σ is $\{g_1, X_1^4\}$, while the reduced Gröbner basis of ${}^hI$ with respect to τ is $\{{}^hg_1, {}^hg_2, X_1^2X_2, X_1^4\}$. Again this happens since $u_1(\sigma) \neq d$.

4. The Gröbner Fan

Let now I be an ideal of A, $B := k[X_0, X_1, \ldots, X_n]$, $d := (d_1, \ldots, d_n) \in (\mathbb{N}^+)^n$, $\Delta := (1, d) = {}^hd$ and let ${}^hI$ be the homogenisation of I with respect to d. We consider the bijective map of 3.6

$$H_\Delta \colon \mathbf{O}_{\mathrm{tot}}(\mathbb{Z}_n) \longrightarrow T_\Delta = \{\tau \in \mathbf{O}_{\mathrm{tot}}(+, \mathbb{N}^{n+1}) / u_1(\tau) = \Delta\}.$$

Using 3.6 and 3.7, we get

$$\mathrm{Mon}^+({}^hI) = \{\mathbf{m}_1, \ldots, \mathbf{m}_r\}$$

and we get Δ-homogeneous reduced Gröbner bases $\mathbf{G}_1, \ldots, \mathbf{G}_r$ (see 3.7), sets $\mathbf{D}_i = D(\mathbf{G}_i)$ and cones $\mathbf{D}_i^* \cap (\mathbb{R}^{n+1})^+$. We recall that $p: \mathbb{R}^{n+1} \longrightarrow \mathbb{R}^n$ is the projection along Δ.

PROPOSITION 4.1. *Mon(I) is finite.*

PROOF: If m is in Mon(I), then $\mathsf{m} = {}^a\mathbf{m}_i$ for some i (see 3.9(c)). This implies that Mon(I) is finite too, by 2.6. □

LEMMA 4.2.

(a) $p(\mathbf{D}_i^*) = ({}^a\mathbf{D}_i)^*$

(b) *If* $\mathbf{D}_i^*$ *is defined by the system*

$$(^\circ) \quad \sum_j a_{ij}X_j + a_iX_0 \geqslant 0 \quad i = 1, \ldots, N$$

then $p(\mathbf{D}_i^*)$ *is defined by the system*

$$(^{\circ\circ}) \quad \sum_j a_{ij}X_j \geqslant 0 \quad i = 1, \ldots, N.$$

PROOF.

(a) Put $\mathbf{D} := \mathbf{D}_i$; it is then sufficient to show that

$$v \in \mathbf{D}^* \Leftrightarrow p(v) \in ({}^a\mathbf{D})^*.$$

And indeed we write $v = \lambda\Delta + i(p(v))$. Since $\mathbf{D}$ is Δ-homogeneous, we get $v \in \mathbf{D}^* \Leftrightarrow i(p(v)) \in \mathbf{D}^* \Leftrightarrow p(v) \in ({}^a\mathbf{D})^*$.

(b) Follows from (a). □

THEOREM 4.3. *Let I be an ideal of $A = k[X_1, \ldots, X_n]$, $d \in (\mathbb{N}^+)^n$. With the preceding notations we have*

(a) $\dim(p(\mathbf{D}_i^*)) = n$ *for* $i = 1 \ldots r$

(b) $\mathbb{R}^n = \bigcup_i(p(\mathbf{D}_i^*)) = \bigcup_i({}^a\mathbf{D}_i)^*$

(c) *For every* i, j $i \neq j$, $p(\mathbf{D}_i^*) \cap p(\mathbf{D}_j^*)$ *is part of a proper face of* $p(\mathbf{D}_i^*)$ *and* $p(\mathbf{D}_j^*)$.

(d) *Given an ordering* $\sigma = \mathbf{ord}(\mathbf{u})$, *where* $\mathbf{u} = (u_1, \ldots, u_s)$, *consider the following conditions*

(i) $\sigma \in \hat{\mathbf{O}}(+, {}^a(D_i))$

(ii) $\sum_k \varepsilon^{k-1}u_k \in (p(D_i^*))^\circ$ *for every small* $\varepsilon > 0$

(iii) $M_\sigma(I) = {}^a(\mathbf{m}_i)$ *and* ${}^a\mathbf{G}_i$ *is a standard set of* I *with respect to* σ.

Then (i) ⇔ (ii) ⇔ (iii).

PROOF.

(a) $\dim(\mathbf{D}_i^*) = n + 1$ by 2.7(a), hence $\dim(p(\mathbf{D}_i^*)) = n$

(b) From 4.2 we get the second equality. Let $u \in \mathbb{R}^n$ $u \neq 0$. Then $u = u_1(\sigma)$ for some $\sigma \in \mathbf{O}(\mathbb{Z}^n)$. Let $\tau := H_\Delta(\sigma)$; being τ a term-ordering, $M_\tau({}^hI) = \mathbf{m}_i$. By 2.5 $\tau = (\Delta, i(\sigma)) \in \mathbf{O}(+, \mathbf{D}_i)$ and since $\mathbf{D}_i$ is Δ-homogeneous, we get $i(\sigma) \in \mathbf{O}(+, \mathbf{D}_i)$, hence $i(u) \in D_i^*$, hence $u \in ({}^a\mathbf{D}_i)^*$.

(c) As in 2.7(c) it is sufficient to show that $\dim(p(\mathbf{D}_i^*) \cap p(\mathbf{D}_j^*)) < n$. And indeed, being $\mathbf{D}_i$ and $\mathbf{D}_j$ Δ-homogeneous, the cones $\mathbf{D}_i^*$ have the 1-dimensional space L generated by Δ in their vertex. Now $p(\mathbf{D}_i^*) \cap p(\mathbf{D}_j^*) = p(\mathbf{D}_i^* \cap \mathbf{D}_j^*)$ and $\dim(\mathbf{D}_i^* \cap \mathbf{D}_j^*) \leqslant n$ by 2.7; being $\mathbf{D}_i^* \cap \mathbf{D}_j^* \supseteq L$, one gets $\dim(p(\mathbf{D}_i^*) \cap p(\mathbf{D}_j^*)) < n$.

(d) (i) $\Leftrightarrow$ (ii) follows from 2.1(d) and Lemma 4.2.
(i) $\Leftrightarrow$ (iii) If $\sigma \in \mathbf{O}(+, {}^a(\mathbf{D}_i))$ then $\tau := H_\Delta(\sigma) \in \mathbf{O}(+, \mathbf{D}_i)$ hence $M_\tau({}^hI) = \mathbf{m}_i$. Therefore $M_\sigma(I) = {}^a(M_\tau({}^hI))$ by 3.9(c). We also get that ${}^a\mathbf{G}_i$ is a standard set of I by 3.9(d). □

Corollary 4.4. *Let I be an ideal, $d \in (\mathbb{N}^+)$, $\Delta := {}^hd$, σ is a term-ordering such that $u_1(\sigma) = d$, $\mathrm{m} := M_\sigma(I)$; $\tau := H_\Delta(\sigma)$, $\mathbf{m} := M_\tau({}^hI)$. Let G be the reduced Gröbner basis of I with respect to σ; let $\mathbf{G}$ be the reduced Gröbner basis of ${}^hI$ with respect to τ and let* DIFF(G), DIFF$(\mathbf{G})$ *be defined as before. Then*

(*a*) $G = {}^a\mathbf{G}$ *hence* $\mathrm{m} = {}^a\mathbf{m}$.
(*b*) DIFF(G) $= {}^a($ DIFF$(\mathbf{G}))$
(*c*) DIFF(G))* $= ({}^a($DIFF$(\mathbf{G}))^* = p(($DIFF$(\mathbf{G})^*)$.

Proof.
(a) From 3.10(b)
(b) From (a)
(c) From (b) and 4.2. □

Definition 3. The set of polyhedral cones

$$\mathbf{F}(I) := \{D_1^*, \ldots, D_r^*\}$$

of Theorem 2.7 will be called the (*extended*) *Gröbner Fan* of I.

Corollary 4.5. *The sets* $\mathbf{F}(I)$, $G(I) := \bigcup_{D^* \in \mathbf{F}(I)} D^*$, $\bigcup_{i=1\ldots r} \mathbf{O}(+, D_i)$ *can be obtained by any homogenisation of I with respect to $d \in (\mathbb{N})^n$. Namely the cones of* $\mathbf{F}(I)$ *are those* $p(D_i^*)(=({}^aD_i)^*)$ *of Theorem* 4.3, *which intersect the interior of* $(\mathbb{R}^n)^+$.

Proof. It is a consequence of 4.4. □

The following example shows instead that outside $G(I)$ the $p(D_i^*)$'s coming from Theorem 4.3 *depend* on d.

Example 2. Let A, f_1, f_2, I, f_3 be as in Example 1, and let $B := k[X_0, X_1, X_2]$. Let $d := (1, 1)$ so that ${}^hf_1 = X_1^2 - X_0X_1$, ${}^hf_2 = f_2$, ${}^hI = ({}^hf_1, {}^hf_2)$, ${}^hf_3 = X_2^2 - X_0X_2$. Let

$$\mathbf{C}_1 := \{(t, x, y) \in \mathbb{R}^3 : x \geqslant y \geqslant t\}, \quad \mathbf{C}_2 := \{(t, x, y) \in \mathbb{R}^3 : y \geqslant x \geqslant t\},$$

$$\mathbf{C}_3 := \{(t, x, y) \in \mathbb{R}^3 : x \geqslant y, t \geqslant y\}, \quad \mathbf{C}_4 := \{(t, x, y) \in \mathbb{R}^3 : y \geqslant x, t \geqslant x\};$$

$$\mathbf{m}_1 := (X_1, X_2^2), \quad \mathbf{m}_2 := (X_1^2, X_2), \quad \mathbf{m}_3 := (X_1, X_0, X_2), \quad \mathbf{m}_4 := (X_0, X_1, X_2);$$

$$\mathbf{G}_1 := \{-{}^hf_2, {}^hf_3\}, \quad \mathbf{G}_2 := \{{}^hf_1, {}^hf_2\},$$

$$\mathbf{G}_3 := \{-({}^hf_2), -({}^hf_3)\}, \quad \mathbf{G}_4 := \{-({}^hf_1), {}^hf_2\}; \ \mathbf{D}_i := \mathrm{DIFF}(\mathbf{G}_i).$$

It is easy to verify that $\mathbf{C}_i = \mathbf{D}_i^*$, $\mathbf{F}({}^hI) = \{\mathbf{C}_1, \mathbf{C}_2, \mathbf{C}_3, \mathbf{C}_4\}$ and that $\mathbf{D}_i, \mathbf{G}_i, \mathbf{m}_i$ satisfy Theorem 4.3. By affinisation we get a partition of $\mathbb{R}^2$ into polyhedral cones

$\{C_1, C_2, C_3, C_4\}$, where $C_i := p(\mathbf{D}_i)^*$ for all i, and $\mathsf{m}_i := {}^a\mathbf{m}_i = M_\sigma(I)$ for all $\sigma \in \mathbf{O}(+, p(\mathbf{D}_i))$; we remark however that $\mathsf{m}_3 = \mathsf{m}_4$:

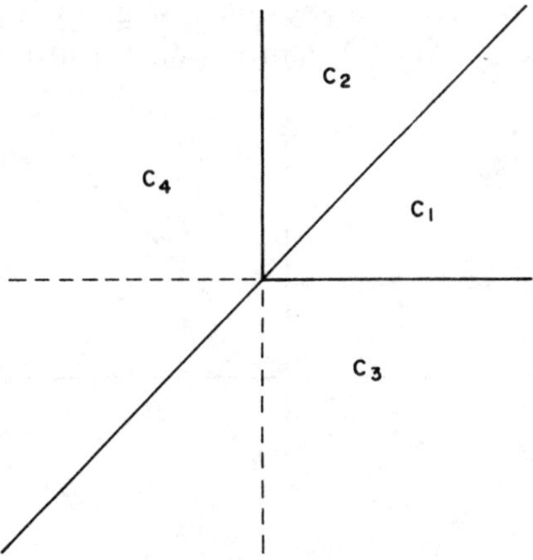

Let now $d := (2, 1)$ so that ${}^hf_1 = X_1^2 - X_0^2X_1$, ${}^hf_2 = X_0X_2 - X_1$, ${}^hf_3 = X_2^2 - X_0X_2$, ${}^hI = ({}^hf_2, {}^hf_3)$; let also $f_4 := X_1X_2 - X_1, f_5 := X_2^2 - X_1$, so that ${}^hf_4 = X_1X_2 - X_0X_1$ and ${}^hf_5 = f_5$. Let

$$\mathbf{C}_1 := \{(t, x, y) \in \mathbb{R}^3 \colon x \geqslant t + y, y \geqslant t\}, \quad \mathbf{C}_2 := \{(t, x, y) \in \mathbb{R}^3 \colon t + y \geqslant x, x \geqslant 2t\},$$

$$\mathbf{C}_3 := \{(t, x, y) \in \mathbb{R}^3 \colon x \geqslant 2y, t \geqslant y\}, \quad \mathbf{C}_4 := \{(t, x, y) \in \mathbb{R}^3 \colon t \geqslant y, 2y \geqslant x\},$$

$$\mathbf{C}_5 := \{t, x, y) \in \mathbb{R}^3 \colon y \geqslant t, 2t \geqslant x\};$$

$$\mathbf{m}_1 := (X_1, X_2^2), \quad \mathbf{m}_2 := (X_1^2, X_0X_2, X_1X_2, X_2^2),$$

$$\mathbf{m}_3 := (X_1, X_0X_2), \quad \mathbf{m}_4 := (X_0X_1, X_0X_2, X_1^2), \quad \mathbf{m}_5 := (X_0^2X_1, X_0X_2, X_1X_2, X_2^2);$$

$$\mathbf{G}_1 := \{-({}^hf_2), {}^hf_3\}, \quad \mathbf{G}_2 := \{{}^hf_1, {}^hf_2, {}^hf_4, {}^hf_5\}, \quad \mathbf{G}_3 := \{-({}^hf_5), {}^hf_3\},$$

$$\mathbf{G}_4 := \{-({}^hf_4), {}^hf_2, {}^hf_5\}, \quad \mathbf{G}_5 := \{-({}^hf^1), {}^hf_2, {}^hf_4, {}^hf_5\};$$

$$\mathbf{D}_i := \mathrm{DIFF}(\mathbf{G}_i).$$

As above, $\mathbf{C}_i = \mathbf{D}_i^*$, $\mathbf{F}(I) = \{\mathbf{C}_1, \mathbf{C}_2, \mathbf{C}_3, \mathbf{C}_4, \mathbf{C}_5\}$ and $\mathbf{D}_i, \mathbf{G}_i, \mathbf{m}_i$ satisfy Theorem 4.3. By affinisation we get a partition of $\mathbb{R}^2$ into polyhedral cones $\{C_1, C_2, C_3, C_4, C_5\}$, where $C_i := p(\mathbf{D}_i)^*$ for all i, and $\mathsf{m}_i := {}^a\mathbf{m}_i = M_\sigma(I)$ for all $\sigma \in \mathbf{O}(+, p(\mathbf{D}_i))$; we remark again that $\mathsf{m}_3 = \mathsf{m}_4 = \mathsf{m}_5$:

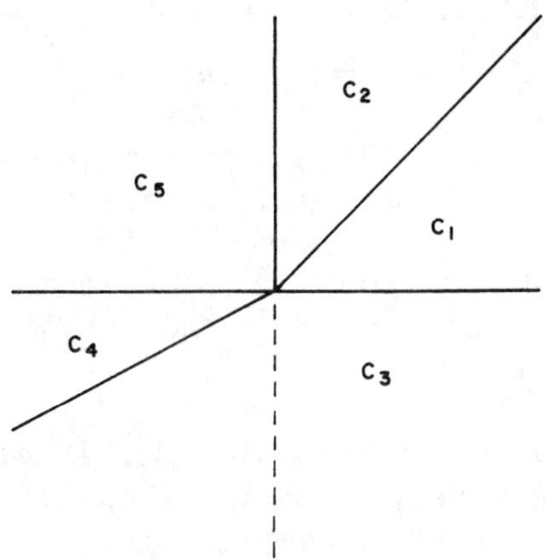

The example shows that outside $G(I)$ the partition depends on d; we remark also that the same monomial ideal can be obtained as $M_\sigma(I)$, even if $L(\sigma)$ belongs to different cones. Actually $\mathbb{R}^2$ can be covered by three regions C_1, C_2, C_3, such that for $L(\sigma) \in C_i$, $M_\sigma(I) = \mathrm{m}_i$, where $\mathrm{m}_i := (X_1, X_2^2)$, $\mathrm{m}_2 := (X_1^2, X_2)$, $\mathrm{m}_3 := (X_1, X_2)$:

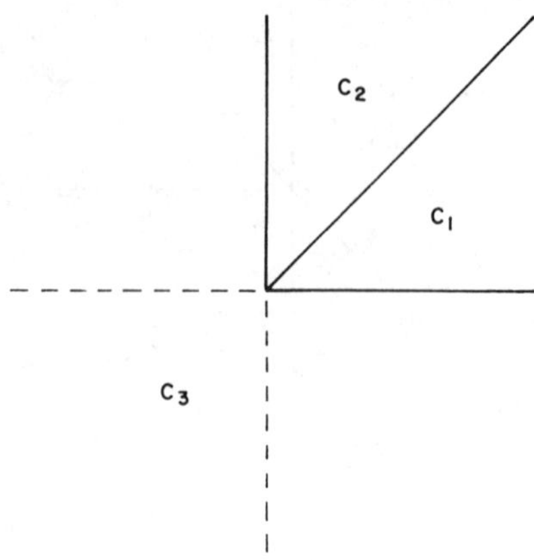

5. The Gröbner Region

The main computational property of a term-ordering σ is that, given an ideal I and a standard set G of I with respect to σ, Buchberger reduction with respect to G terminates when applied to any polynomial f, and it returns a canonical representative for its class $\bar{f}$ in A/I. However, 2.4(d), implies that, once an ideal I is fixed, the same property is shared by a larger set of orderings. So it is natural to pose the following:

DEFINITION 1. An ordering σ is called an *almost term-ordering* with respect to I if there exists a standard set $\{g_1, \ldots, g_r\}$ of I with respect to σ such that $-\wedge\rightarrow^*$ is noetherian, where $\wedge := (g_1, \ldots, g_r; M_\sigma(g_1), \ldots, M_\sigma(g_r)\}$. We will denote by $\mathbf{ATO}(I)$ the set of almost term-orderings with respect to I.

LEMMA 5.1. *Let C be an n-dimensional polyhedral cone. Then*:

$$\dim(C^* \cap (\mathbb{R}^n)^+) < n \Leftrightarrow C \cap (\mathbb{R}^n)^- \neq \{0\}$$

where $(\mathbb{R}^n)^- := \{(a_1, \ldots, a_n) \in \mathbb{R}^n / a_i \leqslant 0 \; i = 1, \ldots, n\}$.

PROOF. $\Rightarrow$ Since $C^* \cap (\mathbb{R}^n)^+ = (C \cup (\mathbb{R}^n)^+)^*$, the cone generated by $C \cup (\mathbb{R}^n)^+$ has a vertex with positive dimension. So there are $v_1, \ldots, v_r \in C$, $w_1, \ldots, w_s \in (\mathbb{R}^n)^-$ and non negative real numbers $\lambda_1, \ldots, \lambda_r, \mu_1, \ldots, \mu_s$ such that $\Sigma \lambda_i v_i = \Sigma \mu_j w_j \neq 0$.
$\Leftarrow$ Let $w \in C \cap (\mathbb{R}^n)^-$, $w \neq 0$, and let $v \in C^* \cap (\mathbb{R}^n)^+$. Then $v \,.\, w \geqslant 0$, since $v \in C$ and $w \in C^*$; however $v \,.\, w \leqslant 0$, since $v \in (\mathbb{R}^n)^+$ and $w \in (\mathbb{R}^n)^-$. So $C^* \cap (\mathbb{R}^n)^+ \subset \{v : v \,.\, w = 0\}$ has dimension less than n. □

LEMMA 5.2. *Let σ be an ordering such that $L(\sigma) \notin \bigcup_{D^* \in \mathbf{F}(I)} D^*$ and let G be a standard set of I with respect to σ. Then there are $g_1, \ldots, g_s \in G$, not necessarily distinct, $n_j \in \operatorname{Supp}(g_j)$ for $j = 1 \ldots s$, and $t \in T_A - \{1\}$ such that, denoting $m_i := M_\sigma(g_i)$, then $n_1 \ldots n_s = t\, m_1 \ldots m_s$.*

PROOF. Let $E := \{\log(M_\sigma(g)) - \log(n): n \in \mathrm{Supp}(g), g \in G\}$ and let C be the cone generated by E. If $\dim(C^* \cap (\mathbb{R}^n)^+) = n$, then there is $w \in C^*$, which is also in the interior of $(R^n)^+$. This implies that there exists a term-ordering τ, such that $w = u_1(\tau)$ is in C^*; therefore $L(\sigma) \in \bigcup_{D^* \in \mathbf{F}(I)} D^*$, a contradiction. Since $\dim(C^* \cap (\mathbb{R}^n)^+) < n$, there are $e_1, \ldots, e_s \in E$, not necessarily distinct, such that $0 \neq \sum_{j=1 \ldots s} e_j \in C \cap (\mathbb{R}^n)^-$. Let $w := -\sum_{j=1 \ldots s} e_j \in (\mathbb{N}^n)^+$ and $t := \log^{-1}(w)$. Since, for $j = 1 \ldots s$, there is $g_j \in G$, $n_j \in \mathrm{Supp}(g_j)$ such that $e_j = \log(M_\sigma(g_j)) - \log(n_j)$, the conclusion follows. □

LEMMA 5.3. *Let G be a finite set of polynomials, and for each $g \in G$ let $M(g) \in \mathrm{Supp}(g)$ be such that if we write $E := \{\log(M(g)) - \log(n): n \in \mathrm{Supp}(g), g \in G\}$, then $O(+, E) \neq \emptyset$. Let moreover $g_1, \ldots, g_s$ be elements, not necessarily distinct, of G; $n_j \in \mathrm{Supp}(g_j)$ for $j = 1 \ldots s$; and $t \in T_a - \{1\}$ be such that, if we write $m_i := M(g_i)$, then $n_1 \ldots n_s = t\, m_1 \ldots m_s$. Then if $\wedge := (g_1, \ldots, g_s; m_1, \ldots, m_s)$, the relation $-\wedge \rightarrow^*$ is not noetherian.*

PROOF. For each $r \in \mathbb{N}$ and each k, $0 \leqslant k < s$, let $t_{rs+k} := t^r n_1 \ldots n_k m_{k+1} \ldots m_s$. It is clear that for each $\sigma \in \mathrm{O}(+, E)$, $t_i > t_{i+1}$ in σ, for each i. Let then h be a polynomial such that $\mathrm{Supp}(h) \cap \{t_i: i \in \mathbb{N}\} \neq \emptyset$ and let $j := \max\{i: t_i \in \mathrm{Supp}(h)\}, j := rs + k$, with $0 \leqslant k < s$, $r \in \mathbb{N}$. Let $r(h) := h - c(h, t_j)\,(c(g_{k+1}, m_{k+1}))^{-1}\, t_j/m_{k+1}\, g_{k+1}$. It is easy to see that $t_{j+1} \in \mathrm{Supp}(r(h))$, so that $r(h) \neq 0$ and $j' := \max\{i: t_i \in \mathrm{Supp}(r(h))\} > j$. Let then $h_0 := t_0$ and, for $i \geqslant 1$, $h_i := r(h_{i-1})$. Then for each i, we have $h_i -\wedge\rightarrow h_{i+1}$. □

We are able now to characterise $\mathbf{ATO}(I)$ in terms of $\mathbf{F}(I)$; let therefore $\{\mathrm{m}_1, \ldots, \mathrm{m}_r\} := \mathrm{Mon}^+(I)$, G_i be the reduced Gröbner basis of I with respect to any $\sigma \in \mathbf{TO}(I, \mathrm{m}_i)$, $D_i = \mathrm{DIFF}(G_i)$.

THEOREM 5.4. $\mathbf{ATO}(I) = \bigcup_{i=1 \ldots r} \mathbf{O}(+, D_i)$.

PROOF. $\mathbf{ATO}(I) \supseteq \bigcup_{i=1 \ldots r} \mathbf{O}(+, D_i)$ follows from 2.5(a) and (c). If $\sigma \notin \bigcup_{i=1 \ldots r} \mathbf{O}(+, D_i)$ then there is τ such that $M_\sigma(I) = M_\tau(I)$ and $L(\tau) \notin \bigcup_{D^* \in \mathbf{F}(I)} D^*$ (cf. 4.3(d)). To prove that $\mathbf{ATO}(I) \subseteq \bigcup_{i=1 \ldots r} \mathbf{O}(+, D_i)$ it is then clearly sufficient to prove that if σ is such that $L(\sigma) \notin \bigcup_{D^* \in \mathbf{F}(I)} D^*$, then $\sigma \notin \mathbf{ATO}(I)$.

Let then σ be such that $L(\sigma) \notin \bigcup_{D^* \in \mathbf{F}(I)} D^*$ and let G be a standard set of I with respect to σ; then Lemma 5.2 implies that G satisfies the assumptions of Lemma 5.3, which in turn implies $\sigma \notin \mathbf{ATO}(I)$. □

DEFINITION 2. $\mathbf{G}(I) := \bigcup_{D^* \in \mathbf{F}(I)} D^*$ will be called the *Gröbner region* of I.

COROLLARY 5.5.

(*a*) $\mathbf{ATO}(I) = \{\sigma = \mathbf{ord}(\mathbf{u}), \mathbf{u} = (u_1, \ldots, u_s),\ \ \sigma \in \mathbf{O}(\mathbb{Z}^n)/\Sigma\, \varepsilon^{k-1} u_k \in (\mathbf{G}(I))^\circ$ for every small $\varepsilon > 0.\}$.

(*b*) $C(\mathbf{ATO}(I)) = \mathbf{G}(I)$.

PROOF. It follows from 2.1. and 5.4. □

The following examples show that $\mathbf{G}(I)$ can be *bigger* than $(\mathbb{R}^n)^+$.

EXAMPLE 1. Let $f = Y^2 - XY - X^4$; $I = (f)$. Then $\mathbf{G}(I)$ is the shaded part of the picture.

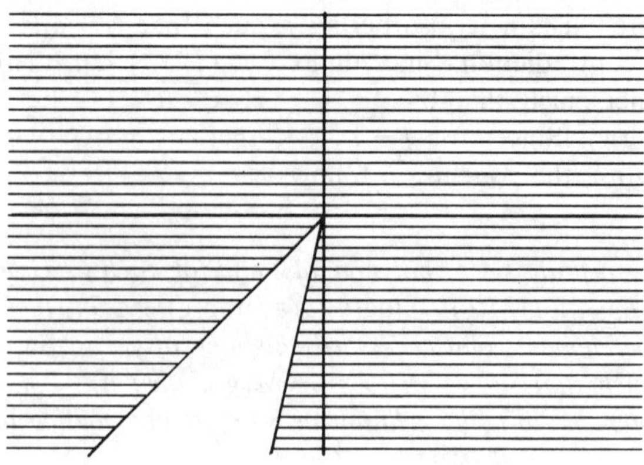

EXAMPLE 2. Let $f = X - X^2Y$; $I = (f)$. Then $\mathbf{G}(I)$ is the shaded part of the picture.

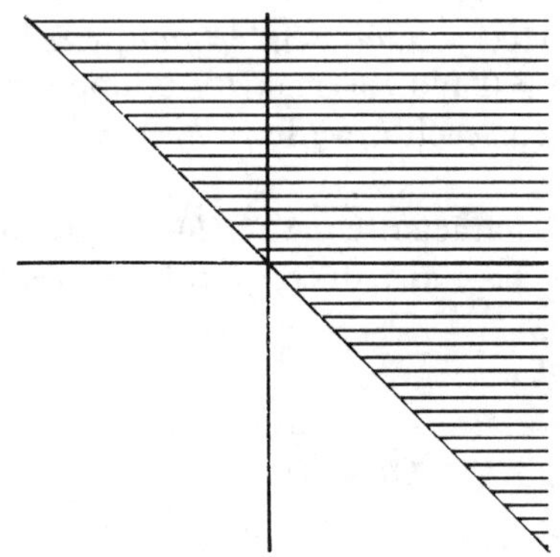

We note that I is $(1, -1)$-homogeneous.

PROPOSITION 5.6. *Let I be an ideal and let $W(I) := \{u \in \mathbb{R}^n / I$ is u-homogeneous$\}$. Then*

(a) *$W(I)$ is an $\mathbb{R}$-subvectorspace of $\mathbb{R}^n$ of type $W_{\mathbb{Q}} \otimes \mathbb{R}$.*
(b) *Given $d \in (\mathbb{N}^+)^n$ and with the notations of Theorem 4.3, $W(I)$ is contained in every $p(D_i^*)$.*

PROOF.
(a) This is an easy exercise.
(b) Let $0 \neq u \in W(I)$, $L(u)$ the line generated by u. If $\mathbf{G}_i$ is a reduced Gröbner basis of ${}^hI$, then $\mathbf{G}_i$ is $i(u)$-homogeneous since ${}^hI$ is such. Then D_i is $i(u)$-homogeneous, whence ${}^aD_i$ is u-homogeneous. Therefore $({}^aD_i)^* = p(D_i^*) \supseteq L(u)$. □

We also remark that $W(I)$ is in the vertex of each D_i^*, so that the orthogonal projection of each D_i^* onto the orthogonal complement of $W(I)$ is sufficient to reconstruct D_i^*.

COROLLARY 5.7. *Let $d \in (\mathbb{N}^+)^n$ and assume that I be d-homogeneous. Then $\mathbf{G}(I) = \mathbb{R}^n$.*

PROOF. From 4.3 we know that $\mathbb{R}^n = \bigcup_i p(D_i^*)$; now every $p(D_i^*) \supseteq L(d)$ by 5.6, hence every $p(D_i^*)$ intersects $(\mathbb{R}^n)^+$, hence it is contained in $\mathbf{G}(I)$ by 4.5. Another direct proof can be given by using 3.8. □

While Example 2 shows that 5.7 cannot be extended to *any* d, the next example shows that $\mathbf{G}(I) = \mathbb{R}^n$ can happen even if I is not d-homogeneous.

EXAMPLE 3. Let $A := k[X_1, X_2, X_3]$, $f := X_1^4X_3 + X_1^3X_2^3 + X_2^4 + X_1^2X_2^2X_3$, $I := (f)$. For I to be (d_1, d_2, d_3)-homogeneous, one should have $4d_1 + d_3 = 3d_1 + 3d_2 = 4d_2 + 2d_1 + 2d_2 + d_3$, while this system has only the trivial solution. However if $u_1(\sigma) = (1, 0, 1)$ then $M_\sigma(f) = X_1^4X_3$; if $u_1(\sigma) = (1, 1, 1)$ then $M_\sigma(f) = X_1^3X_2^3$; if $u_1(\sigma) = (0, 1, 1)$ then $M_\sigma(f) = X_2^4$; if $u_1(\sigma) = (0, 1, 3)$ then $M_\sigma(f) = X_1^2X_2^2X_3$. So the set of orderings is partitioned in four sets, in each of which a term ordering is contained, such that $M_\sigma(I)$ is constant when σ varies in each set.

6. An Algorithm

We conclude by presenting an algorithm that, given a basis $F = \{f_1, \ldots, f_r\}$ of an ideal I, returns all the reduced Gröbner bases and the Gröbner region of I.

In an initialization phase, the algorithm gives a partition of $(\mathbb{R}^n)^+$ into cones, in each of which the maximal terms of the elements of the input basis are constant.

The algorithm then works essentially by running several Buchberger's algorithms in parallel, one for each cone in the partition: every time that in one of the parallel computations, a new element f is added to the basis, a finer partition of the cone is done, in such a way that the maximal term of f is constant in each sub-cone.

So, the algorithm produces a partition of $(\mathbb{R}^n)^+$ into distinct cones, and, for each cone, a basis G which is a (not necessarily reduced) Gröbner basis for all the term-orderings "inside" the cone. Then, a final step is performed in which these bases are reduced and the corresponding cone is computed.

More precisely, the algorithm returns a set $L_{\text{output}} := \{(G_i, M_i, E_i, \Psi_i)/i = 1 \ldots s\}$ where $G_i =: \{g_{i1}, \ldots, g_{ir(i)}\}$ is a finite set of polynomials, $M_i =: \{m_{i1}, \ldots, m_{ir(i)}\}$ is a finite set of terms with $m_{ij} \in \text{Supp}(g_{ij})$, $E_i = \{\log(m_{ij}) - \log(n)/n \in \text{Supp}(g_{ij}), j = 1 \ldots r(i)\}$, $\Psi_i = \{(g_{ij}, m_{ij})/j = 1 \ldots r(i)\}$. (G_i, M_i, E_i, Ψ_i) is such that for all the orderings σ in $\mathbf{O}(+, E_i)$, G_i is a standard set of I with respect to σ, $m_{ij} = M_\sigma(g_{ij})$ for all j, and $(M_i) = M_\sigma(I)$. Moreover, for all the term-orderings σ in $\mathbf{O}(+, E_i)$, G_i is a reduced Gröbner basis of I with respect to σ, $\mathbf{F}(I) = \{E_i^*/i = 1 \ldots s\}$ and $\mathbf{G}(I) = \bigcup_{i=1 \ldots s} E_i^*$.

To prove the correctness of the algorithm, one has to remark, that at each step the union of the cones contains $(\mathbb{R}^n)^+$, so that at the end of the second part of the algorithm, not necessarily reduced Gröbner bases have been obtained for all term-orderings. Moreover, to get the Gröbner region of an ideal, one has to list the elements of $\text{Mon}^+(I)$, and for each of them to produce the corresponding reduced Gröbner basis and the corresponding polyhedral cone. It is easy to verify that this is done by the final part of the algorithm.

We remark that the usual improvements to the Buchberger algorithm can be applied also to this generalisation. We remark also that, given a finite set of polynomials G, it is possible to decide if $\mathbf{TO}(+, \text{DIFF}(G)) = \emptyset$ by linear programming techniques. In the course of the algorithm Ψ denotes a finite subset of $A \times T_A$; when it is convenient, the unique m such that $(g, m) \in \Psi$ will be denoted $\Psi(g)$. By a slight abuse of notations, we will write red(G, M), where G and M are sets (and not arrays) of polynomials and terms, respectively.

ALGORITHM

```
[INITIALISATION]
L := (Ø, Ø, Ø, Ø, Ø)
L_new := {L}
For i = 1 ... r do
  L_old := L_new
  L_new := Ø
  For (G, M, E, Ψ, B) ∈ L_old do
    For m ∈ Supp(f_i) do
      G' := G ∪ {f_i}
      M' := M ∪ {m}
      E' := E ∪ {log(m) − log(n)/n ∈ Supp(f_i) − {m}}
      Ψ' := Ψ ∪ {(f_i, m)}
      B' := B ∪ {{g, f_i}/g ∈ G}
      If TO(+, E') ≠ Ø then
        L := (G', M', E', Ψ', B')
        L_new := L_new ∪ {L}

[COMPUTATION OF THE GRÖBNER BASES]
L_work := L_new
L_partial := Ø
While L_work ≠ Ø do
  Choose (G, M, E, Ψ, B) ∈ L_work
  L_work := L_work − {(G, M, E, Ψ, B)}
  Choose {f, g} ∈ B
  B := B − {{f, g}}
  T := lcm(Ψ(f), Ψ(g))
  h := c(g, Ψ(g))T/Ψ(f)f − c(f, Ψ(f))T/Ψ(g)g
  While there is t ∈ Supp(h), g ∈ G such that Ψ(g) divides t do
    h := h − c(h, t)(c(g, Ψ(g)))^(−1) t/Ψ(g)g
  If h = 0 then
    if B = Ø then
      L_partial := L_partial ∪ {(G, M, E, Ψ)}
    else
      L_work := L_work ∪ {(G, M, E, Ψ, B)}
  else
    for m ∈ Supp(h) do
      G' := G ∪ {h}
      M' := M ∪ {m}
      E' := E ∪ {log(m) − log(n)/n ∈ Supp(h) − {m}}
      Ψ' := Ψ ∪ {(h, m)}
      B' := B ∪ {{g, h}/g ∈ G}
      If TO(+, E') ≠ Ø then
        L := (G', M', E', Ψ', B')
        L_work := L_work ∪ {L}
```

[COMPUTATION OF THE REDUCED GRÖBNER BASES AND OF THE GRÖBNER REGION]

$L_{\text{output}} := \emptyset$
Mon$:= \emptyset$
While $L_{\text{partial}} \neq \emptyset$ do
  Choose $(G, M, E, \Psi) \in L_{\text{partial}}$
  $L_{\text{partial}} := L_{\text{partial}} - \{(G, M, E, \Psi)\}$
  If $(M) \notin$ Mon then
    While there is $g \in G, m \in M - \{\Psi(g)\}, t \in T_A$ such that $\Psi(g) = t\,m$ do
      $G := G - \{g\}$
      $M := M - \{\Psi(g)\}$
      $\Psi := \Psi - \{(g, \Psi(g))\}$
    For all $g \in G$ do
      $g' := \text{red}(G - \{g\}, M - \{\Psi(g)\}, g)$
      $g' := g'/c(g', \Psi(g))$
      $G := G - \{g\} \cup \{g'\}$
      $\Psi := \Psi - \{(g, \Psi(g)\} \cup \{(g', \Psi(g)\}$
    $E := \bigcup_{g \in G} \{\log(\Psi(g)) - \log(n)/n \in \text{Supp}(g) - \{\Psi(g)\}\}$
    $L_{\text{output}} := L_{\text{output}} \cup \{(G, M, E, \Psi)\}$
    Mon$:=$ Mon $\cup \{(M)\}$

Nomenclature

$d(\mathbf{u})$	Rational dimension of $\mathbf{u}$.
$W_{\mathbb{Q}}(\mathbf{u})$	$\mathbb{Q}$-vectorspace generated by the "rational" components of $\mathbf{u}$.
$\mathbf{ord}(\mathbf{u})$	The ordering associated with $(\mathbf{u})$.
$\mathbf{O}(V)$	The set of compatible orderings of linear type on V.
$\mathbf{O}_{\text{tot}}(V)$	The set of compatible total orderings on V.
$\mathbf{O}(+, E)$	The set of orderings for which E is nonnegative.
$C(\mathbf{O}(+, E))$	The cone of "half-lines" of $\mathbf{O}(+, E)$.
C^*	The polar cone of C.
T_A	The monoid of terms in $A := k[X_1, \ldots, X_n]$.
$c(f, m)$	The coefficient of m in the representation of f.
$\text{Supp}(f)$	$\{m/c(f, m) \neq 0\}$.
$M_\sigma(f)$	The maximal monomial of f w.r. to the total ordering σ.
$M_\sigma(I)$	The maximal monomial ideal of I w.r. to the total ordering σ.
$g - \wedge \rightarrow h$	g reduces to h with respect to $\wedge := (f_1, \ldots, f_r; m_1, \ldots, m_r)$.
$- \wedge \rightarrow^*$	Reflexive-transitive closure of $- \wedge \rightarrow$.
$\mathbf{O}$	$\mathbf{O}_{\text{tot}}(\mathbb{Z}^n)$
$\mathbf{TO}$	$\mathbf{O}_{\text{tot}}(+, \mathbb{N}^n)$
$\text{Mon}(I)$	$\{\mathsf{m}/\mathsf{m} = M_\sigma(I), \sigma \in \mathbf{O}\}$,
$\text{Mon}^+(I)$	$\{\mathsf{m}/\mathsf{m} = M_\sigma(I), \sigma \in \mathbf{TO}\}$,
$\mathbf{O}(I, \mathsf{m})$	$\{\sigma \in \mathbf{O}/M_\sigma(I) = \mathsf{m}\}$
$\mathbf{TO}(I, \mathsf{m})$	$\{\sigma \in \mathbf{TO}/M_\sigma(I) = \mathsf{m}\}$
$\mathbf{F}^+(I)$	Restricted Gröbner Fan of I.
$\text{EXT}(\sigma, \Delta)$	Δ-extensions of σ.
$\mathbf{F}(I)$	Gröbner Fan of I.

ATO(I) Almost term-orderings with respect to I.
G(I) Gröbner region of I.

References

Bayer, D. A., Morrison, I. (1986). Standard bases and geometric invariant theory I, Initial ideals and state polytopes. This volume.
Bayer, D. A., Stillman, M. (1987). A theorem on refining division orders by the reverse lexicographical order. *Duke Math. J.* **55,** 321–328.
Buchberger, B. (1965). *Ein Algorithmus zum Auffinden der Basiselemente des Restklassenringes nach einem nulldimensionalen Polynomidealen.* Dissertation, Universität Innsbruck.
Buchberger, B. (1970). Ein algorithmisches Kriterium für die Lösbarkeit eines algebraischen Gleichungssystems. *Aequationes Matematicae,* **4,** 374–383.
Buchberger, B. (1976a). A theoretical basis for the reduction of polynomials to canonical form. *ACM SIGSAM Bull.* **10**(3), 19–29.
Buchberger, B. (1976b). Some properties of Gröbner bases for polynomial ideals. *ACM SIGSAM Bull.* **10**(4), 19–24.
Dress, A., Schiffels, G. (1987). Noetherian quasi orders and canonical ideal bases. Preprint.
Kolchin, E. R. (1973). *Differential Algebra and Algebraic Groups.* Academic Press.
Kollreider, C. (1978). Polynomial reduction: the influence of the ordering of terms on a reduction algorithm. Camp. Linz. Bericht Nr. 124.
Kuhn, H. W., Tucker, A. W. (eds.) (1956). *Linear Inequalities and Related Systems.* Princeton Univ. Press.
Lazard, D. (1983). Gröbner bases, Gaussian elimination and resolution of systems of algebraic equations. Proceeding of EUROCAL 83. *Springer Lec. Notes Comp. Sci.* **162,** 146–156.
Logar, A. (1988). Algorithm, in Algebra Commutativa, Ph.D. Thesis.
Möller, H. M., Mora, F. (1984). Upper and lower bounds for the degree of Gröbner bases. Proceedings of EUROSAM 84. *Springer Lec. Notes Comp. Sci.* **174,** 172–183.
Möller, H. M., Mora, F. (1986). New constructive methods in classical ideal theory *J. Algebra,* **100,** 138–178.
Robbiano, L. (1985). Term orderings on the polynomial ring. Proceedings of EUROCAL 85. *Springer Lec. Notes Comp. Sci.* **204,** 513–517.
Robbiano, L. (1986). On the theory of graded structures. *J. Symb. Comp.* **2,** 139–170.
Schemmel, K. P. (1987). Communication at EUROCAL 87 (Leipzig).
Schwartz, N. (1986). Stability of Gröbner bases. Preprint.
Trevisan, G. (1953). Classificazione dei semplici ordinamenti di un gruppo libero commutativo con N generatori. *Rend. Sem. Mat. Padova,* **22,** 143–156.
Weispfenning, V. (1987). Constructing universal Gröbner Bases. *Proc. AAECC 5 (Menorca).*

Standard Bases and Geometric Invariant Theory
I. Initial Ideals and State Polytopes

DAVID BAYER AND IAN MORRISON†

Department of Mathematics,
Columbia University, New York, NY 10027, U.S.A.

(*Received* 15 *October* 1986)

We characterise the set of monomial ideals that can occur as the initial ideals of a given homogeneous ideal Q in a polynomial ring, as one varies the monomial order used within a fixed coordinate system. This set is canonically bijective to the set of vertices of the *state polytope* of Q, a convex polytope arising in the geometric invariant theory of the Hilbert point of Q.

1. Introduction

Let k be a field of characteristic zero, let V be an n-dimensional k-vector space and let $S := \mathrm{Sym}(V)$. If $X_1, \ldots, X_n$ are a set of coordinates on V we identify S with the polynomial ring $k[X_1, \ldots, X_n]$ and if $I = (i_1, \ldots, i_n)$ is a multi-index, we write as usual X^I to denote the monomial

$$X_1^{i_1} \ldots X_n^{i_n}.$$

We will denote by $>$ the choice of a total ordering on the monomials of *each degree* in S; this is the homogeneous version of what in the affine case is often called a term-ordering. If f is a homogeneous element of S we denote by $in_>(f)$ the $>$-greatest monomial whose coefficient in f is non-zero and if Q is a homogeneous ideal of S we denote by $in_>(Q)$ the span in S of the set $\{in_>(f) \mid f \in Q\}$. The ordering $>$ is called *multiplicative* if for every pair (I, I') such that $X^I > X^{I'}$ and for every J we have $X^{I+J} > X^{I'+J}$. These notions were introduced by Macaulay (1927) who also proved the fundamental result

THEOREM 1.1. *If $>$ is a multiplicative total order and Q is a homogeneous ideal in S, then $in_>(Q)$ is an ideal of S.* ■

REMARK. Macaulay used a particular multiplicative order in his paper, but his proof is equally applicable to any other.

Macaulay's theorem justifies calling $in_>(Q)$ the $>$-*initial ideal* of Q. The importance of these monomial ideals was brought into new relief in 1964 with the pioneering work of Buchberger (1965, 1976) on standard bases (also known as Gröbner bases). His algorithm for computing standard bases through the use of such initial ideals has since served as the foundation for a host of other algorithms in commutative algebra and algebraic geometry

† Research supported by NSF Grant No. DMS–85–03743.

and has spurred considerable work on these ideals. (The authors, in collaboration with Michael Stillman, are preparing a brief survey of this work, entitled "The standard basis algorithm and computations in algebraic geometry.")

Our purpose in this series of articles is to study the connection between the standard basis theory a given Q and the geometric invariant theory of the Hilbert point $H(Q)$ of Q. The main result of this first note—theorem 3.1—is a characterisation of the collection of *all* monomial ideals which occur as $in_>(Q)$ for a given homogeneous ideal Q in terms of what we call the state polytope of Q and denote $\mathscr{P}(Q)$. This state polytope is a combinatorial invariant of the Hilbert point of Q and will be defined in section 2.

THEOREM 3.1. *There is a canonical bijection between the set of ideals $in_>(Q)$ which arise as $>$ varies over all multiplicative total orders on S and the set of vertices of the state polytope $\mathscr{P}(Q)$ of $H(Q)$.* ∎

After developing necessary background material about Hilbert points and their state polytopes in section 2, we give, in section 3, the precise statement of our characterisation (theorem 3.1) and its proof. We also connect these ideas with the work of Mora & Robbiano (1988) on Gröbner fans showing that the two approaches are essentially dual to each other (cf. remark 3.7).

That the computational complexity of finding the standard basis of an ideal Q may be very sensitive to the multiplicative total order used has been widely observed (Bayer & Stillman, 1987*a*, *b*; Lazard, 1983; Giusti, 1984). The desire to better understand this dependence was what led us to theorem 3.1. Implicit in theorem 3.1, however, is the choice of a coordinate system on V and both computational and theoretical experience suggest that the dependence of the performance of standard basis algorithms on this choice may be equally important (Bayer & Stillman, 1987*a*, *b*). We conjecture, but have so far been unable to prove, that there are also connections with geometric invariant theory in this direction. In section 4, we outline these conjectures and illustrate them with a computational example. The deeper study of these conjectures from both computational and theoretical perspectives will be the subject of the next paper in this series.

The work which is described in sections 2 and 3 dates from 1981–82. At the time we put it aside, briefly—or so we thought—planning to go on to further investigate the ideas discussed in section 4 before writing up the result. The already lengthy postponement of these plans might well have become permanent had a new impetus not been provided to follow them up by our invitation to discuss this work at the conference "Meeting on Computer and Commutative Algebra", Genova, 20–23 May 1986. It is therefore doubly a pleasure to thank the organisers, D. Arezzo, T. Mora, G. Niesi and L. Robbiano, not only for the invitation itself but for redirecting our attention towards the ideas outlined here.

2. Hilbert Points and State Polytopes

In this section we recall some facts about Hilbert points and the geometric invariant theory associated to them. All the notions we will need are standard in algebraic geometry but many may be unfamiliar to those whose interest in standard bases comes from computer science. In order to make this paper more accessible to this audience, therefore, we have treated in some detail the principal ideas connected with theorem 3.1. None of the objects we consider here change if we replace the ground field k by its algebraic closure so we will henceforth assume that k is algebraically closed.

Fix as before a homogeneous ideal Q in $S := k[X_1, \ldots, X_n]$ and let $P(m) = P_Q(m)$ be the Hilbert polynomial of Q and $\pi(m)$ that of S. Standard techniques—(cf. Bayer, 1982)—guarantee the existence of a degree m such that for *any* ideal Q' with $P_{Q'} = P$

(a) the saturation of Q' is determined by its degree m part Q'_m;
(b) $P(m) = \dim_k(Q'_m)$; (2.1)
(c) for all multiplicative total orders $>$, $in_>(Q')$ is generated by

$$\{in_>(f) \mid f \in Q', \deg(f) \geqslant m\}.$$

We fix such an m once and for all.

The ideal Q determines via its degree m part Q_m a point $H(Q)$ in $Gr(P(m), S_m)$, the Grassmanian of $P(m)$-dimensional subspaces of S_m. Conversely, in view of (2.1)(a), $H(Q)$ or equivalently Q_m determine the saturation of Q. We call $H(Q)$ the (mth) Hilbert point of the ideal Q. The set of Hilbert points of all ideals Q' of S with Hilbert polynomial P form a closed subscheme $\mathbf{H} := \mathbf{H}_P$ of the Grassmanian called the Pth-Hilbert scheme.

The Hilbert point has a matrix form which is convenient for computational purposes. First, choose an (ordered) basis $B = \{Y_j \mid j = 1, \ldots, P(m)\}$ of Q_m and a total order $>$ on the monomials of degree m. Then let $A = A_{Q,B,>}$ be the $P(m) \times \pi(m)$ matrix whose (j, I)th-entry is the coefficient of the monomial X_I in the basis element Y_j. We call A a *$>$-Hilbert matrix* for Q. If A is a $>$-Hilbert matrix for Q and A' is obtained from A by elementary row operations then A' is also a $>$-Hilbert matrix for Q. More precisely, suppose A is the matrix associated to a choice of basis B and $A' = CA$ for some invertible matrix C. Then there is a unique basis B' for which C is the change of coordinate matrix from B-coordinates into B'-coordinates and A' is the Hilbert matrix associated to B'. By Gaussian elimination, therefore, we may always choose A to be in row-echelon form.

The scheme $\mathbf{H}$ has a natural projective model in the space $\mathbf{P}(T)$ where $T := \Lambda^{P(m)}(S_m)$. In fact, $\mathbf{P}(T)$ even comes equipped with a natural set of homogeneous coordinates—the Plücker coordinates—whose definition we now recall. From our choice of a system of coordinates $\{X_i \mid i = 1, \ldots, n\}$ on the k-vector space V we obtain in the usual way monomial coordinates $\{X_I \mid I \text{ a multi-index s.t. } |I| = m\}$ on S_m. The Plücker coordinates $X_{\mathcal{I}}$ on T are indexed by sets $\mathcal{I}$ consisting of $P(m)$ of the multi-indices I and $X_{\mathcal{I}}(Q)$ is computed as follows. For each set $\mathcal{I}$, $X_{\mathcal{I}}(Q)$ is then the principal $P(m) \times P(m)$ minor of A corresponding to the choice of the $P(m)$-columns I in $\mathcal{I}$. Individually, each $X_{\mathcal{I}}(Q)$ depends on the choice of the basis B giving A. If B' is another basis and C is the change of basis matrix from B to B' coordinates then $A' = CA$. Hence, all the minors $X_{\mathcal{I}}(Q)$ change by the same non-zero factor $\det(C)$. This means that as a set of homogeneous coordinates in $\mathbf{P}(T)$ the collection $\{X_{\mathcal{I}}(Q)\}$ is well defined. Of greater importance in the sequel is the observation that for each individual Plücker coordinate the statements $X_{\mathcal{I}}(Q) = 0$ or $X_{\mathcal{I}}(Q) \neq 0$ make intrinsic sense.

We will also need to use a coarser decomposition of the space T which brings in the connection with geometric invariant theory. Via the choice of the basis $\{X_i\}$ of V, the group $G := GL(n, k)$ acts linearly on V. This action naturally induces linear representations of G on $S_m = \mathrm{Sym}_m(V)$ and on T, and this last descends to an action on $\mathbf{P}(T)$ which clearly leaves invariant the Grassmanian and the Hilbert scheme $\mathbf{H}$. It is the G-orbit of the Hilbert point $H(Q)$ which we will be concerned with in the sequel. For now we wish to focus on the action of the subgroup D of G consisting of the diagonal matrices.

The group D is abelian so the representation of D on T induces a decomposition of T as the direct sum of subspaces T_χ where χ runs over the characters of T in k^*. The characters of

D are indexed by n-tuples $R=(r_1,\ldots,r_n)$ of integers and the character χ_R corresponding to R is given by the formula

$$\chi_R\begin{pmatrix} d_1 & & 0 \\ & \ddots & \\ 0 & & d_n \end{pmatrix} = \prod_{i=1}^{n} d_i^{r_i}$$

A vector $t \in T$ lies in the subspace T_χ if and only if for all matrices $A \in D$, $A \cdot t$ is obtained by scalar multiplying t by $\chi(A)$. Corresponding to this direct sum decomposition of T, we get a decomposition of each Hilbert point

$$H(Q) = \sum_{\chi} H_\chi(Q)$$

into components $H_\chi(Q) \in T_\chi$ well defined up to simultaneous rescaling. Note that the basis of T dual to the Plücker coordinates is compatible with this eigenspace decomposition. Indeed, let $\chi = \chi_R$. Then the element dual to $X_{\mathcal{J}}$ lies in T_χ if and only if for each k between 1 and n,

$$\sum_{I \in \mathcal{J}} i_k(I) = r_k. \tag{2.2}$$

It will be convenient to view each character $\chi = \chi_R$ as a vector in $\mathbf{R}^n$ with coordinates $(r_1,\ldots,r_n)$. We then define the *state polytope* $\mathcal{P}(Q)$ of Q to be the (closed) convex hull in $\mathbf{R}^n$ of the set of vectors $\{\chi \mid H_\chi(Q) \neq 0\}$. The motivation for this terminology is that in representation theory the set $St(Q) = \{\chi \mid H_\chi(Q) \neq 0\}$ whose hull gives $\mathcal{P}(Q)$ is called the *state* of $H(Q)$. It is this polytope which links the geometric invariant theory of the Hilbert point of Q to the standard basis theory of Q.

3. The Main Theorem

Let us define $in(Q)$ to be the set of all ideals $in(Q)$ which arise as $>$ ranges over all multiplicative total orders on the monomials in S. This section is devoted to the proof of

THEOREM 3.1. *There is a canonical bijection between* $in_>(Q)$ *and the set of vertices of the state polytope* $\mathcal{P}(Q)$.

Before we embark on the proof we need one more notion from geometric invariant theory. A group homomorphism $\lambda: k^* \to D$ (or G) is called a *one-parameter subgroup* of D (or G). The one-parameter subgroups of D all have the form

$$\lambda(t) = \begin{pmatrix} t^{w_1} & & 0 \\ & \ddots & \\ 0 & & t^{w_n} \end{pmatrix},$$

where $w_1,\ldots,w_n$ are integers called the weights of λ on $X_1,\ldots,X_n$. In an obvious way, we can also assign weights to each of the monomial coordinates—if $I=(i_1,\ldots,i_n)$, then $w_I = w_1 i_1 + \ldots + w_n i_n$—and to each of the Plücker coordinates:

$$w_{\mathcal{J}} = \sum_{I \in \mathcal{J}} w_I.$$

We obtain in this way a *non-strict* order $\geq_\lambda$ on the monomials in S:

$$I \geq_\lambda I' \Leftrightarrow w_I \geq w_{I'}.$$

Since the weights w_i are integers there will always be ties in large degrees. However, for all λ not on a finite collection of hyperplanes, the partial order $\geq_\lambda$ will give a total order on the monomials of degree at most m. We will call such one-parameter subgroups m-generic. Conversely,

PROPOSITION 3.2. (*Bayer*, 1982) *For any given m, and any multiplicative total order* $>$, *there is an m-generic one parameter subgroup* λ *such that the orders* $>$ *and,* $\geq_\lambda$ *agree on monomials of degree at most m.* ■

REMARK. The definition of $\geq_\lambda$ makes sense starting from any real weights w_i, and Robbiano (1986) has shown that using a sequence of such weight vectors, it is possible to obtain any multiplicative total order in all degrees. We prefer to work in the fixed large degree m as we wish to emphasise later the action of the one-parameter subgroup λ which exists only when the weights are integral.

The connection between one-parameter subgroups and the ideas of section 2 is provided by the following easy but crucial

LEMMA 3.3. *Fix an ideal Q and an m-generic one-parameter subgroup* λ. *Then there is a unique Plücker coordinate* $\mathscr{I} = \mathscr{I}_\lambda(Q)$ *satisfying*

1. $X_{\mathscr{I}}(Q) \neq 0$.
2. *For any other* $\mathscr{I}'$ *such that* $X_{\mathscr{I}'}(Q) \neq 0$, $w_{\mathscr{I}'} < w_{\mathscr{I}}$.

REMARK. We at first found this surprising: m-genericity means that the w_I are distinct, but the $w_{\mathscr{I}}$'s are sums of $P(m)$ of these; the point is that condition 1 imposes very strong restrictions on $\mathscr{I}$. As the subsequent corollaries show, this lemma is essentially equivalent to the existence and uniqueness theorems for standard bases. Now to the easy

PROOF. It will be convenient to order the $P(m)$ monomials in each Plücker set $\mathscr{I}$ using $\geq_\lambda$ and to let I_j denote the jth largest. We may test condition 1 on any $\geq_\lambda$-Hillert matrix A of Q. Choose one in row echelon form and let $\mathscr{I}$ be the set of monomials I such that for some j the first non-zero entry in the jth row of A occurs in the Ith column. Since A is in echelon form, $\mathscr{I}$ has order $P(m)$ and $X_{\mathscr{I}}(Q) \neq 0$. Suppose $\mathscr{I}'$ is any other set of $P(m)$-monomials with the property:

$$(*_j) \qquad I'_j >_\lambda I_j.$$

Then, again since A is in row echelon form $X_{\mathscr{I}'}(Q) = 0$. On the other hand, if $(*_j)$ fails for every $j \leq P(m)$, then either $w_{\mathscr{I}'} < w_{\mathscr{I}}$ or $\mathscr{I}' = \mathscr{I}$. Hence, $\mathscr{I}$ also satisfies condition 2.

It will be convenient to have at hand two other characterisations of the coordinate $\mathscr{I}$ of the lemma which are easy corollaries of its proof. The first is algebraic. Let A be a Hilbert matrix for Q and $B = \{Y_1, \ldots, Y_{P(m)}\}$ be the corresponding basis of Q_m. For each j, $in_>(Y_j) = X_I$ if and only if the first non-zero entry in the jth row of A occurs in the Ith column. If A is in echelon form, then these I's are distinct and therefore

COROLLARY 3.4. *If A is a Hilbert matrix in echelon form and B is the corresponding basis, then*

(i) *B is a* $>$-*standard basis of* Q_m.
(ii) *The monomials in* $\mathscr{I}(Q)$ *span* $in_>(Q_m)$. ■

The second characterisation is more geometric and clarifies the role of the one-parameter subgroup λ. Let us denote by H^* the limit

$$\lim_{t \to 0} \lambda(t) \cdot H(Q)$$

taken in $\mathbf{P}(T)$. Since the Hilbert scheme $\mathbf{H}$ is projective, the point H^* lies in $\mathbf{H}$ and hence corresponds to some ideal Q^* in S with Hilbert polynomial P. The lemma says that $X_{\mathscr{I}(Q)}$ is the unique Plücker coordinate which does not vanish at H^* and so determines H^* as a point of $\mathbf{P}(T)$. But the monomial ideal spanned by the set $\{X_I | I \in \mathscr{I}\}$ has these Plücker coordinates. Therefore,

Corollary 3.5. $Q^* = in_>(Q)$.

In other words, we may view the λ-orbit of $H(Q)$ in $\mathbf{H}$ as a path from $H(Q)$ to the Hilbert point H^* of the $>$-initial ideal Q^* of Q.

Proof of theorem 3.1. A character $\chi = \chi_R$ in the state $St(Q)$ of Q is a vertex of the state polytope $\mathscr{P}(Q)$ if and only if there is an oriented affine hyperplane L in $\mathbf{R}^n$ with equation

$$\sum_{k=1}^{n} w_k r_k = b \tag{3.6}$$

such that χ lies on L but such that every other character in $St(Q)$ lies in the negative half-space determined by L. We take the negative half-space here in order to avoid minus signs in later formulas. We say that L is a *supporting hyperplane* for χ. We will use this observation and the lemma to describe maps in each direction between the set of monomial ideals arising as initial forms of Q and the set of vertices of $\mathscr{P}(Q)$, leaving to the reader the easy verification that these are inverse bijections.

Given a multiplicative total order $>$, let $Q^* = in_>(Q)$ and let H^* be the Hilbert point of Q^*. Choose a one-parameter subgroup λ with weights $w_1, \ldots, w_n$ such that $>$ and $>_\lambda$ agree on monomials of degree m and let $\mathscr{I}$ be the Plücker coordinate of the lemma. Use (3.6) to define L taking $b = w_{\mathscr{I}}$, and let $\chi = \chi_R$ be the character whose eigenspace T_χ contains the basis element dual to $\mathscr{I}$. Condition 1 of the lemma says that $H_\chi(Q) \neq 0$, hence that $\chi \in St(Q)$. Equation (2.2) implies that χ lies on L. Condition 2 of the lemma combined with equation (2.2) says that if χ' is any other character whose eigenspace contains the dual to a Plücker coordinate non-zero at $H(Q)$—i.e. if χ' is any other element of $St(Q)$—then χ' lies on the negative side of L. Therefore, χ is a vertex of $\mathscr{P}(Q)$. We send Q^* to χ.

Conversely, let χ be a vertex of $\mathscr{P}(Q)$ and L be a supporting hyperplane for χ. By perturbing L slightly we can assume that the equation of L has rational coefficients and then by a rescaling that these coefficients are integral. The coefficients w_k therefore determine a one-parameter subgroup λ of D and, by perturbing L again if necessary, we can assume that λ is m-generic. Let $>$ denote the corresponding order, let H^* be the limit of $H(Q)$ under λ and let Q^* be the corresponding monomial ideal. Corollary 3.5 says that $Q^* = in_>(Q)$, and we send χ to Q^*. ■

Remark 3.7. After this work was completed, we learned that Mora & Robbiano (1988) had obtained independently a result which is essentially the dual of theorem 3.1. They associate to each initial form of the ideal Q not the vertex χ of $\mathscr{P}(Q)$ but the cone in $\mathbf{R}^n$ spanned by the coefficient vectors $(w_1, \ldots, w_n)$ of *all* the supporting hyperplanes L of χ.

4. Stability and Complexity

In this section, we would like to introduce a conjectural connection between the ideas from geometric invariant theory introduced above and the theory of standard bases and to illustrate it with one example. A full discussion of these ideas will form the topic of Part II.

The question which first led us to theorem 3.1 was: for what choice(s) of order $>$ will be computation of the standard basis of a given ideal Q have small time and space requirements? This topic is taken up by Bayer & Stillman (1987*a*, *b*) and we wish to examine here the question of optimising in a different direction. Implicit in (3.1) is the choice of a coordinate system on the underlying vector space V of the polynomial algebra S. Both experience with various examples using the Macaulay computer algebra system, and theory (Bayer & Stillman, 1987*a*, *b*) show that for some Q this choice has a very important effect on the complexity (both in time and space) of computing a standard basis and that for others this effect is much less. We ask:

(1) For which ideals Q does the choice of a coordinate system critically effect the complexity of computing standard bases of Q?
(2) For Q of this type, how can optimal sets of coordinates be chosen?

To state our conjectured answer we will need a few more standard facts from geometric invariant theory. The ideal Q is called *semistable* if the origin does not lie in the closure of the $Gl(n, k)$-orbit of (any) lifting of $H(Q)$ from $\mathbf{P}(T)$ to T, and is called *unstable* if the origin does lie in this closure. (For more details on these definitions and the material which follows, see Mumford, 1977.) Empirically, "nice" varieties tend to be semistable, "nasty" ones to be unstable. For example, the ideals of most (but not all—see Morrison, 1980) smooth curves and surfaces are semistable while conversely the ideals of most varieties with singularities of high multiplicity or with embedded nilpotents are unstable. Two theorems relate the state polytope of Q to and its semistability or instability.

THEOREM 4.1 (HILBERT–MUMFORD NUMERICAL CRITERION). (*Mumford & Fogarty*, 1980). *The following conditions on an ideal Q in S are equivalent.*

(i) Q is semistable.
(ii) For any choice of coordinates on V and any one parameter subgroup λ of the corresponding group D of diagonal matrices, there are Plücker coordinates $\mathscr{I}$ and $\mathscr{J}$ which are non-zero at $H(Q)$ and for which $w_{\mathscr{I}} \leq 0 \leq w_{\mathscr{J}}$.
(iii) For any choice of coordinates on V the state polytope $\mathscr{P}(Q)$ contains the origin in $\mathbf{R}^n$. ■

The equivalence of (ii) and (iii) is clear; the point of the theorem is the connection between these and condition (i). The second result is due to Kempf. Reformulated in the language of the preceding sections it is,

THEOREM 4.2. (*Kempf*, 1978). *If the ideal Q is unstable, there is a filtration $\mathscr{F}$ of*

$$\mathscr{F} : \{0\} = U_0 \subset U_1 \subset \ldots \subset U_f = V$$

with the property that the distance from the state polytope $\mathscr{P}(Q)$ associated to a choice of coordinates to the origin in $\mathbf{R}^n$ is maximal if and only if this choice of coordinates is compatible with $\mathscr{F}$. ■

We would like to suggest that the computational complexity of standard basis calculations for an ideal Q is roughly independent of the choice of coordinate system when Q is semistable, but that it is sensitive to this choice when Q is unstable. Moreover, in the latter case the computationally optimal coordinate systems will be those compatible with the Kempf filtration. Our evidence, still fairly weak, for these conjectures comes from examples we have computed by hand and using the Macaulay computer algebra system. If the nature of the connection between the two theories is still not completely clear, we are nonetheless convinced of its existence and feel that better understanding it will add to our insight into both. We shall postpone a detailed discussion of these conjectures to Part II, contenting ourselves here with whetting the reader's appetite with an example.

The equation $y^2z - x^3 - xz^2 = 0$ defines an elliptic curve in $\mathbf{P}^2$. Embed this curve into $\mathbf{P}^3$ via the linear system $a = z^2$, $b = xy$, $c = y^2$, $d = yz$. The image is a degree 6 elliptic curve C, with equations

$$ab - c^2,\ abd - b^2c - d^3,$$

which has a triple point at $(1, 0, 0, 0)$: $(-1, 0, 1)$, $(0, 0, 1)$, and $(1, 0, 1)$ all map to $(1, 0, 0, 0)$ via this linear system. C is unstable, and the filtration

$$\mathscr{F} : \{0\} = U_0 \subset U_1 = \{a = 0\} \subset U_2 = V$$

is the filtration associated with C by theorem 4.2.

We computed standard bases for C with respect to the optimal elimination order which eliminates the variable a (see Bayer & Stillman, 1987*b*) after several linear changes of coordinates on $\mathbf{P}^3$. First, when a completely generic change of coordinates is used, the standard basis has a single element in degrees 2, 3, 4, 5, and 6 and in these generators a total of 104 monomials occur with non-zero coefficients. If we instead make a generic change of coordinates on the three variables a, b, c, a similar standard basis is obtained: the standard basis has a single element in degrees 2, 3, 4, 5, and 6 and a total of 90 monomials occur. However, if we make a generic change of coordinates on the three variables b, c, d, the standard basis has a single element in degrees 2, 3, and 4 and only 35 monomials are needed. This last change of coordinates is the most general change of coordinates which respects the filtration $\mathscr{F}$.

In this example, the generators of C are originally given in a sparse form which yields more efficient computations than any of the above computations. What the above computations and other examples show is that geometric invariant theory may be used to find a coordinate system in which a standard basis computation is relatively sparse. Had we started with a generic set of equations completely lacking in sparseness, the component of the sparseness represented by the filtration $\mathscr{F}$ could have been isolated purely by considerations of instability. Moreover, the flag $\mathscr{F}$ also suggested a computationally efficient order with which to calculate, that which preferentially eliminates the variable a. Hence, geometric invariant theory might permit us to obtain, in a "mechanical" way, the savings illustrated above. We plan to explore these ideas in greater depth in a subsequent paper.

References

Bayer, D. (1982). The division algorithm and the Hilbert scheme. Thesis, Harvard University, Order number 82-22588, University Microfilms International, 300 N. Zeeb Rd., Ann Arbor, MI 48106.

Bayer, D., Stillman, M. (1987*a*). A criterion for detecting *m*-regularity. *Invent. Math.* **87**, 1–11.

Bayer, D., Stillman, M. (1987*b*). A theorem on refining division orders by the reverse lexicographic order. *Duke Math. J.* **55,** No. 2.
Buchberger, B. (1965). Thesis, Univ. Innsbrück.
Buchberger, B. (1976). A theoretical basis for the reduction of polynomials to canonical forms. *ACM SIGSAM Bull.* **39,** 19–29.
Giusti, M. (1984). Some effectivity problems in polynomial ideal theory. In: *Proc. EUROSAM '84', Springer Lect. Notes Comp. Sci.*, **17,** 59–171.
Kempf, G. (1978). Instability in invariant theory. *Ann. Math.* **108,** 299–316.
Lazard, D. (1983). Gröbner bases, Gaussian elimination and resolution of systems of algebraic equations. In: *Computer Algebra—Proc. EUROCAL '83', Springer Lect. Notes Comp. Sci.* **162,** 146–156.
Macaulay, F. S. (1927). Some properties of enumeration in the theory of modular systems. *Proc. Lond. Math. Soc.* **26,** 531–555.
Morrison, I. (1980). Projective stability of ruled surfaces. *Invent. Math.* **56,** 269–304.
Mora, F., Robbiano, L. (1988). The Gröbner fan of an ideal. *J. Symb. Computat.* **6,** 183–208.
Mumford, D., Fogarty, J. (1980). *Geometric Invariant Theory* (2nd edn), *Ergebnisse der Mathematik und ihrer Grenzgebeite* **34.** New York: Springer-Verlag.
Mumford, D. (1977). Stability of projective varieties. *L'Ens. Math.* **23,** 39–110.
Robbiano, L. (1986). On the theory of graded structures. *J. Symb. Computat.* **2,** 139–170.

Gröbner Bases and Hilbert Schemes. I.

GIUSEPPA CARRÀ FERRO

Dipartimento di Matematica, Università di Catania Viale A. Doria 6, 95125 Catania, Italy

Let $<_\sigma$ be a sequential term ordering of the set T of all monomials in the variables $x_1, \ldots, x_n$. The author studies the family A of all ideals in $K[x_1, \ldots, x_n]$, that have the reduced Gröbner basis with respect to $<_\sigma$ with the same associated monomial ideal. It is shown that all such ideals have the same dimension and they are parametrized by an affine scheme V_A over K. Furthermore if $<_\sigma$ is a degree preserving term ordering on T, then all such ideals have the same Hilbert function. The author also shows that V_A is connected and the set of all prime ideals in A and the set of all smooth ideals in A are in one to one correspondence with open subsets of V_A. Finally it is shown that if $J, J' \in A$, then Top(J) and Top(J') can have different associated monomial ideals. Since it is possible to find the Top of the monomial ideal associated with J, then it is possible to decide if this ideal is the same as the monomial ideal associated to Top(J).

Let K be a field of char 0 and let $R = K[x_1, \ldots, x_n]$. Let $T = \{x_1^{a_1} \ldots x_n^{a_n} : (a_1, \ldots, a_n) \in N^n\}$ and let $<_\sigma$ be a term ordering on T i.e. a semigroup total ordering on T with $1 <_\sigma m$ for all $m \in T$. Let $K^* = K - \{0\}$ and let log: $T \to N^n$ be defined by $\log(x_1^{a_1}, \ldots, x_n^{a_n}) = (a_1, \ldots, a_n)$.

If $f \in R - \{0\}$ then it can be written in an unique way as

$$f = \sum_{i=1,\ldots,s} c_i m_i, \qquad c_i \in K^*, \quad m_i \in T, \quad m_1 >_\sigma m_2 >_\sigma \cdots >_\sigma m_s.$$

Denote $T(f) = m_1$, $l_c(f) = c_1$, $M(f) = c_1 m_1$.

DEFINITION. Let $<_\sigma$ be a term ordering on T. Let $f, g \in R$ with $f = \sum_{i,\ldots,s} c_i m_i$ and $g = \sum_{j=1,\ldots,r} c'_j m'_j$. f is called reduced with respect to g iff $m_i \neq m m'_1$ for all i and $m \in T$. A finite subset $\{f_1, \ldots, f_r\}$ of R is called autoreduced iff each f_i is reduced with respect to f_j for all $j \neq i$.

DEFINITION. Let $<_\sigma$ be a term ordering on T and let $f, g \in R$ as above. If f is reduced with respect to g, then we say that f reduces to f modulo g. If f is not reduced in a finite number of steps it is possible to find $h \in R$ reduced with respect to g and we will say that f reduces to h modulo g.

REMARK 1. Given $<_\sigma$ and $f, g \in R$ as above h is obtained in the following way. Suppose that there are $m_{i1}, \ldots, m_{ih}$ such that $m_{ij} = n_j m'_1$ for all $j = 1, \ldots, h$, $n_j \in T$. Let $f_1 = f - \sum_{j=1,\ldots,h} c_{ij}(c_1'^{-1}) n_j g$. If f_1 is reduced with respect to g, then $h = f_1$. If f_1 is not reduced, then we can find $f_2 = (f_1)_1$. If f_2 is reduced with respect to g then $h = f_2$. If f_2 is not reduced we can find $(f_2)_1 = f_3$. Let $f_r = (f_{r-1})_1$, $r \in N$. By Dickson Lemma, after a finite number of steps we can find f_r reduced with respect to g and then $h = f_r$.

DEFINITION. Let $<_\sigma$ be a term ordering on T and let $f, f_1, \ldots, f_r \in R$. We say that f reduces to h modulo $f_1, \ldots, f_r$ if h is obtained by f reducing with respect to each f_i, $i = 1, \ldots, r$.

Let I be an ideal in R. $M(I) = (M(f): f \in I)$ is called the monomial ideal associated with I with respect to $<_\sigma$.

DEFINITION. Let $<_\sigma$ be a term ordering on T. Let I be an ideal in R and let $f_1, \ldots, f_r \in I$ such that $\{f_1, \ldots, f_r\}$ is autoreduced and $M(f_i) = T(f_i)$ for all $i = 1, \ldots, r$. $\{f_1, \ldots, f_r\}$ is called a reduced Gröbner basis of I with respect to $<_\sigma$ iff any of the following conditions is verified:

(i) $M(I) = (M(f_1), \ldots, M(f_r))$
(ii) $f \in I$ iff f reduces to zero modulo $f_1, \ldots, f_r$.

DEFINITION. Let $f, g \in R$. $S(f, g) = (l_c(f))^{-1}(M/T(f))f - (l_c(g))^{-1}(M/T(g))g$ where $M = l.c.m.(T(f), T(g))$.

LEMMA 1. *Let $<_\sigma$ be a term ordering on T. Let I be an ideal in R and let $f_1, \ldots, f_r \in I$ such that $M(f_i) = T(f_i)$ for all $i = 1, \ldots, r$. $\{f_1, \ldots, f_r\}$ is a reduced Gröbner basis of I with respect to $<_\sigma$ iff $\{f_1, \ldots, f_r\}$ is autoreduced and $S(f_i, f_j)$ reduces to zero modulo $f_1, \ldots, f_r$ for all $i, j = 1, \ldots, r$.*

PROOF. It follows by the algorithm of Buchberger in order to construct a Gröbner basis when a finite set of generators of I is given.

LEMMA 2. *Let $<_\sigma$ be a term ordering on T and let $f_1, \ldots, f_r \in I$ such that $\{f_1, \ldots, f_r\}$ is autoreduced and $M(f_i) = T(f_i)$ for all $i = 1, \ldots, r$. $\{f_1, \ldots, f_r\}$ is a reduced Gröbner basis of I with respect to $<_\sigma$ if* G.C.D. $(T(f_i), T(f_j)) = 1$ *for all $i, j = 1, \ldots, r, i \neq j$.*

PROOF. If G.C.D. $(T(f_i), T(f_j)) = 1$ for some $i, j, i \neq j$, then it is well known that $S(f_i, f_j) = (T(f_j)f_i - T(f_i)f_j)$ reduces to zero modulo $f_1, \ldots, f_r$. Now proof follows by Lemma 1.

DEFINITION. Let $<_\sigma$ be a term ordering on T. $<_\sigma$ is called sequential iff for each $m \in T$ there are only finitely many monomials $n \in T$ such that $n <_\sigma m$. $<_\sigma$ is called degree compatible if $m <_\sigma n$ whenever degree of m is less than degree of n.

REMARK 2. The lexicographic order and the reverse lexicographic order on T are not sequential.

LEMMA 3. *Let $<_\sigma$ be a term ordering on T. If $<_\sigma$ is degree compatible then it is sequential.*

PROOF. The number of monomials in T of degree either less than or equal to a given s is equal to $\binom{s+n}{n}$.

REMARK 3. There are sequential term orderings on T that are distinct from the degree compatible term ordering as it is shown in (Robbiano, 1985). As example fixed a set of weights $w = (q_1, \ldots, q_n)$ of the variables such that $q \geqslant 1\ \forall\, i = 1, \ldots, n$, $q_i \in N$, we can define degree of $m = x_1^{a_1}, \ldots, x_n^{a_n} \in T$ as $\deg(m) = \sum_{i=1,\ldots,n} a_i q_i$. Each degree preserving term ordering $<_\sigma$ on T is sequential.

DEFINITION. Let $m \in T$ and let $<_\sigma$ be a sequential term ordering on T. $l_\sigma(m) = \#\{n \in T : n <_\sigma m\}$. Obviously $l_\sigma(1) = 0$ and we define $l_\sigma(0) = -1$. If $f \in R$, then $l_\sigma(f) = l_\sigma(T(f))$. If $f \neq 0$ and $f = \sum_{i=1,\ldots,s} c_i m_i$, $c_i \in K^*$, $m_1 >_\sigma m_2 >_\sigma \cdots >_\sigma m_s$, then $h(f) = s = \mathrm{card}(\mathrm{Supp}(f))$. Obviously $h(f) \leqslant l_\sigma(f) + 1$.

DEFINITION. Let $<_\sigma$ be a sequential term ordering on T. Let $\{m_1, \ldots, m_r\}$ be an autoreduced subset of T. Let $n_{m_1,\ldots,m_r,\sigma}(m_1) = \#\{n \in T : n <_\sigma m_i$ and $n = mm_j$ for some $m \in T$ and $j = 1, \ldots, r\}$. Let $p_\sigma(m_i) = l_\sigma(m_i) - n_{m_1,\ldots,m_r,\sigma}(m_i)$.

Let $\{m_1, \ldots, m_r\}$ be an autoreduced subset of T. Let $f_1, \ldots, f_r \in R$ such that $T(f_i) = M(f_i) = m_i$ for all $i = 1, \ldots, r$. If $\{f_1, \ldots, f_r\}$ is autoreduced, then $h(f_i) \leqslant p_\sigma(m_i) + 1$ for all $i = 1, \ldots, r$. Let $p = p_{m_1,\ldots,m_r,\sigma} = \sum_{i=1,\ldots,r} p_\sigma(m_i)$.

Let $A = A_{m_1,\ldots,m_r,\sigma} = \{$ideal J in $R : J = (f_1, \ldots, f_r)$ with $M(f_i) = T(f_i) = m_i$ for all $i = 1, \ldots, r$ and $\{f_1, \ldots, f_r\}$ reduced Gröbner basis of J with respect to $<_\sigma\}$.

REMARK 4. If J is an ideal in R, then J has an unique reduced Gröbner basis with respect to a fixed term ordering $<_\sigma$ on T (Buchberger, 1976).

LEMMA 4. *Let $<_\sigma$ be a sequential term ordering on T and let $m_1, \ldots, m_r \in T$ such that $\{m_1, \ldots, m_r\}$ is autoreduced. Let $p = \sum_{i=1,\ldots,r} p_\sigma(m_i)$. There is a one to one correspondence between the ideals in A and the points of an affine scheme $V = V_{m_1,\ldots,m_r,\sigma}$ in A_k^p.*

PROOF. Let $J \in A$. By definition there exist $f_1, \ldots, f_r$ such that $J = (f_1, \ldots, f_r)$ and $M(f_i) = T(f_i) = m_i\ \forall\, i = 1, \ldots, r$. So $f_i = m_i + \sum_{j=1,\ldots,p_\sigma(m_i)} c_{ij} n_j$, $c_{ij} \in K$ and $n_j \neq mm_i$ for all $m \in T$ and $i = 1, \ldots, r$. Let $\varphi: A \to A_k^p$ be the map defined by $\varphi(J) = (c_{11}, \ldots, c_{1p_\sigma(m_1)}, \ldots, c_{r1}, \ldots, c_{rp_\sigma(m_r)})$. φ is well defined, because J has an unique reduced Gröbner basis with respect to $<_\sigma$. $S(f_i, f_j)$ reduces to a polynomial $h_{i,j} = \sum_{h=1,\ldots,v(i,j)} b_h n_h$ for all $i, j = 1, \ldots, r$, where $n_h \in T$ and b_h is a polynomial function of the c_{ij}'s for all h, i, j because $T(f_i) = M(f_i)$ for all $i = 1, \ldots, r$. Since $\{f_1, \ldots, f_r\}$ is a Gröbner basis of J, by Lemma 1 $h_{ij} = 0\ \ \forall\, i, j = 1, \ldots, r$ and then $b_h = 0$ $\forall\, h = 1, \ldots, v(i, j)$ and for each $i, j = 1, \ldots, r$. In other words $\varphi(J)$ is a point of the affine scheme V in A_k^p defined by the ideal $I(V) = (b_h : h = 1, \ldots, v(i, j), i, j = 1, \ldots, r)$. The map φ is injective by its own definition. φ is also onto V because each point of V determines by Lemma 1 the reduced Gröbner basis of an ideal $J \in A$ with respect to $<_\sigma$.

COROLLARY 1. *Let $<_\sigma$ be a sequential term ordering on T and let $m_1, \ldots, m_r \in T$ such that $\{m_1, \ldots, m_r\}$ is autoreduced. If* G.C.D. $(m_1, m_j) = 1$ *for all* $i, j = 1, \ldots, r$ *then* $V_{m_1,\ldots,m_r,\sigma} = A_k^p$.

PROOF. It follows by Lemma 4 and by Lemma 2.

Let

$$R' = K[y_{11}, \ldots, y_{1p_\sigma(m_1)}, \ldots, y_{r1}, \ldots, y_{rp_\sigma(m_r)}]$$

and let

$$R'' = K[y_{11}, \ldots, y_{1p_\sigma(m_1)}, \ldots, y_{r1}, \ldots, y_{r,p_\sigma m(r)}, x_1, \ldots, x_n].$$

$$I(W) = \left(I(V), m_i + \sum_{j=1,\ldots,p_\sigma(m_i)} y_{ij} n_j, i = 1, \ldots, r \right)$$

is an ideal in R''.

PROPOSITION 1. *Let $S' = R'/I(V)$ and let $S'' = R''/I(W)$. Then S'' is a free S'-module.*

PROOF. Let $\pi: R' \to S'$ be the canonical ring homomorphism and let $y_{ij}^* = \pi(y_{ij})$ for each $i = 1, \ldots, r$ and $j = 1, \ldots, p_\sigma(m_i)$. $S'' \simeq S'[x_1, \ldots, x_n]/(f_1^*, \ldots, f_r^*)$ where $f_i^* = m_i + \sum_{j=1,\ldots,p_\sigma(m_i)} y_{ij}^* n_j$. Let π': $S'[x_1, \ldots, x_n] \to S''$ be the canonical ring homomorphism and let $n_j^* = \pi'(n_j)$ for all monomials $n_j \in T$. If $f \in S'[x_1, \ldots, x_n]$, then $f = M(f) + \Sigma\, d_j n_j$ with $d_j \in S'$ and $n_j <_\sigma M(f)$, $n_j \in T$. f reduces to a polynomial $h \in S''$ modulo $f_1^*, \ldots, f_r^*$ where $h = M(h) + \Sigma\, l_j n_j$ and $M(h), n_j \neq mm_i$ for all $m \in T$ and all $i = 1, \ldots, r$. In other words $f = g + h$ with $g \in (f_1^*, \ldots, f_r^*)$ and $\pi'(f) = \pi'(h)$. So S'' as S'-module is generated by all monomials n^* such that $n^* \neq m^* m_i^*$ for all $m \in T$ and $i = 1, \ldots, r$. In S'' all such monomials are S'-independent. In fact if $\Sigma\, l_j n_j^* = 0$ in S'', $l_j \in S'$, then $f = \Sigma\, l_j n_j \in (f_1^*, \ldots, f_r^*) \subset S'[x_1, \ldots, x_n]$. Since f is reduced with respect to $f_1^*, \ldots, f_r^*$, then $f = 0$ i.e. $l_j = 0$ in S' for all j, being $\{f_1^*, \ldots, f_r^*\}$ the reduced Gröbner basis in $S'[x_1, \ldots, x_n]$ for the ideal $(f_1^*, \ldots, f_r^*)$ by Zacharias (1978).

COROLLARY 2. *Let $<_\sigma$ be a sequential term ordering on T. The family of affine schemes $\{V(J): J \in A_{m_1,\ldots,m_r,\sigma}\}$ is flat on the affine scheme $V_{m_1,\ldots,m_r,\sigma}$ in A_k^p.*

PROOF. It follows by Proposition 1 and by definition of flat morphism between affine schemes (Hartshorne, 1977, p. 254).

PROPOSITION 2. *Let $\{f_1, \ldots, f_r\}$ be the reduced Gröbner basis of an ideal $I \subset R$ with respect to some term ordering $<_\sigma$ on T. Let $m_i = M(f_i)$ for all $i = 1, \ldots, r$ and let $h = \sup\{k: m_i \in K[x_{jk+1}, \ldots, x_{jn}]$ and $m_i \notin K[x_{jk+2}, \ldots, x_{jn}]$ for all $i = 1, \ldots, r\}$. Let $S = k[x_{j1}, \ldots, x_{jh}]$ and let $S' = R/J$. Then S' is a free S-module.*

PROOF. It is known that $S' \simeq S[x_{jh+1}, \ldots, x_{jn}]/(f_1, \ldots, f_r)$. Let $\pi: R \to S'$ be the canonical ring homomorphism and let $n^* = \pi(x)$ for all $n \in T$. If $f \in R$, then f reduces to a polynomial h modulo $f_1, \ldots, f_r$ where $h = \Sigma\, c_j n_j$, $c_j \in K$ and $n_j \neq mm_i$ for all i, j and all $m \in T$. Furthermore $f = f' + h$, where $f' \in (f_1, \ldots, f_r)$. So $\pi(f) = \pi(h) = \Sigma\, c_j n_j^*$. Since $h \in S[x_{jh+1}, \ldots, x_{jn}]$, then $h = \Sigma\, p_j n_j'$ where $p_j \in S$ and n_j' is a monomial in $x_{jh+1}, \ldots, x_{jn}$ So $\pi(h) = \Sigma\, \pi(p_j) n_j'^*$ and S' as S-module is generated by all monomials n_j^* in $\pi(x_{jh+1}), \ldots, \pi(x_{jn})$ such that $n_j \neq mm_i$ for all $i = 1, \ldots, r$ and all monomials m in $x_{jh+1}, \ldots, x_{jn}$. If $\pi(h) = 0$, then $h \in (f_1, \ldots, f_r)$. So $c_j = 0$ for all j by definition of Gröbner basis (Zacharias, 1978), i.e. $h = 0$ in $R = S[x_{jh+1}, \ldots, x_{jn}]$. We have $p_j = 0$ for all j.

COROLLARY 3. *Let I be an ideal in R and let $\{f_1, \ldots, f_r\}$ be the reduced Gröbner basis of I with respect to some term ordering $<_\sigma$ on T. Let $m_i = M(f_i)$ for all $i = 1, \ldots, r$ and let $h = \sup\{k: m_i \in K[x_{jk+2}, \ldots, x_{jn}]$ and $m_i \notin K[x_{jk}, \ldots, x_{jn}]$ for all $i = 1, \ldots, r\}$. The affine* *scheme* $V(I)$ *is flat on the affine scheme in* A_k^n *defined by the equations* $x_{jh+1}=\cdots=x_{jn}=0$.

PROOF. It follows by Proposition 2 and by definition of flat morphism between affine schemes (Hartshorne, 1977, p. 254).

EXAMPLE. Let $I=(x_1x_2-x_3)\subset K[x_1, x_2, x_3]$. If we take the reverse lexicographic order and assume $x_1>x_2>x_3$ then $M(x_1x_2-x_3)=-x_3$ and $V(I)$ is flat over the line $x_1=x_2=0$. If we take the lexicographic order and assume $x_1>x_2>x_3$ then $M(x_1x_2-x_3)=x_1x_2$ and $V(I)$ is flat over the plane $x_3=0$.

DEFINITION. Let I be an ideal in R. $\dim_k(I)=\sup\{r\colon I\cap K[x_{i1},\ldots,x_{ir}]=(0)\}$.

DEFINITION. Let $\{m_1,\ldots,m_r\}$ be an autoreduced subset of T. $E_{m_1,\ldots,m_r}=E=\{\log(m_i)\colon i=1,\ldots,r\}\in N^n$.

In Carrà Ferro (1987) there are given the definition of dimension of the set E and properties of E related to its dimension.

LEMMA 5. *Let* I *be an ideal in* R *and let* $\{f_1,\ldots,f_r\}$ *be the reduced Gröbner basis of* I *with respect to some term ordering* $<_\sigma$ *on* T. *Let* $m_i=M(f_i)$ *for all* $i=1,\ldots,r$. *Let* $d=\dim(E_{m_1,\ldots,m_r})$. *Then* $d=\dim_K I=\dim_K M(I)$.

PROOF. By (Carrà Ferro, 1987, Theorem 3.1, p. 2629) $d=\sup\{h\colon H(I_{n-h})\cap E_{m_1,\ldots,m_r}=\emptyset$ for some $I_{n-h}=\{i_{h+1},\ldots,i_n\}$ with $1\leqslant i_{h+1}<\cdots<i_n\leqslant n\}$, where $H(I_{n-h})=\{(a_1,\ldots,a_n)\in N^n\colon a_{ij}=0\ \forall j=h+1,\ldots,n\}$. Let $d'=\dim_k M(I)$. By definition $M(I)\cap K[x_{i1},\ldots,x_{id'}]=(0)$, so $d'\leqslant d$. On the other hand $m_i\notin K[x_{i1},\ldots,x_{id}]$ for all $i=1,\ldots,r$ and then $d=d'$. Furthermore $I\cap K[x_{i1},\ldots,x_{id}]=(0)$. In fact if $f\in I\cap K[x_{i1},\ldots,x_{id}]$ then $M(f)=nM(f_i)$ for some $n\in T$ and $i=1,\ldots,r$, i.e. $m_i=M(f_i)\in K[x_{i1},\ldots,x_{id}]$. So $\dim_K(M(I))\leqslant\dim_K(I)$. Let $R'=K[x_0,\ldots,x_n]$ and let T' be the set of all monomials in R'. It is possible to define a degree compatible term ordering $<_\tau$ on T' in the following way: $x_0^{a_0}x_1^{a_1}\ldots x_n^{a_n}<_\tau x_0^{b_0}\ldots x_n^{b_n}$ iff either $\Sigma\, a_i<\Sigma\, b_i$ or $\Sigma\, a_i=\Sigma\, b_i$ and $x_1^{a_1}\ldots x_n^{a_n}<_\sigma x_1^{b_1}\ldots x_n^{b_n}$. Let ${}^hI=({}^hf_1,\ldots,{}^hf_r)$ in R', where ${}^hf_i=x_0^{\deg f_i}f_i(x_1/x_0,\ldots,x_n/x_0)$ for all i. Let $\{g_1,\ldots,g_s\}$ be the reduced Gröbner basis of ${}^hI$ with respect to $<_\tau$ in R'. By Möller (1984) $\{{}^ag_1,\ldots,{}^ag_s\}$ is a Gröbner basis of I with respect to $<_\sigma$, where ${}^ag_i=g_i(1,x_1,\ldots,x_n)$ for all i. Since $\{f_1,\ldots,f_r\}$ is the reduced Gröbner basis of I with respect to $<_\sigma$ we can suppose that ${}^ag_i=f_i$ for all $i=1,\ldots,r$. Since ${}^hI$ is a homogeneous ideal in R', then $d'=\dim_K M({}^hI)=\dim_K({}^hI)$. If $0\neq ij$ for all $j=1,\ldots,d'$ and $0\neq hj$ for all $j=1,\ldots,d'$ whenever $M({}^hI)\cap K[x_{i1},\ldots,x_{id'}]=(0)$ and ${}^hI\cap K[x_{h1},\ldots,x_{hd'}]=(0)$ then $\dim_K(M(I))=\dim_K I=d'$. If $0=ij$ for some $j=1,\ldots,d'$ and $0=hj$ for some $j=1,\ldots,d'$ whenever $M({}^hI)\cap K[x_{i1},\ldots,x_{id'}]=(0)$ and ${}^hI\cap K[x_{h1},\ldots,x_{hd'}]=(0)$, then $\dim_K(M(I))=\dim_K(I)=d'-1$. If $0\neq ij$ for all $j=1,\ldots,d'$ and $0=hj$ for some $j=1,\ldots,d'$ whenever $M({}^hI)\cap K[x_{i1},\ldots,x_{id'}]=(0)$ and ${}^hI\cap K[x_{h1},\ldots,x_{hd'}]=(0)$, then $\dim_K(M(I))=d'$, while $\dim_K I=d'-1$. We have a contradiction, because $\dim_K(M(I))\leqslant\dim_K I$. If $0=ij$ for some $j=1,\ldots,d'$ and $0\neq hj$ for all $j=1,\ldots,d'$, whenever $M({}^hI)\cap K[x_{i1},\ldots,x_{id'}]=(0)$ and ${}^hI\cap K[x_{h1},\ldots,x_{hd'}]=(0)$. Then ${}^hI\cap K[x_{h1},\ldots,\widehat{x_{hj}},x_{hj+1},\ldots,x_{hd'},x_0]\neq(0)$ for all $j=1,\ldots,d'$. So $I\cap K[x_{h1},\ldots,x_{hd'}]=(0)$, while $I\cap K[x_{h1},\ldots,x_{hj},\ldots,\widehat{x_{hd'}},1]\neq(0)$ and we have a contradiction.

COROLLARY 4. *Let $<_\sigma$ be a degree preserving term ordering on T and let $m_1, \ldots, m_r \in T$ such that $\{m_1, \ldots, m_r\}$ is autoreduced. If $J \in A_{m_1,\ldots,m_r,\sigma}$, then the Hilbert function of J is equal to the Hilbert function of $(m_1, \ldots, m_r)$.*

PROOF. It follows by definition of $A_{m_1,\ldots,m_r,\sigma}$ and by Lemma 4.

PROPOSITION 3. *Let $<_\sigma$ be a sequential term ordering on T and let $B = B_{m_1,\ldots,m_r,\sigma} = \{J \in A_{m_1,\ldots,m_r,\sigma}: J$ is prime$\}$. There is a one to one correspondence between the elements of B and the points of an open subset of $V_{m_1,\ldots,m_r,\sigma}$.*

PROOF. Let $J \in A_{m_1,\ldots,m_r,\sigma}$. $J = (f_1, \ldots, f_r)$, where $\{f, \ldots, f_r\}$ is the reduced Gröbner basis with respect to $<_\sigma$ and $f_i = m_i + \sum_{j=1,\ldots,p_\sigma(m_i)} c_{ij}n_j$, $i = 1, \ldots, r$. By using the Ritt algorithm (Ritt, 1950, p. 95) we can find polynomials $g_1, \ldots, g_s \in R$ such that $g_j = \Sigma\, l_{h_j} n_h$ where $n_h \in T$ and l_{h_j} are polynomial functions of the c_{ij}'s and $(g_1, \ldots, g_s) \subseteq J \subseteq (g_1, \ldots, g_s): g^\infty$. It is known (Ritt, 1950, p. 97) that J is prime iff each g_i is irreducible. Furthermore it is known by (Perron, 1951, Theor. 140, p. 282) that each g_j is reducible iff the l_{h_j}'s are solutions of a finite number of homogeneous polynomial equations with coefficients in K. So J is not prime iff the c_{ij}'s satisfy some homogeneous polynomial equations $d_s(c_{ij}: i = 1, \ldots, r, j = 1, \ldots, p_\sigma(m_i)) = 0$, $s \in S$ where S is a finite set. So the set of all $J \in A_{m_1,\ldots,m_r,\sigma}$ that are not prime is in one to one correspondence with the closed subset $Z \cap V_{m_1,\ldots,m_r,\sigma}$ of $V_{m_1,\ldots,m_r,\sigma}$, being $I(Z) = (d_s(y_{ij}): s \in S) \subseteq K[y_{ij}: i = 1, \ldots, r, j = 1, \ldots, p_\sigma(m)]$.

PROPOSITION 4. *Let $<_\sigma$ be a sequential term ordering on T and let $C_{m_1,\ldots,m_r,\sigma} = \{J \in A_{m_1,\ldots,m_2,\sigma}$: the affine scheme $V(J)$ is smooth$\}$. There is a one to one correspondence between the elements of $C_{m_1,\ldots,m_r,\sigma}$ and the points of an open subset of $V_{m_1,\ldots,m_r,\sigma}$.*

PROOF. Let $I \in A_{m_1,\ldots,m_r,\sigma}$. $I = (f_1, \ldots, f_r)$ where $\{f_1, \ldots, f_r\}$ i.e. the reduced Gröbner basis with respect to $<_\sigma$ and $f_i = m_i + \sum_{j=1,\ldots,p_\sigma(m_i)} c_{ij}n_j$ for all $i = 1, \ldots, r$. Let $J(I) = \|\partial f_i/\partial x_j\|$ be the jacobian matrix of I. The singular locus of $V(I)$ is the intersection of $V(I)$ with the affine scheme W defined by the minors of maximal rank in $J(I)$. Each polynomial $\partial f_i/\partial x_j$ is equal to $\Sigma\, a_h n_h$ where $n_h \in T$ and a_h's are polynomial functions of the c_{ij}'s for all h. So each minor of maximal rank in $J(I)$ is defined by a polynomial $\Sigma\, b_k r_k$ where $r_k \in T$ and b_k's are polynomial functions of the c_{ij}'s for all k. Let $I(W) = (a_b: b \in B, B$ finite set$) \subset R$ be the defining ideal of W. Let $I' = I(V(I) \cap W) = (f_1, \ldots, f_r, a_b: b \in B)$and let $\{g_1, \ldots, g_s\}$ be the reduced Gröbner basis of I' with respect to $<_\sigma$. $g_j = \Sigma\, l_{h,j} n_h$ where $n_h \in T$ and the $l_{h,j}$ are again polynomial functions of the c_{ij}'s for all j. $V(I)$ is smooth iff $W \cap V(I) = \emptyset$, i.e. iff $1 \in I'$. Now $1 \in I'$ only when some of the g_j's as polynomial in $K[x_1, \ldots, x_n, y_{ij}: i = 1, \ldots, r\, j = 1, \ldots, p_\sigma(m_i)]$ is a non zero polynomial in the y_{ij}'s. Let $g_1, \ldots, g_k$ be such polynomials in the y_{ij}'s and let Z be the affine scheme in A_k^p such that $I(Z) = (g_1, \ldots, g_k)$. So there is a one to one correspondence between the points of $C_{m_1,\ldots,m_r,\sigma}$ and the points of $V_{m_1,\ldots,m_r,\sigma} \cap (A_k^p - Z)$.

Let J be an ideal in R and let $<_\sigma$ be a term ordering on T. By Bayer (1980—Theorem 2.12) there exists an ideal $J(t)$ in $k[x_1, \ldots, x_n, t]$ such that $J(0) = M(J)$ and $J(1) = J$. Furthermore $R/M(J)$ is a special fibre of a flat family parametrized by $K[t]$ and all fibres over closed points except $J(0)$ are isomorphic to J. The ideal $J(t)$ is given in the following way. Choose a set of weights $w = (d_1, \ldots, d_n) \in Z_+^n$ for the variables $x_1, \ldots, x_n$ and weight $d_0 = 1$ for t. Let $\{f_1, \ldots, f_r\}$ be the reduced Gröbner basis of J with respect to $<_\sigma$ and let $f_i = M(f_i) + \Sigma\, c_{ij} n_{ij}$ where $c_{ij} \in K^*$ and $n_{ij} <_\sigma M(f_i)$ for all j. Let $w(n)$ be the weight of n with respect to the set w for all $n \in T$. We must choose w in such way as $w(M(f_i)) > w(n_{ij})$ for all i and j. Let $\alpha_{ij} = w(M(f_i)) - w(n_{ij})$ and let $f_i(t) = M(f_i) + \Sigma\, c_{ij} t^{\alpha_{ij}} n_{ij}$ for all $i = 1, \ldots, r$. $J(t) = (f_1(t), \ldots, f_r(t))$ and $\{f_1(t), \ldots, f_r(t)\}$ is the reduced Gröbner basis of $J(t)$ by Bayer (1980—Theorem 2.6).

Now let $<_\sigma$ be a sequential term ordering on T and let $J \in A_{m_1, \ldots, m_r, \sigma}$ for some autoreduced subset $\{m_1, \ldots, m_r\}$ of T. All points $t^{\alpha_{ij}} c_{ij}$: $i = 1, \ldots, r$ $j = 1, \ldots, p_\sigma(m_i)$ are in $V_{m_1, \ldots, m_r, \sigma}$ and they are all the points of the affine one-dimensional subscheme of $V_{m_1, \ldots, m_r, \sigma}$ defined by the equations $y_{ij} = t^{\alpha_{ij}} c_{ij}$, $i = 1, \ldots, r$ $j = 1, \ldots, p_\sigma(m_i)$. Let $L_{c_{ij},w}$ be such subscheme. By Bayer (1980—Theorem 2.12) all $J \in A_{m_1, \ldots, m_r, \sigma}$ that correspond to the points of the open subset $L_{c_{ij},w} - \{(0, \ldots, 0)\}$ of $V_{m_1, \ldots, m_r, \sigma}$ are isomorphic.

Let $w_h = (d_{1_h}, \ldots, d_{n_h})$ be a set of weights for the variables $x_1, \ldots, x_n$, that satisfies conditions $w_h(M(f_i)) > w_h(n_{ij})$ for all i and j and some $h \in N$.

COROLLARY 5. *Let $<_\sigma$ be a sequential term ordering on T and let $J, J' \in A_{m_1, \ldots, m_r, \sigma}$ such that $J, J' \neq M(J)$. J and J' are isomorphic if there exists $L_{c^{(1)}_{ij},w_1}, \ldots, L_{c^{(s)}_{ij},w_s}$ such that $L_{c^{(i)}_{ij},w_i} \cap L_{c^{(i+1)}_{ij},w_{i+1}} \neq \{(0, \ldots, 0)\}$ for all $i = 1, \ldots, s-1$, $(c_{ij}) \in L_{c^{(1)}_{ij},w_1}$, and $(c'_{ij}) \in L_{c^{(s)}_{ij},w_s}$.*

PROOF. It follows by definition of isomorphism and by the properties of the product of morphisms.

COROLLARY 6. *Let $<_\sigma$ be a sequential term ordering on T and let $\{m_1, \ldots, m_r\}$ be an autoreduced subset of T. Then $V = V_{m_1, \ldots, m_r, \sigma}$ is connected.*

PROOF. Let $w = (d_1, \ldots, d_n)$ be a set of weights for which $w(m_i) > w(n_{ij})$ for all i and j such that $n_{ij} <_\sigma m_i$. First of all we shall prove that $L_{c_{ij},w}$ is connected for each $(c_{ij}) \in V$. In fact $K[y_{ij}: i = 1, \ldots, r, j = 1, \ldots, p_\sigma(m_{ij})]/(y_{ij} - t^{\alpha_{ij}} c_{ij}: i = 1, \ldots, r, j = 1, \ldots, p_\sigma(m_i)) \sim K[t^{\alpha_{ij}} c_{ij}: i = 1, \ldots, r, j = 1, \ldots, p_\sigma(m_i)] = B$, which is a subring of $K[t]$. If $f \in B$, then $f = \sum_{i=1,\ldots,t} b_t M_t$, $b_i \in K$ and M_i monomial in the $t^{\alpha_{ij}} c_{ij}$'s. So $f = \sum_{j=0,\ldots,d} \beta_j t^j$, where the β_j's are polynomial functions of the c_{ij}'s and $\beta_d \neq 0$. If $f^2 = f$, then $\beta_d^2 = 0$, i.e. $\beta_d = 0$ because $\beta d \in K$. So Spec(B) is connected. Since $L_{c_{ij},w}$ can be identified with max(B), suppose that max$(B) = (C_1 \cap \text{max}(B)) \cup (C_2 \cap \text{max}(B))$, with $C_1 \cap C_2 \cap \text{max}(B) = \emptyset$ being C_1, C_2 nonempty closed subsets of Spec(B). So either $C_1 \cup C_2 \neq$ Spec(B) or $C_1 \cup C_2 =$ Spec(B) and $C_1 \cap C_2 \neq \emptyset$. In the first case $C_1 \cup C_2 = V(\text{max}(B)) = \text{Spec}(B)$, because the radical of Jacobson of B is zero. In the second case there exists $m \in \text{max}(B)$ in $C_1 \cap C_2$. Now suppose that $V = C_1 \cup C_2$, where C_1 and C_2 are nonempty closed subsets of V with $C_1 \cap C_2 = \emptyset$. Since $(0, \ldots, 0) \in V$ then we can suppose that $(0, \ldots, 0) \in C_1$. Let $(c_{ij}) \in C_2$. $L_{c_{ij},w} \ni (0, \ldots, 0)$, (c_{ij}) so $L_{c_{ij},w} = (L_{c_{ij},w} \cap C_1) \cup (L_{c_{ij},w} \cap C_2)$ would be disconnected.

REMARK 5. Fixed a set of weights $w = (d_1, \ldots, d_n)$ such that $w(m_i) > w(n_{ij})$ for all i and j, then $L_{c_{ij},w}$ can be reducible. Let $L_{c_{ij},w}$ be defined by the equations $y_{ij} = t^{\alpha_{ij}} c_{ij}$. We can give a new order to the variables y_{ij} in such way as $\alpha_{11} = \min(\alpha_{ij}: j = 1, \ldots, p_\sigma(m_i))$. So by eliminating the parameter t we have the equations $y_{ij}^{b_{ij}} c_{11}^{d_{ij}} = y_{11}^{d_{ij}} c_{ij}^{b_{ij}}$, where $d_{ij}\alpha_{11} = b_{ij}\alpha_{ij} = \text{l.c.m.}(\alpha_{11}, \alpha_{ij})$. If the polynomial $y_{ij}^{b_{ij}} c_{11}^{d_{ij}} - c_{ij}^{b_{ij}} y_{11}^{d_{ij}}$ is reducible, then it factorizes in polynomials h_j such that the system $h_j = 0, j \in J$ has only $(0, \ldots, 0)$ as solution. So if $L_{c_{ij},w}$ is reducible, then it is union of irreducible one dimensional affine schemes, that meet only at the point $(0, \ldots, 0)$.

COROLLARY 7. *Let* $<_\sigma$ *be a sequential term ordering on* T. *Let* $\{m_1, \ldots, m_r\}$ *be an autoreduced subset of* T. *If* $V = V_{m_1,\ldots,m_r,\sigma}$ *is reducible, then* $(0, \ldots, 0)$ *is in the intersection of all irreducible components.*

PROOF. Let $V = V_1 \cup \cdots \cup V_t$, where V_j is an irreducible component for all $j = 1, \ldots, t$. Suppose that $(0, \ldots, 0) \in \bigcap_{j=1,\ldots,s} V_j$ when $s < t$. Let $(c_{ij}) \in V_{s+1}$. There exists an irreducible component L of $L_{c_{ij},w}$ containing $(0, \ldots, 0)$ and (c_{ij}) for each set of weights $w = (d_1, \ldots, d_n)$ as above. Since $L \subset V$, then $L \subset V_j$ for some j. $L \subset V_i$ for all $i = 1, \ldots, s+1$ by hypothesis. If $t = s+1$ we have a contradiction. If $t > s+1$, then $L \subset V_i$, $s+2 \leqslant i \leqslant t$. So $(0, \ldots, 0) \in V_1 \cap \cdots \cap V_s \cap V_i$ and we have again a contradiction.

COROLLARY 8. *Let* $<_\sigma$ *be a sequential term ordering on* T. *Let* $\{m_1, \ldots, m_r\}$ *be an autoreduced subset of* T. *Then the point* $(0, \ldots, 0) \in V_{m_1,\ldots,m_r,\sigma} = V$ *is in the closure of the nonempty open subsets* A *of* V, *that corresponds respectively to the irreducible affine schemes* $V(J)$ *and to the smooth affine schemes* $V(J)$, *when* $J \in A_{m_1,\ldots,m_r,\sigma}$.

PROOF. It follows by Bayer (1980). In fact if $(c_{ij}) \in A$ then each point of $L_{c_{ij},w} - \{(0, \ldots, 0)\} \subset A$ and $(0, \ldots, 0)$ is in the closure of each such $L_{c_{ij},w} - \{(0, \ldots, 0)\}$. If $A = \emptyset$ there is nothing to prove.

Now let $<_\sigma$ be a term ordering on T and let $\{m_1, \ldots, m_r\}$ be an autoreduced subset of T. For each i let $n_{ij} \in T$ such that $n_{ij} <_\sigma m_i$ and $n_{ij} \neq mm_i$ for all $m \in T$ and $i = 1, \ldots, r$. Let $f_i = m_i + \sum_{j=1,\ldots,\lambda(i)} c_{ij}n_{ij}$, $c_{ij} \in K$ and n_{ij} as above with $n_{i1} >_\sigma n_{i2} >_\sigma \cdots >_\sigma n_{i\lambda(i)}$ for all i. Let $A' = A_{m_1,\ldots,m_r,n_{ij},\sigma} = \{J$ ideal in R: $\{f_1, \ldots, f_r\}$ is the reduced Gröbner basis of J with respect to $<_\sigma$ for some $(c_{ij}$: $i = 1, \ldots, r\; j = 1, \ldots, \lambda(i))\}$. $A' \neq \emptyset$ because $J = (m_1, \ldots, m_r) \in A'$. Let $p' = \sum_{i=1,\ldots,r} \lambda(i)$. Lemma 4, Propositions 1, 3 and 4, Corollaries 1, 2, 5, 6, 7 and 8 are still true when we change the family A with A' and p with p'.

Let $<_\sigma$ be a term ordering on T, let $\{m_1, \ldots, m_r\}$ be an autoreduced subset of T and let $E = \{\log(m_i)$: $i = 1, \ldots, r\}$. Let $V(E) = \{(a_i, \ldots, a_n) \in N^n$: $(a_1, \ldots, a_n) \geqslant \log(m_i)$ for all i with respect to the product order$\}$. It is known (Kolchin, 1973, Lemma 16, p. 51) that the number of $(v_1, \ldots, v_n) \in V(E)$ such that $\sum_{i=1,\ldots,n} v_i \leqslant s$ is equal to a numerical polynomial $\omega_E(s)$ when $s \gg 0$. Let

$$\omega_E = \sum_{j=0,\ldots,h} a_j \binom{s+j}{j}.$$

Let $N_n = \{1, \ldots, n\}$. Given a subset $K \subset N_n$ and $a = (a_1, \ldots, a_n) \in N^n$ we define $a_K = \{(b_1, \ldots, b_n) \in N^n$: $b_k = a_k \; \forall\, k \notin K\}$. A subset $W \subset N^n$ is said to be r-dimensional if $W = a_K$ for some $a \in N^n$ and $K \subset N_n$ with $|K| = r$. A subset W of N^n is properly r-dimensional in $V(E)$ if W is an r-dimensional subset of $V(E)$ and it is not contained in any $(r+1)$-dimensional subset of $V(E)$. Let $d_r(V(E))$ be the number of subsets of N^n properly r-dimensional in $V(E)$. Note that $d_n(V(E)) \leqslant 1$ and equality holds iff $V(E) = N^n$, in which case $d_r(V(E)) = 0$ for all $r \neq n$. It is known (Sit, 1975, Prop. 1, p. 38) that $d_r(V(E))$ is finite for all r. Let $\alpha(E) = (d_0(V(E)), \ldots, d_n(V(E)))$, which is called the dimension sequence of $V(E)$. It is known (Sit, 1975, Coroll. 3, p. 40) that $d_r(V(E)) = 0$ when $r > h$ and $d_h(V(E)) = a_h$.

PROPOSITION 5. *Given* E *and* $V(E)$ *with dimension sequence* $\alpha(E) =$

$(d_0(V(E)), \ldots, d_h(V(E)), 0, \ldots, 0)$, *there exist subsets* $V_h = V(E) \supseteq V_{h-1} \supseteq \cdots \supseteq V_0$ *such that* $d_k(V_{h-r}) = d_k(V(E))$ *for all* $r \leqslant k \leqslant h$, *while* $d_k(V_{h-r}) = 0$ *when* $0 \leqslant k \leqslant r-1$. *Furthermore for all* $r = 0, \ldots, h$ *we have:*

(a) $V_{h-r} = \{(v_1, \ldots, v_n) \in V(E) \colon (v_1, \ldots, v_n)$ *is in some* r*-dimensional subset of* $V(E)\}$
(b) *there exists a numerical polynomial* $\beta_r(s)$ *of degree less than* r *such that when* $s \gg 0$ *the number of* $(v_1, \ldots, v_n) \in V_{h-r} - V_{h-r-1}$ *with* $\sum_{j=1,\ldots,n} v_j \leqslant s$ is equal to

$$d_r(V(E))\binom{s+r}{r} - \beta_r(s).$$

Proof. It follows by (Sit, 1975, Prop. 3, p. 34) and by (Sit, 1975, Coroll. 2, p. 40).

Given $V(E)$, let E_{h-r} be the set of minimal points in $N^n - V_{h-r}$ with respect to the product order. So $V_{h-r} = V(E_{h-r})$ for all $r = 0, \ldots, h$.

Remark 6. When $<_\sigma$ is a degree compatible term ordering on T and $J \in A_{m_1,\ldots,m_r,\sigma}$, then $\omega_E = H(J, s)$ for all $s \in N$ and all $J \in A_{m_1,\ldots,m_r,\sigma}$.

Definition. Let J be an ideal in R and let $h = \dim_K J$. Then J_r is the greatest ideal in R containing J such that $H(J_r/J, s)$ has degree either less than or equal to $r-1$. Obviously $r = 0, \ldots, h = \dim_K J$. If $h = r$, then J_h is denoted also by $\mathrm{Top}(J)$.

Lemma 6. *Let* J *be an ideal in* R. *Then* $J_h \supseteq J_{h-1} \supseteq \cdots \supseteq J_1 \supseteq J_0 = J$. *Furthermore* $(J_r)_{r'} = J_{r'}$ *for all* $r \leqslant r' \leqslant h$ *and* $(J_r)_{r'} = J_r$ *for all* $0 \leqslant r' \leqslant r-1$.

Proof. Suppose that $J_r \nsubseteq J_{r'}$ for some $r < r' \leqslant h$. Let $H_{r'} = (J_r)_{r'}$. $H_{r'} \supseteq J_r \supseteq J$. We have $H(H_{r'}, s) = H(J_r, s) - H(H_{r'}/J_r, s) = H(J, s) - H(J_r/J, s) - H(H_{r'}/J_{r,s}) = H(J, s) - H(H_{r'}/J, s)$. So $\deg H(H_{r'}/J, s)$ is either less than or equal to $r'-1$ and then $H_{r'} \subset J_{r'}$, i.e. $J_r \subset (J_r)_{r'} \subset J_{r'}$. On the other hand $H(J_{r'}/J_r, s) = H(J_{r'}/J, s) - H(J_r/J, s)$ i.e. $\deg H(J_{r'}/J_r, s) \leqslant r'-1$, so $J_{r'} \subseteq (J_r)_{r'}$. Now let $0 \leqslant r' < r$. $(J_r)_{r'} \supseteq J_r$ by its own definition and $J_r \supseteq (J_r)_{r'}$, because $H((J_r)_{r'}/J, s) = H((J_r)_{r'}/J_r, s) + H(J_r/J, s)$ has degree either less than or equal to $r-1$.

Fixed a degree preserving term ordering $<_\sigma$ on T, let $\{f_1, \ldots, f_r\}$ be the reduced Gröbner basis of the ideal J with respect to $<_\sigma$ and let $m_i = M(f_i) = T(f_i)$ for all $i = 1, \ldots, r$. Let $E = E_{m_1,\ldots,m_r,\sigma}$ and let E_{h-r} be the subsets of N^n defined as before when $r = 0, \ldots, h$, being $h = \dim_K J$. Each E_{h-r} defines the autoreduced set of monomials in $T\ \{n_j \colon \log(n_j) \in E_{h-r}, j = 1, \ldots, \gamma(r)\}$. Let $\{h_1, \ldots, h_{\varepsilon(r)}\}$ be the reduced Gröbner basis of J_r with respect to σ and let $s_i = M(h_i) = T(h_i)$ for all $i = 1, \ldots, \varepsilon(r)$, $r = 0, \ldots, h$. Let $E'_r = \{\log(s_i) \colon i = 1, \ldots, \varepsilon(r)\}$.

Proposition 6. *Let* J *be an ideal in* R *and let* $<_\sigma$ *be a degree compatible term ordering on* T. $V(E'_r) \supseteq V_{h-r}$ for all $r = 0,, \ldots, h$, *being* $h = \dim_K J$.

Proof. When $r = 0$, then $V_h = V(E)$ and $V(E'_0) = V(E)$, being $J_0 = J$. Now let $r = h$.

Since $V(E'_h) \subseteq V(E)$, then $d_h(V(E'_h)) \leqslant d_h(V(E))$.

$$H(J, s) = \omega_E(s) = \sum_{j=0,\ldots,h} d_j(V(E))\binom{s+j}{j} - \sum_{j=0,\ldots,h} \beta_j(s)$$

with $\deg \beta_j \leqslant j-1$ by Proposition 5. If $d_h(V(E'_h)) < d_h(V(E))$, then $H(J, s) - H(J_h, s)$ would have degree h and we have a contradiction. Suppose that $H(J, s) - H(J_h, s) > \sum_{j=0,\ldots,h-1} d_j(V(E)) - \sum_{j=0,\ldots,h-1} \beta_j(s)$. Then there exist subsets $K' \subset K$ with $0 \leqslant |K'| \leqslant h-1$, $|K| = h$ such that for some $a \in N^n$, $a_K \subset V(E)$ and $b \in a_K$ with $a_K - b_{K'} \subset V(E'_h)$. If $a_K = \{(c_1, \ldots, c_n) \in N^n \colon c_k = a_k \ \forall\, k \notin K\}$, then $b_{K'} = \{(c_1, \ldots, c_n) \in N^n \colon c_k = a_k \ \forall\, k \notin K$ and $c_k = b_k \ \forall\, k \in K - K'\}$. So there exists $(c'_1, \ldots, c'_n) \in a_K - b_{K'}$ with $c'_k = a_k$ when $k \notin K$ and $c'_k = b_k + 1$ when $k \in K - K'$. Furthermore each element of b'_K is less than some of these $(c'_1, \ldots, c'_n)$ with respect to the product order and then it belongs to $V(E'_h)$ by its own definition. If we take $K'' \subseteq K$ with $|K''| \leqslant h-1$ and suppose that $a_K - (b_{K'} \cup b'_{K''}) \subset V(E'_h)$, we can repeat the proof as before by taking $c'_k = \max(b_k, b'_k) + 1$ when $k \in K - (K' \cup K'')$. So $V_0 \subseteq V(E'_h)$. Now suppose that $r = h-1$. Since $(J_{h-1})_h = J_h$, then $d_h(V(E'_{h-1})) = d_h(V(E)) = d_h(V_1) = d_h(V_0) = d_h(V(E'_h))$. So

$$H(J, s) - H(J_{h-1}, s) = \sum_{j=0,\ldots,h-1} d_j(V(E))\binom{s+j}{j} - \sum_{j=0,\ldots,h-1} \beta_j(s) - \sum_{j=0,\ldots,h-1} d_j(V(E'_{h-1})) + \sum_{j=0,\ldots,h-1} \beta'_j(s)$$

by Proposition 5. It follows that $d_{h-1}(V(E'_{h-1})) = d_{h-1}(V(E)) = d_h(V_1)$. If

$$H(J, s) - H(J_{h-1}, s) > \sum_{j=0,\ldots,h-2} d_j(V(E))\binom{s+j}{j} - \sum_{j=0,\ldots,h-2} \beta_j(s)$$

we have a contradiction by repeating the proof as above. So $V_1 \subseteq V(E'_{h-1})$. Now by induction on $h-r$ we have $V_{h-r} \subseteq V(E'_r)$ for all $r = 0, \ldots, h$.

Definition. Let $\{m_1, \ldots, m_r\}$ be an autoreduced subset of T and let $J_0 = (m_1, \ldots, m_r)$ in R. Let $K_r = \{K \colon K \subset N_n$ and $|K| = r\}$ and let $M_K = (x_k - 1 \colon k \in K, K \in K_r)$. Let $P_K = (J_0, M_K) \cap K[x_k \colon k \notin K]$ when $K \in K_r$ and let $P_r = \cap\{P_K \colon K \in K_r\}$.

Lemma 7. *Let $I = (m_1, \ldots, m_r)$ and $J = (n_1, \ldots, n_s)$ be ideals in R such that $m_i, n_j \in T$ for all $i = 1, \ldots, r$ and $j = 1, \ldots, s$. Then $I \cap J = (v_{ij} \colon v_{ij} = \text{l.c.m.}(m_i, n_j), i = 1, \ldots, r$ and $j = 1, \ldots, s)$.*

Proof. Let t be an indeterminate. $I \cap J = (tI, (t-1)J)R[t] \cap R$ by Buchberger (1976). So $I \cap J = (tm_1, \ldots, tm_r, tn_1 - n_1, \ldots, tn_s - n_s)R[t] \cap R$. We can always suppose that $\{m_1, \ldots, m_r\}$ and $\{n_1, \ldots, n_s\}$ are autoreduced subsets of T. So $SP(m_i, m_j) = SP(n_i, n_j) = 0$. Now $SP(tm_i, tn_j - n_j) = \text{l.c.m.}(m_i, n_j)/n_j = v_{ij}$ for all $i = 1, \ldots, r$ and $j = 1, \ldots, s$. So $SP(tm_{i'}, v_{ij}) = 0 \ \forall\, i, i' = 1, \ldots, r$ and $j = 1, \ldots, s$ while $SP(tn_{j'} - n_{j'}, v_{ij}) = \text{l.c.m.}(n_{j'}, v_{ij})/n_{j'} \,.\, n_{j'} = \text{l.c.m.}(n_{j'}, v_{ij}) = a_{j'}v_{ij'}, a_{j'} \in T$. So $(tm_1, \ldots, tm_2, tn_1 - n_1, \ldots, tn_0, v_{ij})$ is a Gröbner basis of the ideal $(tI, (t-1)J)$ and then the lemma is proved.

Proposition 7. *Let $\{m_1, \ldots, m_r\}$ be an autoreduced subset of T and let $J_0 = (m_1, \ldots, m_r)$. Let $h = \dim_K J_0$. Then $P_r = (J_0)_r$ for all $r = 0, \ldots, h$.*

PROOF. Let $K \in K_r$. By definition $P_K = (s_{1K}, \ldots, s_{rK})$, where $s_{iK} m'_i = m_i$. s_{iK}, $m'_i \in T$ and m'_i monomial in the only x_k's, $k \in K$, while s_{iK} is a monomial in the only x_k's, $k \notin K$. Let $E_K = E_{s_{1K}, \ldots, s_{rK}}$ and let $V_K = V(E_K)$. Since $P_K \supseteq J_0$, then $V_K \subseteq V(E)$. Let $(a_{i_1}, \ldots, a_{i_n}) = \log(s_{iK})$ for all $i = 1, \ldots, r$. $a_{i_k} = 0$ when $k \in K$ and $(v_1, \ldots, v_n) \in V_K$ if some $v_k < a_{i_k}$ when $k \notin K$ and for all $i = 1, \ldots, r$. By (a) of Proposition 5 $V_K \subseteq V_{h-r}$. Let $P_r = (d_1, \ldots, d_{\lambda(r)})$. By Lemma 7 $d_l \in T$ for all $l = 1, \ldots, \lambda(r)$. Let $E''_r = E_{d_1, \ldots, d_{\lambda(r)}}$. Since $J_0 \subseteq P_r \subseteq P_K$ for all $K \in K_r$, then $V_K \subseteq V(E''_r) \subseteq V(E)$. Furthermore $\cup \{V_K: K \in K_r\} \subseteq V(E''_r)$. If $a \in V(E''_r)$ and $a \notin V_K$ for all $K \in K_r$, then $a \geqslant \log(s_{i_k})$ for all $K \in K_r$ and some $i = 1, \ldots, r$, i.e. $a \in N^n - V(E''_r)$, being $P_r \subset P_K$. So $V(E''_r) = \cup \{V_K: K \in K_r\} \subseteq V_{h-r}$. Let W_K be the union of all r-dimensional spaces a_K, that are contained in V_{h-r}. $V_K \subseteq W_K$. Let L_K be the set of minimal points in $N^n - W_K$ and let $N_K = \{t \in T: \log(t) \in L_K\}$. We have $J_0 \subseteq N_K \subseteq P_K$. On the other hand $(J_0, M_K) \cap K[x_k: k \notin K] \subseteq (N_K, M_K) \cap K[x_k: k \notin K] \subseteq (P_K, M_K) \cap K[x_k: k \notin K]$. So $P_K \subseteq (N_K, M_K) \cap K[x_k: k \notin K] \subseteq P_K$. But $(N_K, M_K) \cap K[x_k: k \notin K] = N_K$ for all $K \in K_r$ and then $N_K = P_K$, i.e. $V_K = W_K$. By Proposition 5 we have $V_{h-r} = \cup \{W_K: K \in K_r\}$ and then $V_{h-r} = V(E''_r)$. Since $P_r \supset J_0$ and $H(J_0, s) - H(P_r, s)$ has degree less than r by definition of V_{h-r}, we have $P_r = (J_0)_r$ by Proposition 6.

Unfortunately there are cases in which $V(E'_r) \neq V_{h-r}$ for some r. In other words given an ideal J in R and a degree compatible term ordering $<_\sigma$ on T, we have only $M(J_r) \subseteq M(J)_r$ for all $r = 0, \ldots, \dim_K J$.

EXAMPLE. Let $R = K[x_1, x_2, x_3]$ and let $<_\sigma$ be defined in the following way: if $m = x_1^{a_1} x_2^{a_2} x_3^{a_3}$ and $n = x_1^{b_1} x_2^{b_2} x_3^{b_3}$, then $m <_\sigma n$ if either $a_1 + a_2 + a_3 < b_1 + b_2 + b_3$ or $a_1 + a_2 + a_3 = b_1 + b_2 + b_3$ and $m < n$ with respect to the reverse lexicographic order. Let $J = (x_3 x_2^2 - x_1, x_3^2 x_2 - 1, x_1 x_3 - x_2, x_2^3 - x_1^2)$. These four polynomials are the reduced Gröbner basis of J with respect to $<_\sigma$. $M(J) = (x_3 x_2^2, x_3^2 x_2, x_1 x_3, x_2^3)$. It is not too hard to show that $H(J, s) = 4s + 1$ so $\dim_K J = 1$. Let $J_0 = M(J)$. By Proposition 7 $M(J)_1 = (x_3 x_2, x_2^3, x_1 x_3)$ and $H(M(J)_1, s) = 4s$. There is no ideal $I \in A_{x_1 x_2, x_2^3, x_1 x_3, \sigma}$ such that $J \subseteq I$. So $J_1 = J$ and $M(J_1) = M(J) \neq M(J)$.

Given an ideal J and the degree compatible term ordering $<_\sigma$ on T by Proposition 7 it is possible to find $M(J)_r$ for all $r = 0, \ldots, \dim_K J$. Let $(n_1, \ldots, n_{\lambda(r)}) = M(J)_r$ with $n_j \in T$ for all $j = 1, \ldots, \lambda(r)$. It is possible to decide if there exists $I \in A_{n_1, \ldots, n_{\lambda(r)}, \sigma}$ such that $J \subset I$. Such $I = (g_1, \ldots, g_{\lambda(r)})$ with $T(g_j) = n_j$ for all $j = 1, \ldots, \lambda(r)$ and $g_j = n_j + \sum_{h=1, \ldots, p_\sigma(n_j)} d_{h_j} r_h$ for all j. By definition of Gröbner basis $J \subseteq I$ if each polynomial of the reduced Gröbner basis of J with respect to $<_\sigma$ reduces to zero modulo $g_1, \ldots, g_{\lambda(r)}$. If this is the case the d_{h_j}'s must satisfy a system of polynomial equations, so $J \subseteq I$ if the system has solutions. Furthermore when such I exists, then $I = J_r$ by Proposition 6.

References

Bayer, D. (1982). The division algorithm and the Hilbert Scheme, Ph.D. Thesis, Harvard.

Buchberger, B. (1976). Some properties of Gröbner bases for polynomial ideals. *ACM SIGSAM Bull.* **10**(4), 19–24.

Carrà Ferro, G. (1987). Some properties of the lattice points and their applications to differential algebra. *Communications in Algebra*, **15**(12), 2625–2632.

Gianni, P., Trager, B., Zacharias, G. (1984). Gröbner Bases and Primary Decomposition of Polynomial Ideals, Preprint.

Hartshorne, R. (1977). *Algebraic Geometry*, Graduate Texts in Mathematics no. 52, Springer-Verlag, New York.
Kolchin, E. R. (1973). *Differential Albegra and Algebraic Groups*. Academic Press, London.
Möller, H. M., Mora, F. (1984). *Upper and Lower Bounds for the Degree of Gröbner Bases*, Proc. Eurosam 1984, Springer L.N.C.S. volume 162, pp. 172–183.
Perron, O. (1951). *Algebra*, vol. I, Die Grundlagen, W. de Gruyter, Berlin.
Ritt, J. F. (1950). *Differential Algebra*. AMS Publications.
Robbiano, L. (1985). *Term Orderings on the Polynomial Rings*. Proceedings of Eurocal 85, Springer, L.N.C.S. volume 204, pp. 513–517.
Sit, W. Y. (1975). Well ordering of certain numerical polynomials. *Trans. Am. Math. Soc.* **212,** 37–45.
Zacharias, G. (1978). Generalized Gröbner Bases in Commutative Polynomial Rings. Bachelor Thesis, M.I.T.

Computing Dimension and Independent Sets for Polynomial Ideals

HEINZ KREDEL* AND VOLKER WEISPFENNING†

Universität Passau, D-8390 Passau, FRG

(*Received* 5 *November* 1986, *and in revised form* 8 *April* 1988)

We present an algorithm that computes the dimension and maximal independent sets for a polynomial ideal I from a Gröbner basis for I with respect to a lexicographical term order, and extends Buchberger's test for zero-dimensionality. The algorithm is (for a given lexicographical Gröbner basis) faster, than the method of Möller & Mora (1983) using the computation of the Hilbert polynomial, and yields important additional information on independent sets and associated prime ideals. Its verification involves the new concept of strong independence modulo an ideal with respect to an admissible term order. The algorithm is implemented in the ALDES/SAC-2 system of Collins & Loos (1980), and has been tested successfully on the examples from Böge *et al.* (1986) and other examples.

Introduction

Among the basic problems of the algorithmic theory of polynomial ideals, the computation of the dimension $dim(I)$ of an ideal I in a polynomial ring $\mathbf{R} = \mathbf{K}[X_1, \ldots, X_n]$ occupies a prominent place. The geometric definition of $dim(I)$ as the maximal dimension of all isolated prime ideals J associated with I is unfavourable for computation, since it involves the primary decomposition of I. Instead, $dim(I)$ can be described more directly as the largest number of elements in $\mathbf{R}$ that are independent modulo I in a natural sense (see Gröbner, 1968/1970). A third approach characterizes $dim(I)$ as the degree of the Hilbert polynomial of I, that can be computed from the vector space dimension of the $\mathbf{K}$-linear spaces $S_m = \{f + I | deg(f) \leqslant m\} \subseteq \mathbf{R}/I$.

This last characterization has been combined successfully with the algorithmic technique of Gröbner bases introduced by Buchberger (1965), Möller & Mora (1983). The resulting algorithms are applicable to non-trivial cases. (The special problem to test whether $dim(I) \leqslant 0$ can be handled more easily by Buchberger's criterion (Method 6.9 in Buchberger, 1985).

The Gröbner basis method can also be combined with the second characterization of $dim(I)$: Consider all pure lexicographical orderings $<_L$ of terms in $\mathbf{R}$ induced by permutations of the variables X_i. Compute a Gröbner base $G = G(<_L)$ of I with respect to $<_L$, and let $S = S(<_L)$ be the largest initial segment of the set of variables, such that no head term of a polynomial in G contains only variables from S. Then each S is independent modulo I and the largest S determines the dimension of I (see Kandri-Rody,

* Formerly: Gesellschaft für Schwerionenforschung, D-6100 Darmstadt, FRG.
† Formerly: Universität Heidelberg, D-6900 Heidelberg, FRG.

1985), compare also Kutzler & Stifter (1985), Lemma 5. The advantage of this method is that it determines besides *dim*(I) also sets of variables independent modulo I. On the other hand, the number of Gröbner basis calculations involved in this method renders it useless for practical purposes in most cases. Further related methods for computing the dimension of a polynomial ideal appear in Giusti (1984) and Carra-Ferro (1986). The application of resultant calculus to this problem has been described in Kredel (1985).

In this paper, we present an algorithm that computes both the dimension of I and maximal independent sets of variables modulo I from a single Gröbner basis of I. We verify the correctness for the case of a pure lexicographical term order and a few other cases. A recent theorem of Carra-Ferro (1987) implies that the algorithm is correct for an arbitrary admissible term order. The algorithm is significantly faster than the algorithms in Möller & Mora (1983) using the Hilbert polynomial. It also provides considerable additional information on independent sets and dimensions of isolated prime ideals associated with I. Thus, it is particularly suitable to the method of geometrical theorem proving developed in Kutzler & Stifter (1985/86). It is also useful as a tool in the primary decomposition of I (compare Gianni *et al.*, 1986 and also Kredel 1987). The method provides parametrizations of affine algebraic sets. Thus it may be helpful in determining the spatial structure of molecules from algebraic equations describing this structure (compare example 4.6).

The verification of the algorithm employs the novel notion of strong independence modulo an ideal I, a concept that correlates well with Gröbner bases and may be of independent interest.

The algorithm is implemented in the ALDES/SAC-2 system of Collins & Loos (1980), and has been tested successfully on substantial examples, including those studies in Böge *et al.* (1986).

The plan of the paper is as follows: **Section 1** introduces strong independence modulo a polynomial ideal I, and studies its properties and relations to the traditional concept of independence modulo I. **Section 2** relates strong independence with Gröbner bases and provides the theoretical basis for our algorithm. **Section 3** presents the algorithm and its implementation. **Section 4** gives an overview of the performance of the algorithm for a number of examples.

1. Independence Modulo an Ideal

Let $\mathbf{K}$ be a field, $\mathbf{R} = \mathbf{K}[X_1, \ldots, X_n]$ a polynomial ring over $\mathbf{K}$, and let $\mathbf{X} = \{X_1, \ldots, X_n\}$. For any subset $S = \{X_{i_1}, \ldots, X_{i_r}\} \subseteq \mathbf{X}$, we let $T(S)$ be the set of all terms (power products) of variables in S, and we let $\mathbf{K}[S] = \mathbf{K}[X_{i_1}, \ldots, X_{i_r}]$. An ideal I of $\mathbf{R}$ is proper, if $I \neq \mathbf{R}$, or equivalently $1 \notin I$. Guided by the concept of algebraic dependence in fields (see Zariski & Samuel, 1958/60), the following notions of independence and dependence (appearing in Gröbner (1968/70)) seem to be the most natural: Let I be a proper ideal in $\mathbf{R}$, let $S \subseteq \mathbf{X}$, $X \in \mathbf{X} \backslash S$. Then S is **independently modulo** I if $\mathbf{K}[S] \cap I = \{0\}$; otherwise S is **dependent modulo** I. X is **dependent on** S **modulo** I if there exists $f \in \mathbf{K}[S][X] \cap I$ such that f has positive degree in X. For the case of a prime ideal I, these concepts coincide essentially with algebraic independence and dependence in the quotient field of the residue ring $\mathbf{R}/I$:

LEMMA 1.1. *Suppose I is a prime ideal in* $\mathbf{R}$. *Denote the residue class $f + I \in \mathbf{R}/I$ by f and let $\mathbf{K}'$ be the quotient field of $\mathbf{R}/I$, so that $\mathbf{K} \subseteq \mathbf{R}/I \subseteq \mathbf{K}'$. Then the following hold for $S \subseteq \mathbf{X}$, $X \in \mathbf{X} \backslash S$.*

(*1*) *S is independent mod I* **iff** $\bar{S} = \{\bar{Y} \mid Y \in S\}$ *is algebraically independent in* $\mathbf{K}'$ *over* $\mathbf{K}$ *and* $|S| = |\bar{S}|$.

(*2*) *If S is independent mod I, then X is dependent on S mod I* **iff** $\bar{X}$ *is algebraically dependent on* $\bar{S}$ *in* $\mathbf{K}'$ *over* $\mathbf{K}$.

PROOF. Obvious □

For an arbitrary ideal I, the following properties hold:

LEMMA 1.2. *Let* $S \subseteq \mathbf{X}$, $X \in \mathbf{X} \setminus S$.

(*1*) $S \cup \{X\}$ *is dependent mod I* **iff** *S is dependent mod I* **or** *X is dependent on S mod I.*

(*2*) *S is independent mod* **iff** *no* $Y \in S$ *is dependent on* $S \setminus \{Y\}$ *mod I.*

(*3*) *If* $S' \subseteq S$ *and X is dependent on* S' *mod I, then X is dependent on S mod I.*

(*4*) (**Steinitz exchange property**) *Let* $Y \in S$, $S' = S \setminus \{Y\}$. *If X is dependent mod I on S but not on* S', *then Y is dependent on* $S' \cup \{X\}$.

PROOF. (1)–(3) are obvious. (4) Let $f \in \mathbf{K}[S'][Y][X] \cap I$ be of positive degree in X. Then f is of positive degree in Y, since otherwise $f \in \mathbf{K}[S'][X]$, contradicting the hypothesis. So we may construe f as a polynomial in $\mathbf{K}[S' \cup \{X\}][Y] \cap I$ of positive degree in Y. □

Recall that the **dimension** of a prime ideal J in $\mathbf{R}$, *dim*(J), is defined as the transcendence degree of the quotient field $\mathbf{K}'$ of any $\mathbf{R}/J$ over $\mathbf{K}$. By lemma 1.1, *dim*(J) is the number of elements of any maximal set S of variables independent mod J. For an arbitrary proper ideal I in $\mathbf{R}$, *dim*(I) is defined as the maximal dimension of an isolated prime ideal associated with I (see Zariski & Samuel, 1958/60). *dim*(I) can be characterized more directly after Gröbner (1968/70) as follows:

LEMMA 1.3. *Let I be a proper ideal in* $\mathbf{R}$. *Then dim(I) is the maximal number d of elements in any set S of variables independent mod I.*

PROOF. Let J be an isolated prime ideal associated with I such that *dim*(J) = *dim*(I) and let $S \subseteq \mathbf{X}$, $|S| = dim(J)$, be independent mod J. Then S is also independent mod I, and so $|S| \leqslant d$. Conversely, if $S' \subseteq \mathbf{X}$, $|S'| = d$, and S' is independent mod I, then the set $M = \mathbf{K}[S'] \setminus \{0\}$ is multiplicatively closed and disjoint to I. So there exists a prime ideal $J' \supseteq I$ disjoint to M. Let J'' be an isolated prime ideal associated with I such that $J'' \subseteq J'$. Then S' is independent mod J'', and so $dim(I) \geqslant dim(J'') \geqslant d$. □

Unfortunately, the transitivity of the dependence modulo an ideal fails, and so this is not an algebraic dependence relation in the axiomatic sense (see Zariski & Samuel, 1958/60). In fact, the most important property of an axiomatic algebraic dependence relation fails: There exist maximal sets of variables independent mod I of different cardinalities.

EXAMPLE 1.4. Let $\mathbf{R} = \mathbf{K}[X, Y, Z]$, $J = (X)$, $J' = (Y, Z)$, $I = J \,.\, J' = (XY, XZ)$. Then $S = \{X\}$ is a maximal subset of $\{X, Y, Z\}$ independent mod I, since XY, $XZ \in I$. On the other hand, $S' = \{Y, Z\}$ is also independent mod I. Thus $|S| = 1 = dim(J') < 2 = |S'| = dim(J) = dim(I)$.

For a **prime** ideal I, however, all maximal sets of variables independent mod I have the same cardinality by lemma 1.1.

It turns out that this natural concept of independence does not correlate well with Gröbner bases for the ideal I. This leads us to a more refined concept of independence mod I that refers in addition to a given admissible order $<_T$ of the set $T = T(\mathbf{X})$. Recall from Buchberger (1985) that a linear order $<_T$ of T is **admissible**, if $1 < X$, and $t <_T t'$ implies $Xt <_T Xt'$ for $X \in \mathbf{X}$ and $t, t' \in T$. As a consequence, multiplication is monotonic with respect to $<_T$. Examples of admissible orders are the **pure lexicographical ordering** ($<_L$) and the **total degree order** ($<_G$) on T induced by some linear order on $\mathbf{X}$. Characterizations of **all** admissible term orders can be found in Robbiano (1985) and Weispfenning (1987). For $S \subseteq T, t \in T$ we write $S <_T t$ if $s <_T t$ for all $s \in S$, $S <_T S'$ is defined similarly. For $f \in \mathbf{R}$, $F \subseteq \mathbf{R}$, we let $HT(f)$ denote the highest term of f with respect to $<_T$, and let $HT(F) = \{HT(f) | f \in F\}$. So we may write $f \in \mathbf{R}$ as $f = a \,.\, HT(f) + r_f$ with $0 \neq a \in \mathbf{K}^*$, $r_f \in \mathbf{R}$, r_f is called the **reduct of** f. We define $f >_T g$ **iff** $HT(f) >_T HT(g)$ or if $HT(f) = HT(g)$ and $r_f >_T r_g$.

The following definitions are fundamental for our study. Let S and A be disjoint subsets of $\mathbf{X}$, and let I be a proper ideal of $\mathbf{R}$. Then $\mathbf{K}[S/A]$ denotes the set of all non-zero polynomials $f \in \mathbf{K}[S \cup A]$ such that $HT(f) \in \mathbf{K}[S]$:

$$\mathbf{K}[S/A] = \{f | 0 \neq f \in \mathbf{K}[S \cup A] \text{ and } HT(f) \in \mathbf{K}[S]\}$$

(so for $S = \emptyset$, $\mathbf{K}[S/A] = \mathbf{K}^*$). We say S is **independent mod** I **with respect to** A, if $\mathbf{K}[S/A] \cap I = \emptyset$. If $X \in \mathbf{X} \backslash (S \cup A)$ we say X is **dependent on** S/A **mod** I, if there exists $f \in \mathbf{K}[S \cup \{X\}/A] \cap I$ such that X occurs in $HT(f)$. This implies that there exists $f \in \mathbf{K}[S \cup A][X] \cap I$ such that the coefficient of some positive power of X in f is in $\mathbf{K}[S/A]$. For $A = \emptyset$, these concepts coincide with those considered above. We say S is **strongly independent modulo** I, if S is independent mod I with respect to $\mathbf{X} \backslash S$. So any strongly independent set S of variables is also independent mod I. The converse fails as the following easy example shows:

EXAMPLE 1.5. Let $\mathbf{R} = \mathbf{K}[X, Y]$, $I = (Y - X)$ and let $X <_T Y$. Then $S = \{Y\}$ is independent but not strongly independent mod I.

A much more delicate problem arises, when we consider **maximal** sets S strongly independent mod I: Clearly, any such S is independent mod I. Is it also **maximal** independent mod I? Using the interrelations between Gröbner bases and strongly independent sets, we are going to verify the following example in the next section.

EXAMPLE 1.6. Let $\mathbf{R} = \mathbf{K}[X, Y, Z]$, where $<_T = <_L$ is the pure lexicographical order with $X <_T Y <_T Z$; let $J = (Z - X)$, $J' = (X, Y)$, $I = J \,.\, J' = (ZX - X^2, ZY - XY)$, and let $S = \{Z\}$. Then S is maximal strongly independent mod I, but $S' = \{X, Y\}$ is independent mod I, and so S is not maximal independent mod I.

Notice that in the example, I is a product of two prime ideals of different dimensions 2 and 1.

In view of the results below we conjecture that for a **prime** ideal I any maximal set of variables strongly independent mod I is also maximal independent mod I and hence determines the dimension of I. Some consequences of this conjecture will be discussed at the end of this section.

The results of this section will be concerned with sufficient conditions on a maximal set S of variables strongly independent mod I to be also maximal independent mod I.

We begin by relating strongly independent sets of variables modulo an arbitrary ideal I to the corresponding sets modulo the isolated prime ideals associated with I.

LEMMA 1.7. *Let I be a proper ideal in* $\mathbf{R}$, *let* $S \subseteq \mathbf{X}$, *and let* $<_T$ *be an arbitrary admissible term order on* T.

(1) *If S is (maximal) strongly independent mod I with respect to $<_T$, then there exists an isolated prime ideal J associated with I such that S is also (maximal) strongly independent mod J with respect to $<_T$.*
(2) *Let $U = \{S \subseteq \mathbf{X} | S$ is strongly independent mod I with respect to $<_T\}$, and $U' = \{S \subseteq \mathbf{X} |$ there exists an isolated prime ideal J associated with I such that S is strongly independent mod J with respect to $<_T\}$ then $U = U'$.*

PROOF.
(1) Let S be strongly independent mod I and let $A = \mathbf{X} \backslash S$. Then $\mathbf{K}[S/A]$ is a multiplicatively closed subset of $\mathbf{R}$ disjoint to I. So there exists a prime ideal $J \supseteq I$ in $\mathbf{R}$ disjoint to $\mathbf{K}[S/A]$. Let J' be an isolated prime ideal associated with I such that $I \subseteq J' \subseteq J$. Then S is strongly independent mod J'. Moreover, if S is maximal strongly independent mod I, then for any $S \subseteq S' \subseteq \mathbf{X}$ with S' strongly independent mod J', S' is strongly independent mod I, and so $S' = S$.
(2) By (1), $U \subseteq U'$, the converse, $U' \subseteq U$, is trivial. □

THEOREM 1.8. *Let J be a prime ideal in* $\mathbf{R}$, *let* $<_T$ *be an arbitrary admissible term order on T and let* $S \subseteq \mathbf{X}$ *be maximal strongly independent mod J with respect to* $<_T$. *If* $|S| \geqslant n-2$, *then S is also maximal independent mod J, and so* $|S| = dim(J)$.

PROOF. If $|S| = n-1$ there is nothing to prove. So we may assume $S = \mathbf{X} \backslash \{X, Y\}$ with $X \neq Y$. Pick $f \in \mathbf{K}[S \cup \{X\}/\{Y\}] \cap J$ and $g \in \mathbf{K}[S \cup \{Y\}/\{X\}] \cap J$. Then any irreducible factor of f is also in $\mathbf{K}\{S \cup \{X\}/\{Y\}] \cap J$, and similar for g. So we assume without restriction that f and g are irreducible. Since S is strongly independent mod J, X occurs in $HT(f)$ and Y occurs in $HT(g)$, and so g is not a multiple of f. Consequently, the prime ideal J' generated by f is properly contained in J, and so by Zariski & Samuel (1958/60) chapter VII, theorem 20, $dim(J) \leqslant n-2$ and so S is maximal independent mod J. □

The following somewhat technical theorem is the central result of this section. It requires the concept of an inessential set of variables: let $S \subseteq \mathbf{X}, f \in \mathbf{R}$, and let $<_T$ be an arbitrary term order on T. Then we denote by f^S the polynomial resulting from f by substituting 1 for all variables from S in f. We say S **is inessential for** f if all terms t occurring in f, $t^S \leqslant HT(f)^S$.

THEOREM 1.9. *Let* $S \subseteq \mathbf{X}$, *I a prime ideal in* $\mathbf{R}$ *and let* $<_T$ *be an arbitrary admissible term order on T. Assume moreover that S is independent mod I and that for any* $X \in \mathbf{X} \backslash S$ *there exists a polynomial* $f_X \in \mathbf{K}[S \cup \{X\}/\mathbf{X} \backslash (S \cup \{X\})] \cap I$ *such that S is inessential for* f_X. *Then S is maximal independent mod I, and so* $|S| = dim(I)$.

PROOF. For $X \in \mathbf{X} \backslash S$, let d_X be the degree of $HT(f_X)$ in X. Then $d_X > 0$; for otherwise $HT(f_X^S) = 1$, and so $t^S = 1$ for all terms t occurring in f, and so $f \in \mathbf{K}[S]$ which contradicts the independence of S mod I. Let T' be the set of all $t \in T$ such that for every $X \in \mathbf{X} \backslash S$, the degree of t in X is $\leqslant d_X$.

CLAIM. For every $t \in T \backslash T'$ there exists $0 \neq p, p_1, \ldots p_m \in \mathbf{K}[S]$, $t_1, \ldots, t_m \in T'$ and $f \in I$ such that $pt = p_1 t_1 + \cdots + p_m t_m + f$.

PROOF OF THE CLAIM. Assume for a contradiction that the claim fails for some $t \in T \backslash T'$ and that t is $>_T$-minimal with this property. Choose $X \in \mathbf{X} \backslash S$ such that the degree d of t in X is greater than d_X, and put $u = t \cdot X^{-d} \in T$. By our hypothesis, f_X may be written in the form $pX^{d_X} = p_1 t_1 + \cdots + p_m t_m$ with $t_i \in T(\mathbf{X} \backslash S)$, $0 \neq p$, $p_i \in \mathbf{K}[S]$, $X^{d_X} > t_i$ for $1 \leqslant i \leqslant m$. So $pt = X^{d-d_X} \cdot u \cdot f_X - p_1 t_1 X^{d-d_X} \cdot u - \cdots - p_m t_m X^{d-d_X} \cdot u$ with $X^{d-d_X} \cdot u \cdot f_X \in I$ and $t_i X^{d-d_X} < X^{d_X} \cdot X^{d-d_X} \cdot u = t$ for $1 \leqslant i < m$. So the claim is valid for all $t_i \cdot X^{d-d_X} . u$, and hence for t as well, a contradiction.

With the notation of lemma 1.1 we may now conclude that $\mathbf{K} \subseteq \mathbf{K}(\bar{S}) \subset \mathbf{K}'$, the quotient field of $\mathbf{R}/I$, and that $\mathbf{R}/I$ is generated as a $\mathbf{K}(\bar{S})$-vector space by the finite set $\bar{T}'$. So each $\bar{X}$ with $X \in \mathbf{X} \backslash S$ is algebraic over $\mathbf{K}(\bar{S})$, and so $\mathbf{K}' = \mathbf{R}/I$ is a finite algebraic extension of $\mathbf{K}(\bar{S})$, and so $dim(I) = |\bar{S}| = |S|$. □

Next we define a special type of maximal strongly independent set, the **left basic set** of an ideal I.

Let $<_T$ be an arbitrary admissible order on T and let I be a proper ideal in $\mathbf{R}$. For $0 \leqslant k \leqslant n$ define $S_k \subseteq \mathbf{X}$ inductively by

$$S_0 = \emptyset$$

$$S_{k+1} = \begin{cases} S_k \cup \{X_k\} & \text{if } S_k \cup \{X_k\} \text{ is strongly independent mod } I \text{ wrt. } <_T, \\ S_k & \text{otherwise.} \end{cases}$$

Then we call $S = S_n$ the **left basic set of I with respect to** $<_T$. Notice that by definition S is maximal strongly independent modulo I with respect to $<_T$. We shall see in section 2 that S can be constructed from $<_T$ and an ideal basis for I.

As an immediate consequence of the definition, we note:

PROPOSITION 1.10. *Let I be an ideal in $\mathbf{R}$, let $<_T$ be an arbitrary admissible term order on T, and let S be the left basic set of I with respect to $<_T$. Then $S = \emptyset$ iff for all $X \in S$ there exists a polynomial $f_X \in \mathbf{K}[\{X\}/(\mathbf{X} \backslash \{X\})] \cap I$.*

Combining 1.10 with 1.9 and 1.7 we obtain:

THEOREM 1.11. *Let I be a proper ideal in $\mathbf{R}$, let $<_T$ be an arbitrary admissible term order on T, and let S be the left basic set of I with respect to $<_T$. Then $S = \emptyset$ iff $dim(I) = 0$.*

Specializing theorem 1.9 to a pure lexicographic term order, we get:

COROLLARY 1.12. *Let I be a prime ideal in $\mathbf{R}$, let $<_L$ be a pure lexicographic term order on T and let S be the left basic set of I with respect to $<_L$. Then S is maximal independent mod I and so $|S| = dim(I)$.*

PROOF. Since S is maximal strongly independent mod I, we find for every $X \in \mathbf{X} \backslash S$ a polynomial $f_X \in \mathbf{K}[S \cup \{X\}/\mathbf{X}(S \cup \{X\})] \cap I$. f_X contains no variable $Y \in \mathbf{X}$ with $Y >_L X$;

moreover, for every term t occurring in f_X, the degree of t in X is smaller or equal to the degree d_X of $HT(f_X)$ in X. Consequently, $HT(f_X)^S = X^{d_X} \geqslant_L t^S$ for all terms t occurring in f_X, and so S is inessential for f_X. This verifies the hypothesis of theorem 1.9. □

The main application of 1.12 is as follows:

THEOREM 1.13. *Let I be a proper ideal in* **R**, *let $<_L$ be a pure lexicographic term order on T, and put*

$$d = max\{|S|: S \subseteq \mathbf{X}, S \text{ is maximal strongly independent mod } I \text{ wrt. } <_L\}.$$

Then $d = dim(I)$.

PROOF. By lemma 1.3, $dim(I) \geqslant d$. Pick an isolated prime ideal J associated with I such that $dim(J) = dim(I)$, and let S be the left basic set of J. Then $|S| = dim(J) = dim(I)$ and S is strongly independent mod J and hence mod I, and so $d \geqslant dim(I)$. □

Recall that a proper ideal I in **R** is **unmixed** if all isolated prime ideals associated with I have the same dimension. We call I **weekly unmixed**, if all isolated prime ideals of positive dimension associated with I have the same dimension.

COROLLARY 1.14. *Let I be a weakly unmixed ideal in* **R**, *let $<_L$ be a pure lexicographic term order on T, and let S be the left basic set of I. Then $|S| = dim(I)$.*

PROOF. By theorem 1.11, it suffices to consider the case that $S \neq \emptyset$. By 1.7 there exists an isolated prime ideal J associated with I such that S is strongly independent mod J. Then S is also the left basic set of J and so by 1.12, $|S| = dim(J) = dim(I)$. □

REMARK 1.15.

(i) In a recent preprint, Carra-Ferro (1987), states a combinatorial theorem (theorem 3.1 in Carro-Ferro, 1987), which together with our theorem 2.1 below implies that theorem 1.13 is valid for an arbitrary admissible term order $<_T$ on T.

(ii) Suppose our conjecture is valid, that for any prime ideal J in **R**, an arbitrary admissible term order $<_T$ on T, and any maximal strongly independent set S mod J, $|S| = \text{dim}(J)$. Then we may conclude from 1.7: If I is a proper ideal in **R** and S is maximal strongly independent mod I, then $S = dim(J)$ for some isolated prime ideal J associated with I.

2. Strong Independence and Gröbner Bases

In this section we establish the connection between Gröbner bases for polynomial ideals I and strong independence modulo I. This will enable us to turn the main results of section 1 into algorithmic methods for computing maximal independent sets and dimensions for polynomial ideals. For the relevant information on Gröbner bases, we refer the reader to Buchberger (1985).

Let $<_T$ be an arbitrary but fixed admissible order on T, and let F be a finite non-empty set of polynomials in **R**. Then $\rightarrow_F$ denotes the reduction on **R** induced by F, and $\rightarrow_F^*$ denotes the reflexive-transitive closure of $\rightarrow_F$. F is a Gröbner basis, **iff** for all $f \in (F), f \rightarrow_F^* 0$.

THEOREM 2.1. *Let $<_T$ be an arbitrary admissible order on T, let $S \subseteq \mathbf{X}$, and let G be a Gröbner basis in* **R** *with respect to $<_T$. Then S is strongly independent modulo (G) with respect to $<_T$* **iff** $T(S) \cap HT(G) = \emptyset$.

PROOF. If $g \in G$ with $HT(g) \in T(S)$, then $g \in \mathbf{K}[S/(\mathbf{X}\backslash S)] \cap (G)$, and so S is not strongly independent modulo (G). Conversely, assume $f \in \mathbf{K}[S/(\mathbf{X}\backslash S)] \cap (G)$. Then $f \rightarrow_G^* 0$ say $f = f_1 \rightarrow_G f_2 \rightarrow_G f_3 \rightarrow_G \ldots \rightarrow_G 0$. Pick k minimal such that $HT(f_k) >_T HT(f_{k+1})$, and pick $g \in G$ such that $f_k \rightarrow_g f_{k+1}$. Then $HT(g)$ divides $HT(f_k) = HT(f)$ in T, and so $HT(g) \in T(S)$. □

Using theorem 2.1, we can now verify the statements in example 1.6: First, one can observe that $G = \{ZX - X^2, ZY - XY\}$ is a Gröbner basis wrt. $<_L$ with $X <_L Y <_L Z$, since the S-polynomial of these two polynomials is zero (see Buchberger, 1985). $S = \{Z\}$ is now obviously a maximal subset of $\mathbf{X}$ satisfying $T(S) \cap HT(G) = \emptyset$, and so by 2.1, S is maximal strongly independent mod I. Next we reorder T by the pure lexicographical order with $Y <_L Z <_L X$. Then $G = \{-X^2 + ZX, -XY + ZY\}$ is again a Gröbner basis with respect to the new order on T. For $S' = \{X, Z\}$ we now have $T(S') \cap HT(G) = \emptyset$. So by 2.1, S' is strongly independent, and hence independent mod I.

COROLLARY 2.2. *Let $<_T$ be an arbitrary admissible order on T, let G be a Gröbner basis in* **R** *with respect to $<_T$, and put $I = (G)$. Then the left basic set S of I can be constructed from G and $<_T$ by the following algorithm LBS*:

Algorithm $S \leftarrow LBS(G)$
Left basic set of polynomial ideal from a Gröbner basis.
Given: $<_T$ an admissible order on T,
$G =$ Gröbner basis for an ideal $I \subseteq \mathbf{K}[X_1, \ldots, X_n]$ with respect to $<_T$.
Find: $S =$ left basic set of (G) with respect to $<_T$.
if $dim(I) = -1$ then S is undefined.
comment Initialise, and check dimension < 0.
if $1 \in G$ **then return**.
$S \leftarrow \emptyset$. $U \leftarrow \{X_1, \ldots, X_n\}$.
repeat select x from U. $U \leftarrow U \backslash \{x\}$.
if $T(S \cup \{x\}) \cap HT(G) = \emptyset$ **then** $S \leftarrow S \cup \{x\}$.
until $U = \emptyset$.
return.

The combination of theorem 2.1 with 1.13 and 1.8 yields the following algorithmic result:

THEOREM 2.3. *Let $<_L$ be a pure lexicographic order on T, let G be a Gröbner basis in* **R** *with respect to $<_L$ such that $G \cap \mathbf{K} = \emptyset$, let $I = (G)$, and let S be a subset of* **X** *such that*

$$(*) \qquad T(S) \cap HT(G) = \emptyset,$$

and such that S has the largest number of elements among all subsets of **X** *satisfying (*). Then S is maximal independent mod I and $|S| = dim(I)$. The same condition holds for an arbitrary admissible term order $<_T$ on T in case $|S| \geqslant |\mathbf{X}| - 2$.*

REMARK 2.4.

(i) The easy special case of theorem 2.3, where S is an initial segment of **X** under the order $<_L$ was noted already in Kutzler & Stifter (1985/86) and in Kandri-Rodi (1985).

(ii) A recent result of Carra-Ferro (1987, theorem 3.1), states that theorem 2.3 is valid for an arbitrary admissible term order $<_T$ on T, provided G is a **reduced** Gröbner basis.

(iii) Let $<_T$ be an arbitrary admissible term order on T, let G be an arbitrary ideal basis in $\mathbf{R}$, and let S and I be as in theorem 2.3. Then G can be extended to a Gröbner basis G' of I, and so $|S| \geqslant dim(I)$. This fact may be used to guess $dim(I)$ from an "approximate" Gröbner basis for I.

Condition (*) depends only on $HT(G)$, which is obviously a Gröbner basis in $\mathbf{R}$. As a consequence, we have:

COROLLARY 2.5. *Let $<_L$ be a pure lexicographic term order on T, and let G be a Gröbner basis in* $\mathbf{R}$ *with respect to $<_L$. Then $dim((G)) = dim((HT(G))$.*

Notice that Buchberger's criterion for deciding, whether a polynomial ideal is zero-dimensional, is an immediate consequence of 2.1 and 1.11.

COROLLARY 2.6 (Buchberger, 1965, 1985). *Let $<_T$ be an admissible order on T, and let G be a Gröbner basis in* $\mathbf{R}$ *with respect to $<_T$, and let $I = (G)$. Then I is zero-dimensional* **iff** *for all $X \in \mathbf{X}$, $T(\{X\}) \cap HT(G) \neq \emptyset$.*

PROOF. By 2.3 and 1.11, I is zero-dimensional **iff** every singleton $S = \{X\}$ violates condition (*). □

We have indicated in 1.15(ii) that if our conjecture on prime ideals holds then **every** maximal set S of variables satisfying $T(S) \cap HT(G) = \emptyset$ for a Gröbner basis G in R is a maximal independent set for some isolated prime ideal J associated with $I = (G)$; this is the case e.g. for $|S| \geqslant |\mathbf{X}| - 2$. So one may ask for a converse: Given some isolated prime ideal J associated with I, and a maximal set S of variables independent mod J, does S occur as a maximal set with $T(S) \cap HT(G) = \emptyset$. The answer is no, for the simple reason that S may be a proper subset of a maximal set S' independent mod J' for another isolated prime ideal J' associated with I:

EXAMPLE 2.7. Let $\mathbf{R} = \mathbf{K}[X, Y, Z]$, $J = (X)$, $J' = (X+1, Y)$, $I = J \cdot J' = (X^2 + X, XY)$, Then $G = \{YX, X^2 + X\}$ is a Gröbner basis of I with respect to the pure lexicographical order on T with $X < Y$. So by 2.1, $S = \{Y, Z\}$ is the only maximal set of variables with $T(S) \cap HT(G) = \emptyset$. Thus the only maximal set $S' = \{Z\}$ of variables independent mod J' is 'hidden' by the larger set S belonging to J.

Finally we give a new, constructive proof of a theorem proved in Giusti (1984) for homogeneous ideals.

THEOREM 2.8. *Let $<_T = <_L$ be the pure lexicographical order on T with $X_1 <_L \cdots <_L X_n$. For any finite ideal basis F of I in* $\mathbf{R}$ *one can construct a non-empty Zariski-open subset M of GL(n) (regarded as acting on $X_1, \ldots, X_n$) such that for all $\varphi \in M$ the left basic set S^φ of the transformed ideal $I^\varphi = (F^\varphi)$ equals some $\{X_1, \ldots, X_s\}$ $(0 \leqslant s \leqslant n)$ and determines the dimension of I: $dim(I) = dim(I^\varphi) = |S^\varphi|$.*

PROOF. First we compute a Gröbner basis G of I with respect to $<_T$ and a strongly independent set $S \subseteq \mathbf{X}$ with maximal number of elements. Then $s = |S| = dim(I)$. Next we determine $\varphi \in GL(n)$ permuting $X_1, \ldots, X_n$, such that $S^\varphi = \{X_1, \ldots, X_s\}$. Then S^φ is independent modulo I^φ. Since $<_T$ is pure lexicographic, S^φ is strongly independent modulo I^φ and hence is the left basic set of I^φ. Compute a Gröbner basis G' of I^φ. Then $HT(G') \cap T(S^\varphi) = \emptyset$. Next we apply theorem 2.1 in Weispfenning (1988) to determine for G' a Zariski-open subset M of $GL(n)$ such that for all $\psi \in M$ the reduced Gröbner basis $G''(\psi)$ of $(I^\varphi)^\psi$ is computed uniformly in ψ; in particular the coefficients of the polynomials in $G''(\psi)$ are rational functions in the components of ψ. We claim that for all $\psi \in M$, $T(S^\varphi) \cap HT(G''(\psi)) = \emptyset$. This will prove the theorem.

Assume for a contradiction that for some $\psi \in M$, $HT(G''(\psi)) \cap T(S^\varphi) \neq \emptyset$. There is a Zariski-open subset M' of $GL(n)$ such that for all $\rho \in M'$, $T(S^\varphi) \cap HT(G''(\psi)^\rho) = \emptyset$. Pick $\sigma \in M \cap M' \circ \psi$. Then $G''(\sigma \circ \psi) = G''(\psi)^\sigma$, and so $T(S^\varphi) \cap HT(G''(\sigma \circ \psi)) = \emptyset$. This contradicts the uniformity of the construction of $G''(\psi)$ and $G''(\sigma \circ \psi)$ from G', by which also $HT(G''(\sigma \circ \psi)) \cap T(S^\varphi) \neq \emptyset$. □

3. Algorithm Description

The following two algorithms *DIMENSION* and *DIMREC* compute the dimension d of a polynomial ideal $I \subseteq \mathbf{K}[\mathbf{X}]$, $\mathbf{X} = \{X_1, \ldots, X_n\}$ from a given Gröbner basis G of I with respect to a pure lexicographic order on T. They compute a set M, whose elements $S \subseteq \mathbf{X}$ are maximal strongly independent sets of variables mod I with respect to $<_T$. Then $d = max\{|S|: S \in M\}$.

The algorithm *DIMENSION* first checks if $1 \in G$. Then it initializes the sets S, U and M and calls the recursive algorithm *DIMREC*. *DIMREC* computes M by testing if there is a headterm in G in the variables of subsets S of $\mathbf{X}$. There after *DIMENSION* determines the element of M with maximal number of variables.

Algorithm *DIMENSION*$(G. d, S, M)$

Dimension of a polynomial ideal from a Gröbner basis.

Given: $G =$ Gröbner basis for an ideal $I \subseteq \mathbf{K}[X_1, \ldots, X_n]$.

Find: $d =$ dimension $ideal(G)$.

$S =$ greatest maximal set of variables with $T(S) \cap HT(G) = \emptyset$, if $d \geqslant 0$.

$M =$ set of maximal sets of variables with $T(S) \cap HT(G) = \emptyset$.

comment Initialize, and check dimension < 0.

$M \leftarrow \emptyset$. $d \leftarrow -1$.

if $1 \in G$ **then return**.

$S \leftarrow \emptyset$. $U \leftarrow \{X_1, \ldots, X_n\}$.

comment Call of the recursive algorithm for the computation of M.

$M \leftarrow DIMREC(G, S, U, M)$.

comment Search for greatest $S \in M$

$M' \leftarrow M$.

repeat select m from M'. $M' \leftarrow M' \setminus \{m\}$.

$d' \leftarrow |m|$.

if $d' > d$ **then begin**

$d \leftarrow d'$. $S \leftarrow m$ **end**

until $M' = \emptyset$

return.

Algorithm $M' \leftarrow DIMREC(G, S, U, M)$

Recursive computation of maximal sets S' with $T(S') \cap HT(G) = \emptyset$.

Given: G = Gröbner basis for an ideal $I \subseteq \mathbf{K}[X_1, \ldots, X_n]$.
S = set of variables with $T(S) \cap HT(G) = \emptyset$.
U = the set of unprocessed variables, $U \subseteq \{X_1, \ldots, X_n\}$.
M = set of already computed maximal sets S' with $T(S') \cap HT(G) = \emptyset$.
Find: M' = updated set of maximal sets S' with $T(S') \cap HT(G) = \emptyset$.
comment Loop until U becomes empty.
$M' \leftarrow M$
while $U \neq \emptyset$ **do begin**
select first u from U. $U \leftarrow U \backslash \{u\}$.
if $T(S \cup \{u\}) \cap HT(G) = \emptyset$
then $M' \leftarrow DIMREC(G, S \cup \{u\}, U, M')$
end
comment Test if S is already contained in some element of M'.
$M'' \leftarrow M'$. $t \leftarrow$ **true**.
while $M'' \neq \emptyset$ and t **do begin**
select m from M''. $M'' \backslash \{m\}$.
if $S \subseteq m$ **then** $t \leftarrow$ **false end**.
if t **then** $M' \leftarrow M' \cup \{S\}$
return.

3.1. NOTES ON THE CORRECTNESS

To test

$$T(S \cup \{u\}) \backslash T(S) \cap HT(G) = \emptyset,$$

as requested by Theorem 2.3, it is sufficient to test

$$T(S \cup \{u\}) \cap HT(G) = \emptyset,$$

since

$$T(S) \cap HT(G) = \emptyset$$

by recursion assumption.

At the end of the while-loop in *DIMREC* at least S is independent mod F with respect to $\mathbf{X} \backslash S$. So if S is not already contained in some element of M, S can correctly be added to M.

Crucial for the correctness of the algorithm is the step when

$$T(S \cup \{u\}) \cap HT(G) = \emptyset$$

and after return from the recursion in the next while-loop a variable v is found with $S <_T u <_T v$ and

$$T(S \cup \{v]) \cap HT(G) = \emptyset.$$

We have to consider the two cases

$$T(S \cup \{u\} \cup \{v\}) \cap HT(G) \begin{cases} = \emptyset \\ \neq \emptyset \end{cases}.$$

Let $A <_T v$ and $A \subseteq \mathbf{X} \backslash \{S \cup \{u\}\}$ in the following.

- the intersection is empty:
  In this case $S' = S \cup \{u\} \cup \{v\}$ is independent mod I with respect to A and so S'

has become subset of an element in M by the previous recursion. Then at the end of the next recursion level it is tested if $S \cup \{v\}$ is a subset of an element in M and since this is true, $S \cup \{v\}$ is not entered into M.

- the intersection is not empty:
  Since $T(S \cup \{v\}) \cap HT(G) = \emptyset$ we know per definition that $S \cup \{v\}$ is independent mod I with respect to $A \cup \{u\}$. And so $S \cup \{v\}$ correctly enters into M as subset of some S'.

For an arbitrary admissible term order $<_T$, Buchberger's test for dimension zero is included, since in this case for all $u \in U$:

$$T(\emptyset \cup \{u\}) \cap HT(G) \neq \emptyset,$$

and so the algorithm *DIMREC* finishes without any further recursion.

3.2. NOTES ON THE COMPLEXITY

The depth of recursion is equal to $d+1$, since only when $T(S \cup \{u\}) \cap HT(G) = \emptyset$, a new recursion level is entered. The while-loop in *DIMREC* is performed at most $|U| = n$ times. When a new recursion level is entered, U is at least one element smaller than before.

By this at most

$$\binom{n}{1} + \cdots + \binom{n}{d+1}$$

subsets of $\mathbf{X}$ are tested if there is a headterm in the Gröbner base. If $d = 0$ only n subsets of $\mathbf{X}$ with one element are tested. If $d = n-1$ then $2^n - 1$ subsets of $\mathbf{X}$ are tested. (The empty set is tested in *DIMENSION*, i.e. the test for $1 \in G$.)

The number of calls of the *DIMREC* algorithm is then at most

$$\binom{n}{0} + \cdots + \binom{n}{d},$$

i.e. if $d = 0$, *DIMREC* is only called once. In the case $d = n-1$, *DIMREC* is called $2^n - 1$ times.

4. Examples for the Computation

In the following examples the ground field $\mathbf{K}$ is always the field of the rational numbers. To improve readability, only the headterms of the computed Gröbner bases are displayed. Listings of the input ideal bases can be found in Böge *et al.* (1986). The algorithms were coded in ALDES, and the computation was done using the SAC-2 computer algebra system by Collins & Loos (1980) and the Buchberger Algorithm System by Gebauer & Kredel (1983), on an IBM 3090-200 mainframe under MVS/XA at GSI. The definitions of the used term orders can be found in Kredel (1988). If the polynomial ring is $\mathbf{K}[X_1, X_2, \ldots, X_n]$ we always assume that $X_1 <_T X_2 <_T \cdots <_T X_n$. Although the algorithms are proved to be correct only in the case of the pure lexicographic term order, we include also examples with different term orders, to support the evidence in these cases.

EXAMPLE 4.1 (HAIRER, RUNGE-KUTTA 1). *See example* 1 *in Boge* et al. (1986).

Polynomial ring: $\mathbf{K}[C_2, C_3, B_3, B_2, B_1, A_{21}, A_{32}, A_{31}]$
Term order is inverse lexicographic $<_L$.

Input: $HT(G) = \{C_3^2B_3, C_2B_2, B_1, A_{21}, C_2^2A_{32}, C_2C_3A_{32}, B_3A_{32}, C_3B_2A_{32}, A_{31}\}$

Output: d = 2
S = $\{C_2, C_3\}$
M = $\{\{C_2, C_3\}, \{C_2, B_3\}, \{C_3, B_2\}, \{C_3, A_{32}\}, \{B_3, B_2\}, \{B_2, A_{32}\}\}$

Statistics: Computing time = 20 ms, Used storage cells = 822

EXAMPLE 4.2 (HAIRER, RUNGE-KUTTA 2). *See example* 2 *in Boge* et al. (1986).

Polynomial ring: $\mathbf{K}[C_2, C_3, C_4, B_4, B_3, A_{43}, A_{32}, A_{42}, B_2, A_{41}, A_{31}, A_{21}, B_1]$
Term order is inverse lexicographic $<_L$.

Input: $HT(G) = \{C_4, C_2C_3B_4, C_3^2B_3, C_2^2B_4B_3, C_2C_3A_{43}, B_4A_{43}, C_2^3B_3A_{43}, C_3B_3A_{43}, C_2^2A_{32}, C_3B_4A_{32}, C_2B_3A_{32}, B_4B_3A_{32}, C_3A_{43}A_{32}, B_3A_{43}A_{32}, A_{42}, B_2, A_{41}, A_{31}, A_{21}, B_1\}$

Output: d = 2
S = $\{C_2, C_3\}$
M = $\{\{C_2, C_3\}, \{C_2, B_4\}, \{C_2, B_3\}, \{C_2, A_{43}\}, \{C_3, B_4\}, \{C_3, A_{43}\}, \{C_3, A_{32}\}, \{B_4, B_3\}, \{B_4, A_{32}\}, \{B_3, A_{32}\}, \{B_3, A_{32}\}, \{A_{43}, A_{32}\}\}$

Statistics: Computing time = 120 ms, Used storage cells = 6552.

EXAMPLE 4.3 (BUTCHER, RUNGE-KUTTA, S = 3, PT = 4). *See example* 5 *in Böge* et al. (1986). *Gröbner base of the extension ideal containing the third factor* $(B + 1)$ *of the univariate polynomial in* B: $B^7 + 7/2B^6 + 14/3B^5 + 23/8B^4 + 97/144B^3 - 17/144B^2 - 13/144B - 1/144$.

Polynomial ring: $\mathbf{K}[B, C_2, C_3, A, B_3, B_2, A_{32}, B_1]$
Term order is inverse lexicographic $<_L$.

Input: $HT(G) = \{B, A, C_3^2B_3, C_2B_2, C_2B_3A_{32}, C_3B_3^2A_{32}, B_1\}$

Output: d = 3
S = $\{C_2, C_3, A_{32}\}$
M = $\{\{C_2, C_3, A_{32}\}, \{C_2, B_3\}, \{C_3, B_2, A_{32}\}, \{B_3, B_2, A_{32}\}\}$

Statistics: Computing time = 20 ms, Used storage cells = 690

With a different lexicographical term order we get:

Polynomial ring: $\mathbf{K}[B_3, B_2, A_{32}, C_2, C_3, B, A, B_1]$
Term order is inverse lexicographic $<_L$.

Input: $HT(G) = \{B_3A_{32}C_2, B_2^2C_2^2, B_2A_{32}C_2^2, B_3C_3, B_2C_2C_3, B, A, B_1\}$

Output: d = 3
S = $\{B_3, B_2, A_{32}\}$
M = $\{\{B_3, B_2, A_{32}\}, \{B_3, C_2\}, \{B_2, A_{32}, C_3\}, \{A_{32}, C_2, C_3\}\}$

Statistics: Computing time = 30 ms, Used Storege cells = 1509
With Buchberger's total degree term order we get:

Polynomial ring: $K[B, C_2, C_3, A, B_3, B_2, A_{32}, B_1]$
Term order is $<_B$.

Input: $HT(G) = \{B, A, B_1, C_3, B_3, C_2C_3B_2, C_2B_3A_{32}, C_2^2B_2^2, C_2^2B_2A_{32}\}$

Output: d = 3
S = $\{C_2, C_3, A_{32}\}$
M = $\{\{C_2, C_3, A_{32}\}, \{C_2, B_3\}, \{C_3, B_2, A_{32}\}, \{B_3, B_2, A_{32}\}\}$

Statistics: Computing time = 20 ms, Used storage cells = 855
With the inverse graduated term order we get:

Polynomial ring: $\mathbf{K}[B, C_2, C_3, A, B_3, B_2, A_{32}, B_1]$
Term order is $<_G$.

Input: $HT(G) = \{B, A, B_1, C_2, B_2, C_3^2B_3, C_2B_3A_{32}, C_3B_3^2A_{32}\}$
Output: d = 3
S = $\{C_2, C_3, A_{32}\}$
M = $\{\{C_2, C_3, A_{32}\}, \{C_2, B_3\}, \{C_3, B_2, A_{32}\}, \{B_3, B_2, A_{32}\}\}$

Statistics: Computing time = 20 ms, Used storage cells = 791

EXAMPLE 4.4. (GERDT). *See special example* 1 *in Böge* et al. (1986).

Polynomial ring: $\mathbf{K}[L_1, L_2, L_4, L_5, L_6, L_3, L_7]$
Term order is inverse lexicographic $<_L$.

Input: $HT(G) = \{L_1^3L_2^3, L_1^2L_2^4, L_1L_2^5, L_1^4L_4, L_1^3L_2L_4, L_1L_2^2L_4, L_1L_4^2, L_1^3L_5, L_1^2 L_2L_5, L_1L_2^2L_5, L_1L_4L_5, L_2L_4^2L_5, L_1L_5^2, L_4L_5^2, L_1L_6, L_4L_6, L_2L_6^2, L_5L_6^2, L_1^3L_3, L_1L_2L_3, L_4L_3, L_1L_5L_3, L_2L_6L_3, L_1L_3^2, L_7\}$
Output: d = 3
S = $\{L_2, L_5, L_3\}$
M = $\{\{L_1\}, \{L_2, L_4\}, \{L_2, L_5, L_3\}, \{L_6, L_3\}\}$

Statistics: Computing time = 50 ms, Used storage cells = 2097
With a different lexicographical term order we get:

Polynomial ring: $\mathbf{K}[L_2, L_3, L_1, L_4, L_5, L_6, L_7]$
Term order is inverse lexicographic $<_L$.

Input: $HT(G) = \{L_2^2L_3^3L_1, L_2L_3^4L_1, L_3^5L_1, L_2^4L_1^2, L_2^3L_3L_1^2, L_2L_3^2L_1^2, L_3^3L_1^2, L_1^3, L_2^2L_3L_4, L_3^2L_4, L_1L_4, L_3L_4^2, L_1L_5, L_3L_4L_5, L_2L_4^2L_5, L_4L_5^2, L_1L_6, L_4L_6, L_5L_6, L_2L_6^2, L_7\}$
Output: d = 3
S = $\{L_2, L_3, L_5\}$
M = $\{\{L_2, L_3, L_5\}, \{L_2, L_4\}, \{L_3, L_6\}\}$

Statistics: Computing time = 30 ms, Used storage cells = 1782
With a different lexicographical term order we get:

Polynomial ring: $\mathbf{K}[L_2, L_3, L_5, L_1, L_4, L_6, L_7]$
Term order is inverse lexicographic $<_L$.

Input: $HT(G) = \{L_2^2L_3^3L_1, L_2L_3^4L_1, L_3^5L_1, L_2^3L_5L_1, L_2^2L_3L_5L_1, L_3^2L_5L_1, L_5^2L_1, L_2^3L_1^2, L_2^2L_3L_1^2, L_3^2L_1^2, L_5L_1^2, L_1^3, L_3L_4, L_1L_4, L_2L_5L_4^2, L_4^3, L_5L_6, L_1L_6, L_4L_6, L_2L_6^2, L_7\}$
Output: d = 3
S = $\{L_2, L_3, L_5\}$
M = $\{\{L_2, L_3, L_5\}, \{L_3, L_6\}\}$

Statistics: Computing time = 30 ms, Used storage cells = 1735

EXAMPLE 4.5 (GEDDES). *See special example* 4 *in Böge* et al. (1986).

Polynomial ring: $\mathbf{K}[B_5, B_4, A_5, B_0, A_4, B_3, A_3, B_1, B_2, A_2, A_0, C_5, C_4, C_3, C_1, C_2, C_0]$
Term order is inverse lexicographic $<_L$.

Input: $HT(G) = \{B_5^2, B_5B_4, B_4^2, B_5A_5, B_4A_5, A_5^2, B_5A_4, B_4A_4, A_5A_4, A_4^2, B_5B_3, A_5B_3, A_4B_3, B_3^2, B_5A_3, B_4A_3, A_4A_3, B_3A_3, B_5B_1, A_5B_1, A_4B_1, A_3B_1, B_5B_2, B_4B_2, A_5B_2, A_4B_2, B_2^2, B_4A_2, A_5A_2, A_4A_2, B_3A_2, A_3A_2, B_1A_2, B_2A_2, C_5, C_4, C_3, C_1, C_2, C_0\}$

Output: d = 3
S = $\{B_0, A_3, A_0\}$
M = $\{\{B_0, A_3, A_0\}, \{B_0, B_1, A_0\}, \{B_0, A_2, A_0\}\}$

Statistics: Computing time = 220 ms, Used storage cells = 8681
With a different lexicographical term order we get:

Polynomial ring: $\mathbf{K}[B_0, A_0, B_5, B_4, A_5, A_4, B_3, A_3, B_1, B_2, A_2, C_5, C_4, C_3, C_1, C_2, C_0]$
Term order is inverse lexicographic $<_L$.

Input: $HT(G) = \{B_5^2, B_5B_4, B_4^2, B_5A_5, B_4A_5, A_5^2, B_5A_4, B_4A_4, A_5A_4, A_4^2, B_5B_3, A_5B_3, A_4B_3, B_3^2, B_5A_3, B_4A_3, A_4A_3, B_3A_3, B_5B_1, A_5B_1, A_4B_1, A_3B_1, B_5B_2, B_4B_2, A_5B_2, A_4B_2, B_2^2, B_4A_2, A_5A_2, A_4A_2, B_3A_2, A_3A_2, B_1A_2, B_2A_2, C_5, C_4, C_3, C_1, C_2, C_0\}$

Output: d = 3
S = $\{B_0, A_0, A_3\}$
M = $\{\{B_0, A_0, A_3\}, \{B_0, A_0, B_1\}, \{B_0, A_0, A_2\}\}$

Statistics: Computing time = 250 ms, Used storage cells = 9545

EXAMPLE 4.6 (HEMION, CYCLOHEXAN C_6H_{12} MOLECULE). *Hemion* (1986) private communication.

Polynomial ring: $\mathbf{K}[X, Y, Z]$
Term order is inverse lexicographic $<_L$.

Input: $HT(G) = \{X^2Y^2, X^4Z, YZ, X^2Z^2\}$
Output: d = 1
S = $\{X\}$
M = $\{\{X\}, \{Y\}, \{Z\}\}$

Statistics: Computing time = 4 ms, Used storage cells = 319

EXAMPLE 4.7 (MACAULAY'S CURVES, $s = 10$). *See the examples in Möller and Mora* (1983).

Polynomial ring: $\mathbf{K}[X_0, X_1, X_2, X_3]$
Term order is inverse lexicographic $<_L$.

Input: $HT(G) = \{X_0^8X_2, X_0X_3, X_1^8X_3, X_1^7X_3^2, X_1^6X_3^3, X_1^5X_3^4, X_1^4X_3^5, X_1^3X_3^6, X_1^2X_3^7, X_1X_3^8\}$
Output: d = 2
S = $\{X_0, X_1\}$
M = $\{\{X_0, X_1\}, \{X_1, X_2\}, \{X_2, X_3\}\}$

Statistics: Computing time = 10 ms, Used storage cells = 484

EXAMPLE 4.8 (MACAULAY'S CURVES, $s = 30$.) *See the examples in Möller and Mora* (1983)
Polynomial ring: $\mathbf{K}[X_0, X_1, X_2, X_3]$
Term order is inverse lexicographic $<_L$.

Input: $HT(G) = \{X_0^{28}X_2, X_0, X_3, X_1^{28}X_3, X_1^{27}X_3^2, X_1^{26}X_3^3, X_1^{25}X_3^4, X_1^{24}X_3^5, X_1^{23}X_3^6, X_1^{22}X_3^7, X_1^{21}X_3^8, X_1^{20}X_3^9, X_1^{19}X_3^{10}, X_1^{18}X_3^{11}, X_1^{17}X_3^{12}, X_1^{16}X_3^{13}, X_1^{15}X_3^{14}, X_1^{14}X_3^{15}, X_1^{13}X_3^{16}, X_1^{12}X_3^{17}, X_1^{11}X_3^{18}, X_1^{10}X_3^{19}, X_1^9X_3^{20}, X_1^8X_3^{21}, X_1^7X_3^{22}, X_1^6X_3^{23}, X_1^5X_3^{24}, X_1^4X_3^{25}, X_1^3X_3^{26}, X_1^2X_3^{27}, X_1X_3^{28}\}$

Output: $d = 2$
$S = \{X_0, X_1\}$
$M = \{\{X_0, X_1\}, \{X_1, X_2\}, \{X_2, X_3\}\}$

Statistics: Computing time = 20 ms, Used storage cells = 764
With different term orders, we get the following maximal strongly independent sets:

Polynomial ring: $\mathbf{K}[X_0, X_1, X_2, X_3]$
Term order is inverse graduated $<_G$.

Output: $d = 2$
$S = \{X_0, X_1\}$
$M = \{\{X_0, X_1\}, \{X_1, X_2\}, \{X_2, X_3\}\}$

Statistics: Computing time = 20 ms, Used storage cells = 764
Polynomial ring: $\mathbf{K}[X_0, X_1, X_2, X_3]$
Term order is inverse total degree $<_B$.

Output: $d = 2$
$S = \{X_0, X_3\}$
$M = \{\{X_0, X_3\}\}$

Statistics: Computing time = 20 ms, Used storage cells = 580
Polynomial ring: $\mathbf{K}[X_0, X_1, X_2, X_3]$
Term order is inverse reverse exponent vector $<_S$.

Output: $d = 2$
$S = \{X_0, X_1\}$
$M = \{\{X_0, X_1\}, \{X_1, X_2\}, \{X_2, X_3\}\}$

Statistics: Computing time = 20 ms, Used storage cells = 762

Research on this paper was carried out at the Gesellschaft für Schwerionenforschung, Darmstadt, and the University of Heidelberg, and was partially supported by a DFG-Heisenberg grant. We are indebted to B. Buchberger and H. Möller for helpful comments on an earlier version of the paper, to A. Kandri-Rody for valuable discussions on the subject, and to R. Loos and G. Hemion for providing us with example 4.6. We would also like to thank the GSI for providing computing time.

References

Böge, W., Gebauer, R., Kredel, H. (1986). Some examples for solving systems of algebraic equations by calculating Gröbner bases. *J. Symbolic Computation*, **1,** 83–98.
Buchberger, B. (1965). *Ein Algorithmus zum Auffinden der Basiselemente des Restklassenringes nach einem nulldimensionalen Polynomideal*, Ph.D. Thesis, University of Insbruck.
Buchberger, B. (1985). Gröbner bases: An algorithmic method in polynomial ideal theory, In (Bose, N. K. ed.)

Progress, Directions and Open Problems in Multidimensional Systems Theory. Dordrecht: Reidel Publ. Comp. Pp. 184–232.

Carra-Ferro, G. (1985). *Some Upper Bounds for the Multiplicity of an Autoreduced Subset of N^m and their Application*, AAECC-3, Grenoble, Springer, LNCS, Vol. 229.

Carra-Ferro, G. (1987). *Some Properties of Lattice Points and Their Application to Differential Algebra*. Preprint.

Collins, G. E., Loos, R. G. (1980). SAC 2—Symbolic and Algebraic Computation version 2, a computer algebra system, ALDES—ALgorithm DEscription language. *ACM SIGSAM Bulletin*, **14,** no. 2.

Gebauer, R., Kredel, H. (1983). Buchberger algorithm system. Technical Report, Institut für Angewandte Mathematik, Universität Heidelberg (see also *ACM SIGSAM Bulletin* **18,** 19).

Gianni, P., Trager, B., Zacharias, G. (1986). *Gröbner bases and Primary Decomposition of Polynomial Ideals*. Preprint.

Giusti, M. (1984). Some Effectivity Problems in Polynomial Ideal Theory, EUROSAM '84, Springer LNCS, Vol. 174 (1984), pp. 159–171.

Gröbner, W. (1970). *Algebraic Geometrie I, II*, Bibliographisches Institut, Mannheim (1968, 1970).

Kandri-Rody, A. (1985). Dimension of ideals in polynomial rings. *Proc. Combinatorial Algorithms in Algebraic Structures*. Otzenhausen 1985, J. Avenhaus, K. Madlener Eds., Fachbereich Informatik, University of Kaiserslautern.

Kredel, H. (1985). *Über die Bestimmung der Dimension von Polynomidealen*. Diplomarbeit, Heidelberg.

Kredel, H. (1987). *Primary ideal decomposition*, presented at EUROCAL '87, Leipzig.

Kredel, H. (1988). Admissible term orderings used in Computer Algebra Systems. *ACM SIGSAM Bulletin* **22,** 28–31.

Kutzler, B., Stifter, S. (1986). Automated geometry theorem proving using Buchberger's algorithm. CAMP-Publ.-Nr. 85-29.0, University of Linz (1986); *Procl SYMSAC '86*, Waterloo, Canada (1986).

Möller, H. M., Mora, F. (1983). The computation of the Hilbert function, in EUROCAL '83, *Computer algebra*, Springer LNCS vol. 162 (1983).

Robbiano, L. (1985), Term orderings on the polynomial ring, EUROCAL '85, Springer LNCS, Vol. 204 (1985), pp. 513–517.

Weispfenning (1987). Admissible Orders and Linear Forms, *ACM SIGSAM Bulletin* **21,** 16–18.

Weispfenning, V. (1988). Some bounds for the construction of Gröbner Bases, Proc. AAECC-4, Karlsruhe 1986, Springer LNCS, Vol. 307, pp. 195–201.

Zariski, O. Samuel, P. (1958/1960). *Commutative Algebra*, vols. I, II. van Nostrand.

Combinatorial Dimension Theory of Algebraic Varieties

MARC GIUSTI†

Centre de Mathématiques, Ecole Polytechnique, 91128 Palaiseau Cedex, Unité associée au CNRS No. 169, France

(*Received* 13 *January* 1988, *and in revised form* 28 *March* 1988)

Several algorithms are already known to compute the dimension of a projective algebraic variety. But they all rely on the construction of the whole standard basis of the defining ideal, leading to non optimal worst case complexities. We present below a new algorithm, based on a truncation idea, with better complexity.

1. Introduction

How can one compute the dimension of a projective algebraic subset, given by a system of equations? There are different definitions, and their equivalence is part of the classical dimension theory in algebraic geometry. The game is then the following: to make some definition of dimension effective enough to get an algorithm.

A first method consists to think *algebraically* and to use the Hilbert function of the quotient ring. It can be easily computed once a particular set of generators (a so called standard basis (= Gröbner basis) of the defining ideal) is known. In fact such a basis yields a monomial ideal with the same Hilbert function, hence the same Hilbert polynomial, whose degree is the dimension. So it is the order at infinity of this numerical function, attained for example through the maximal dimension of a coordinate plane containing no elements of the monomial ideal. Based on this idea there are different algorithms already known in the literature (see Carrà Ferro, 1986; Kandri–Rody, 1985; Kredel & Weispfennig, 1988; Lejeune-Jalabert, 1984–1985). Unfortunately this method needs to construct a whole standard basis from a given set of generators, and leads necessarily to a disastrous upper bound for the worst case complexity, as shown by the explicit examples of Mayr & Meyer (1982), Demazure (1985): there are ideals for which deciding whether a given polynomial belongs to them needs exponential space. As the knowledge of a standard basis easily solves the previous problem, computing such a basis needs also exponential space.

In order to bypass that, why not to determine the Hilbert polynomial by interpolation, if we know an upper bound for the regularity of the Hilbert function? Alas, it happens that the same catastrophic behaviour cannot be avoided, as indicated in Giusti (1984).

Another way is to think *geometrically* and to cut the variety by linear subspaces, as proposed by Lazard (1981; 1982). Once again this dimension can be read on a standard

† Partially supported by a grant from the PRC "Mathématiques et Informatique", and GRECO de Calcul Formel No. 60.

basis, this time for a particular ordering (Giusti, 1984). But it involves the costly introduction of generic coordinates.

We will propose below an approach combining the combinatorial point of view (standard bases) and the geometric ones (section/projection). We shall introduce two new characterizations of the dimension, which come in between the classical approaches in a remarkable way, since this gives a new proof of the dimension theorem through their relationships with the classical definitions.

Furthermore we do not need a whole standard basis, which is a goal much too complicated, in order to extract a very partial useful information. Actually, a part in rather low degree is enough. Using this truncation idea, we will give an algorithm computing the dimension, whose time and space complexity are polynomial in a suitable measure of the input data.

2. Standard Bases

Let k be a field, and R the polynomial algebra $k[x_0, x_1, \ldots, x_n]$, with the usual graduation induced by the total degree. To study homogeneous ideals of R, an essential idea going back to Macaulay (1927), consists in totally ordering the monomials of the polynomial ring, in one-to-one correspondence with the points of $\mathbf{N}^{n+1}$, but in a compatible way with the multiplication of R. We will use the following orderings:

2.1. DEFINITION: LOWER (RESP. UPPER) LEXICOGRAPHIC ORDERING

A point $a = (a_0, \ldots, a_n)$ of $\mathbf{N}^{n+1}$ is smaller than a point $b = (b_0, \ldots, b_n)$ for the lower lexicographic ordering (resp. larger for the upper lexicographic ordering) if and only if there exists an index i $(0 \leqslant i \leqslant n)$ such that:

$$a_0 = b_0, \ldots, a_{i-1} = b_{i-1}, a_i < b_i$$

(resp.

$$a_n = b_n, \ldots, a_{i+1} = b_{i+1}, a_i > b_i).$$

2.2. DEFINITIONS: LOWER (RESP. UPPER) LEXICOGRAPHIC FILTRATION

To every non zero homogeneous polynomial:

$$f = \sum_{a \in \mathbf{N}^{n+1}} f_a x^a$$

is then associated its support $\{a \in \mathbf{N}^{n+1} \mid f_a \neq 0\}$ in $\mathbf{N}^{n+1}$. The smallest (resp. largest) element of the support of f for the lower (resp. upper) lexicographic ordering is called its *leading monomial exp*(f). The *leading* or *initial form in* (f) is the term

$$f_{exp(f)} x^{exp(f)}$$

We shall speak from now on of the lower (resp. upper) filtration of R.

2.3. DEFINITION: MIXED LEXICOGRAPHIC FILTRATION OF ORDER m

Let us fix an integer m between 0 and $n-1$. After Bayer & Stillman (1985), we introduce now the following total ordering: by definition a point $a = (a_0, \ldots, a_n)$ of $\mathbf{N}^{n+1}$

is smaller than a point $b = (b_0, \ldots, b_n)$ if the point $(a_0 + \cdots + a_m, a_0, \ldots, a_{n-1})$ is smaller than the point $(b_0 + \cdots + b_m, b_0, \ldots, b_{n-1})$ for the lower lexicographic ordering.

The leading monomial of a non zero polynomial is then the smallest element of its support.

2.4. DEFINITION: STABLE SUBSETS

Given a non zero homogeneous ideal I, the set of all the leading monomials of its non zero elements form a so called *stable* subset of $\mathbf{N}^{n+1}$, i.e.:

$$a \in E(I) \Rightarrow a + \mathbf{N}^{n+1} \subseteq E(I).$$

By convention, the empty set is associated to the trivial ideal (0).

2.5. DICKSON'S LEMMA

Every stable subset E of $\mathbf{N}^{n+1}$ is finitely generated, and there exists a unique minimally generating finite family $a^{(1)}, \ldots, a^{(p)}$, i.e.:

$$E = \bigcup_{i=1}^{p} (a^{(i)} + \mathbf{N}^{n+1}).$$

The proof is easy by induction on n.

2.6. DEFINITION: STANDARD BASES

Let I be an homogeneous ideal of R. By Dickson's lemma, $E(I)$ is minimally generated by some family, say $a_1, \ldots, a_p$. Following Hironaka (1964), we define a *standard basis* of I as a family of polynomials of I such that their leading monomials form this minimal generating subset.

2.7. NOTATION

If E is a stable subset of $\mathbf{N}^{n+1}$, we denote by $D(E)$ the maximal degree of the elements of a generating family of E, which is minimal for the inclusion.

2.8. HIRONAKA'S DIVISION THEOREM

Let I be an homogeneous ideal of R. Every polynomial of R is equivalent modulo I to a unique polynomial called the remainder of the division by I, either zero or owning all its monomials outside of $E(I)$.

Proof: Actually if the polynomial f to be divided is not zero, it has a leading monomial. If this last monomial belongs to $E(I)$, it can be divided by the leading monomial of some element of I, say g. The input polynomial f is then equivalent to $f - (in(f)/in(g))g$, which, if non zero, owns a strictly smaller leading monomial. Since there is only a finite number of monomials of given degree, the iterated process stops on a polynomial, which if not zero, has a leading monomial outside of $E(I)$. To conclude we apply the same division procedure to the reductum of this last polynomial, i.e. itself minus its initial form.

The uniqueness of the remainder follows trivially from its property of having all its monomials outside of $E(I)$.

2.9. COROLLARY

Every standard basis of an ideal generates this ideal.

Proof: In fact, the remainder of the division of any polynomial of the ideal by a standard basis $f_1, \ldots, f_p$ must be zero: if not, its leading monomial should be simultaneously in $E(I)$ and its complementary.

Note that it proves by the way the noetherianity of polynomial rings.

2.10. COROLLARY

As k-vector space, the quotient R/I is isomorphic to the direct sum of the k-vector subspaces of dimension one generated by the monomials of the complementary of $E(I)$:

$$R/I = \bigoplus_{a \notin E(I)} kx^a.$$

Eventually we will show that the filtrations we introduced before have a nice geometric interpretation:

2.11. PROPOSITION: SECTION BY A LINEAR SUBVARIETY

Fix an integer s between 0 and $n-1$. Let $f_1, \ldots, f_t$ be a standard basis of I, with respect to the lower lexicographic filtration. Then the f_i's with leading monomial not depending on $x_0, \ldots, x_s$, together with $x_0, \ldots, x_s$ form a standard basis of $I + (x_0, \ldots, x_s)$, with respect to the induced lower lexicographic filtration; hence they define the section of $Z(I)$ by the linear subvariety $x_0 = \cdots = x_s = 0$.

Proof: It follows from the remark that if x_0 divides the leading monomial of a polynomial, it actually divides this polynomial. Hence the proposition is true for $s = 0$. The general result is clear by induction.

2.12. PROPOSITION: PROJECTION ON A LINEAR SUBVARIETY

In this paragraph the ground field k will be algebraically closed.

Fix an integer p between 0 and $n-1$. Let $f_1, \ldots, f_t$ be a standard basis of I, with respect to the upper lexicographic filtration. Then the f_i's with leading monomial depending only on the variables $x_0, \ldots, x_p$ form a standard basis of $I \cap k[x_0, \ldots, x_p]$; hence if we assume that the linear subvariety defined by $x_0, \ldots, x_p$ does not intersect $Z(I)$, they define the projection of $Z(I)$ on the linear subvariety $x_{p+1} = \cdots = x_n = 0$.

Proof: Let p be equal to zero. The first part follows from the remark that if the leading monomial of a polynomial does not depend on x_n, the polynomial itself does not. Now let π be the morphism restriction to $Z(I)$ of the projection of $\mathbf{P}^n \backslash (0, \ldots, 0, 1)$ on the hyperplane $x_n = 0$. By abstract non-sense we get the equality of the Zariski closure of $\pi(Z(I))$ with the algebraic subset defined by $I \cap k[x_0, \ldots, x_{n-1}]$. Since the restriction of π to $Z(I)$ is proper, we obtain the wanted equality.

The general result follows easily by induction.

2.13. GEOMETRIC INTERPRETATION OF THE MIXED FILTRATIONS

Fix an integer m between 0 *and* $n-1$. *Let* $f_1, \ldots, f_t$ *be a standard basis of I, with respect to the mixed lexicographic filtration of order m.*

Then the f_i*'s with leading monomial not depending on the variables* $x_0, \ldots, x_m$ *form a standard basis of* $I+(x_0, \ldots, x_m)$; *hence they define the section of* $Z(I)$ *by the linear subvariety* $x_0 = \cdots = x_m = 0$.

On the other hand the f_i*'s with leading monomial depending only on the variables* $x_0, \ldots, x_m$ *form a standard basis of* $I \cap k[x_0, \ldots, x_m]$ *for the induced lower lexicographic filtration. Hence if we assume that linear subvariety defined by* $x_0, \ldots, x_m$ *does not intersect* $Z(I)$, *they define the projection of* $Z(I)$ *on the linear subvariety defined by* $x_{m+1} = \cdots = x_n = 0$.

The proof is similar to the two previous ones.

3. Some Bounds for the Regularity of the Hilbert Function

The *degree* of a point $a = (a_0, \ldots, a_n)$ of $\mathbf{N}^{n+1}$ is the integer $|a| = a_0 + \cdots + a_n$.

3.1. DEFINITION: HILBERT FUNCTION

Let E be a stable subset of $\mathbf{N}^{n+1}$; the function HF_E, which associates to every integer s the number of elements of degree s not belonging to E, is called the *Hilbert function* of the complementary of E:

$$HFE(s) = \#\{a \in \mathbf{N}^{n+1} \mid a \notin E, |a| = s\}.$$

3.2. DEFINITION: HILBERT POLYNOMIAL

It is classical that for s large enough, the Hilbert function HF_E is equal to a polynomial HP_E (the so called *Hilbert polynomial*). Moreover there are integers $c_0, \ldots, c_d$ such that:

$$HP_E(s) = \sum_{i=0}^{d} c_i \binom{s}{d-i}$$

where

$$\binom{s}{r} = s(s-1)(\cdots)(s-r+1)/r!$$

is the binomial coefficient function. By definition, d is the *dimension* of the complementary of E, and c_0 its *degree*. Finally by convention the degree -1 will be associated to the zero polynomial.

3.3. DEFINITION: REGULARITY OF THE HILBERT FUNCTION

The *regularity* $H(E)$ of the Hilbert function is the smallest integer where it becomes equal to the Hilbert polynomial of E.

3.4. PROPOSITION

$$H(E) \leqslant (n+1)(D(E)-1)+1.$$

Proof: By induction on n. The claim is clear for $n=0$. Then from Dickson's lemma, the section E_i of E by the hyperplane $\{x_n = i\}$ is constant for i at least e and not before. Let us consider the following development of the Hilbert function:

$$HF_E(s) = \sum_{i=0}^{s} HF_{E_i}(s-i)$$

which breaks into three pieces as soon as s is large enough:

$$HF_E(s) = \sum_{0 \leqslant i \leqslant e-1} HF_{E_i}(s-i) + \sum_{e \leqslant i \leqslant s-H(E_e)} HF_{E_e}(s-i) + \sum_{s-H(E_e)+1 \leqslant i \leqslant s} HF_{E_e}(s-i)$$

$$= \sum_{0 \leqslant i \leqslant e-1} HF_{E_i}(s-i) + \sum_{H(E_e) \leqslant i \leqslant s-e} HF_{E_e}(i) + \sum_{0 \leqslant i \leqslant H(E_e)-1} HF_{E_e}(i).$$

For s large enough this function becomes:

$$HF_E(s) = \sum_{0 \leqslant i \leqslant e-1} HP_{E_i}(s-i) + \sum_{H(E_e) \leqslant i \leqslant s-e} HP_{E_e}(i) + \sum_{0 \leqslant i \leqslant H(E_e)-1} HF_{E_e}(i)$$

which is the Hilbert polynomial HP_E since $\sum_{H(E_e) \leqslant i \leqslant s-e} HP_{E_e}(i)$ is effectively a polynomial in s. This is a consequence of the following classical lemma:

LEMMA: *Let P be a polynomial of degree d with rational coefficients. For any two integers r and s, $\sum_{r \leqslant i \leqslant s} P(i)$ is a polynomial in s of degree $d+1$, with integral coefficients on the binomial basis if it was the case for P.*

The proof is straightforward: as the binomial coefficient functions

$$\binom{s}{0}, \ldots, \binom{s}{d}$$

form a basis on $\mathbf{Q}$ of the vector space of all polynomials of degree less or equal to d, it is enough to prove the first claim on these particular polynomials, where it becomes trivial.

More precisely this expression is true as soon as s is larger than $Sup\{i + H(E_i) \mid 0 \leqslant i \leqslant e\}$, which occurs through the induction hypothesis if s is bigger than $(n+1)(D(E)-1)+1$.

(Note that we reproved by the way the classical facts on the Hilbert polynomial recalled in 3.2.)

3.5. EXAMPLE

Given a sequence of $n+1$ positive integers $a_0, \ldots, a_n$, consider the ideal $(x_0^{a_0}, \ldots, x_n^{a_n})$. This set of generators form a standard basis, and the associated stable subset E is the complement of a parallelpiped, whose point of maximal degree is $(a_0 - 1, \ldots, a_n - 1)$.

Hence the regularity is $1 + \sum_{i=0}^{n} (a_i - 1)$. Note that it is larger than $D(E)$ which is the maximum of the a_i's; this remark will be developed below.

The particular case where all the a_i's are equal show that the previous bound of the regularity is sharp.

3.6. PROPOSITION

If E is a negative dimensional stable subset, its regularity $H(E)$ is at least $D(E)$.

Proof: First let us discard the trivial case $E = \mathbf{N}^{n+1}$ where $D(E) = H(E) = 0$. Then consider an element $A = (A_0, \ldots, A_n)$ of a minimally generating family of E. Since the degree of A is positive, there exists an index i for which A_i is positive, and the point $(A_0, \ldots, A_i - 1, \ldots, A_n)$ does not belong to E. Thus the Hilbert function is not zero for the argument $|A| - 1$, hence is not equal to the Hilbert polynomial. So the regularity is at least $|A|$, and in particular at least $D(E)$.

3.7. REMARK

If I is an homogeneous ideal of R, the Hilbert function of $E(I)$ is independent of the ordering by 2.10 and is called the *Hilbert function* of R/I.

4. Dimension Theory

From now on, the ground field k is assumed to be algebraically closed. Let I be an homogeneous ideal of R defined by a given system of generators, and $Z(I)$ the projective algebraic subset defined in $\mathbf{P}^n$.

4.1. VARIOUS DEFINITIONS OF DIMENSION

First an *algebraic* definition: the *Hilbert dimension* is the degree d of the Hilbert polynomial of R/I.

Then two *geometrical* definitions: the *section dimension* is the smallest integer s such that there exists a linear subvariety of codimension $s + 1$ not intersecting $Z(I)$.

The *projection dimension* is the biggest integer p such that there exists a linear subvariety of dimension p on which $Z(I)$ can be surjectively projected.

It is classical that the three notions of dimension recalled above coincide, and their common value is called the *dimension* of the projective algebraic subset $Z(I)$. We will now introduce two combinatorial definitions:

4.2. DEFINITIONS: LOWER AND UPPER LEXICOGRAPHIC DIMENSIONS

Let x be a system of coordinates on $\mathbf{P}^n$. We denote by $linf_x$ (resp. $lsup_x$) the smallest (resp. largest) integer e such that $E_x(I)$ relative to the lower (resp. upper) lexicographic filtration intersects the last $n - e$ coordinate axes of $\mathbf{N}^{n+1}$ (resp. does not intersect the plane of the first $e + 1$ coordinates).

The minimum *linf* of all $linf_x$ (resp. the maximum *lsup* of all $lsup_x$) when x ranges through all systems of coordinates of $\mathbf{P}^n$ is called the *lower* (resp. *upper*) *lexicographic dimension*.

4.3. DIMENSION THEOREM

The lower and upper lexicographic dimensions are equal to the dimension of the projective algebraic subset.

The proof goes as follows, showing actually through five successive inequalities that the five notions of dimension coincide, thus reproving the classical dimension theorem for projective algebraic subsets:

4.3.1. $d \geqslant lsup$

Let us assume that there exists coordinates of $\mathbf{P}^n$ such that, with respect to some compatible ordering, for example the upper lexicographic ordering, $E(I)$ does not intersect the plane of the first $e+1$ coordinates of $\mathbf{P}^n$. Then $HF(s)$ is bounded below by the number of monomials of degree s on $e+1$ letters, which is a $O(s^e)$ when s is going to infinity. Hence d is bounded below by $lsup$.

4.3.2. $lsup \geqslant p$

Let us assume that $Z(I)$ can be projected surjectively on a plane P of dimension p. We can choose coordinates such that this plane is defined by the equations $x_{p+1} = \cdots = x_n = 0$. Then the ideal $I \cap k[x_0, \ldots, x_p]$ reduces to (0).

For if there is a non zero polynomial $f(x_0, \ldots, x_p)$ in I, it does not vanish on the whole plane P and the algebraic subset $Z(I)$ contained in the proper cylinder of equation $f = 0$ cannot be projected on the whole plane P.

Let us look now at $E(I)$ for such coordinates with respect to the upper lexicographic filtration. If a polynomial owns a leading form depending only on the first $p+1$ variables, it has the same property; hence $E(I)$ does not intersect the plane of the first $p+1$ variables.

4.3.3. $p \geqslant s$

From the definition of s, there exists a linear subvariety L of codimension $s+1$ not intersecting $Z(I)$. On the other hand it is always possible to choose a linear subvariety B of dimension s avoiding L. The projection of center L sends $Z(I)$ surjectively on B since s is minimal; for if b is a point of B, the linear subvariety generated with L is of codimension s, hence intersects $Z(I)$ in at least one point with image b.

4.3.4. $s \geqslant linf$

From the definition of s, there exists a linear subvariety L of codimension $s+1$ avoiding $Z(I)$. We can choose coordinates such that L is defined by the equations $x_0 = \cdots = x_s = 0$. The ideal $J = I + (x_0, \ldots, x_s)$ defines the empty subvariety, therefore contains a power of the ideal $(x_0, \ldots, x_n)$, according to the Nullstellensatz theorem. For any choice of the ordering, $E(J)$ intersects all coordinate axes. From now on let us consider only the lower lexicographic filtration, and look at a polynomial whose leading monomial lies in one of the axes $x_{s+1}, \ldots, x_n$. This last polynomial is then equivalent to some polynomial g of I modulo $(x_0, \ldots, x_s)$, which cannot be zero, and whose leading monomial is equal to the one of f. Hence the section of $E(I)$ by the plane of the last $n-s$ coordinates has points on all axes.

4.3.5. *linf* $\geqslant d$

Let us assume that there are coordinates of $\mathbf{P}^n$ such that, with respect to some compatible ordering, for example the lower lexicographic one, $E(I)$ intersects the last $n - linf$ axes. In degree u large enough, the complementary of $E(I)$ contains only monomials with last $n - linf$ components bounded above by some fixed integer, hence a fixed integer times the number of monomials of degree u on $linf + 1$ letters. So the Hilbert function $HF(u)$ is bounded above by $O(u^{linf})$ when u goes to infinity.

4.4. NOTE ADDED IN PROOF

Given a stable subset E of $\mathbf{N}^{n+1}$, following several authors (Carrà Ferro, 1986; Kredel & Weispfennig, 1988), let us consider independent variables for E, i.e. such that E does not contain an element in just those variables. Call $m(E)$ the cardinality of a maximal set of such independent variables.

Actually we prove in 4.3.1 that the dimension of the complementary of E is greater or equal to the cardinality of every set of independent variables minus one, hence to $m(E) - 1$. On the other hand, it is easy to prove by double induction on r and n that this dimension is smaller or equal to $m(E) - 1$. The assertion is clear for $m = 0$ and trivial for $n = 0$. Then going back to the proof of 3.4, let us remark that if every r-plane of coordinates contains a point of E, every r-plane of coordinates of $\mathbf{N}^n$ contains a point of any section E_i, and every $(r - 1)$-plane of coordinates of $\mathbf{N}^n$ contains a point of the section at infinity E_e. Hence the degree of the Hilbert polynomial of E is smaller or equal to r using the induction hypotheses.

Hence we reproved a result of Carrà Ferro (1986), see also Kredel & Weispfennig. Remark that this assertion is independent of the ordering chosen.

5. Complexity of Computation of the Dimension

5.1. MEASURE OF THE COMPLEXITY OF THE INPUT DATA

An algebraic projective set $Z = Z(I)$ is defined by some homogeneous ideal I of a polynomial algebra $k[x_0, \ldots, x_n]$, where k is an algebraically closed field. The ideal I is given by generators, as usual. So the complexity of the input data should include their number t and their maximal degree d. So the number

$$\binom{d+n}{n}$$

of monomials of degree d over $n + 1$ letters occurs naturally, through the space required to store one input polynomial with the dense representation. Asymptotically this is $O(d^n)$ if d goes to infinity with fixed n.

If the nature of the ground field k is not specified we shall denote below by *elementary time complexity* the number of elementary operations in k (additions, multiplications, comparisons, each counting for one) needed to perform an algorithm. Furthermore, if we want to consider the cost of each of these elementary operations, we shall speak of the *total time complexity*. For example if the generators are defined over the integers, we need their size or maximal number of bits M needed to write any coefficient.

We will now successively study the time complexity of the naive algorithms computing the Hilbert dimension, show that they all lead to a disastrous lower bound for the worst case complexity, study the complexity of the naive geometric approach, which is not bad but not the best, and introduce a new algorithm based on a truncation idea with subexponential complexity. Actually its time complexity is bounded above by a polynomial in td^{n^2}.

Furthermore, if the input generators are defined over the integers the space complexity is bounded above by a polynomial in $Mt(d^{n^2})$. So even if we keep track of the growth of intermediate computations, at the ultimate level of manipulations of bits the time complexity stays definitely polynomial in Mtd^{n^2}.

5.2. CALCULABILITY OF THE HILBERT DIMENSION

The principle of the computation of the Hilbert function, once a standard basis is known follows easily from 2.10, and goes back at least to Macaulay (1927). A key point is now to construct a standard basis of I from a given set of generators: the first algorithms are due to Buchberger (1965) (see also Galligo, 1983; Giusti, 1984; Lazard, 1983). Actually in the homogeneous case we are concerned by, the computation of $HF(s)$ (as well as the construction of such a basis in degree s) is reduced to a pure linear algebra process, i.e. a triangulation by column operations (see also Lazard, 1983; Giusti, 1984).

It still remains to determine the Hilbert polynomial by interpolation, and this is possible through the upper bound of the regularity given in 3.4; and indeed upper bounds of $D(E(I))$ are known in terms of the input data, at least in some particular case (Giusti, 1984). Anyway, it is shown in Giusti (1984) that the regularity is bounded below by the maximal degree needed to compute the syzygy modules occurring in a free resolution of the quotient ring R/I. But the study of the pathological examples of Mayr & Mayer (1982) yields a double exponential behaviour for this number (Demazure, 1985; Giusti & Lazard, 1986).

The method proposed by Carrà Ferro (1986) presents the same drawback. The worst case complexity of any algorithm computing the dimension based on the construction of a whole standard basis is necessarily catastrophic, since Mayr & Meyer (1982) showed that the space required to obtain such a basis can be exponential in the input data.

5.3. GENERIC COORDINATES

We observe now that *linf* (resp. *lsup*) (4.2) is reached for generic coordinates. To give a precise meaning to this assertion, we must first as usual define a parameter space. Coordinates of the projective space are determined by a simplex configuration of $n + 1$ linearly independent hyperplanes $\check{h}_0, \ldots, \check{h}_n$ (elements of $\check{\mathbf{P}}^n$) and $n + 1$ linearly independent points $h_0, \ldots, h_n$ (elements of $\mathbf{P}^n$); the equation of $\check{h}_i$ is $x_i = 0$, and the point h_i has all coordinates zero except the ith one. The parameter space is then the non empty Zariski open subset $\check{\Omega}$ (resp. Ω), complement of the determinant hypersurface in $(\check{\mathbf{P}}^n)^{n+1}$ (resp. in $(\mathbf{P}^n)^{n+1}$).

5.3.1. DEFINITION

Let $Z(I)$ be an algebraic subset of dimension l. We shall denote by U the non empty Zariski open subset of the Grassmanian $G(n - l - 1, n)$ of all linear subvarieties of codimension $l + 1$, which do not intersect $Z(I)$.

A coordinate system x of the projective space, such that the linear subvariety defined by $x_0 = \cdots = x_l = 0$ belongs to U is called *generic* for this algebraic subset.

5.3.2. PROPOSITION

The subset of coordinates systems where $linf_x(I)$ *(resp.* $lsup_x(I)$*) reaches its minimum* $linf(I)$ *(resp. its maximum* $lsup(I)$*) contains the following non empty Zariski open subset of* $\check{\Omega}$ *(resp.* Ω*):*

$$\check{\Gamma} = \{(\check{h}_0, \ldots, \check{h}_n) \in (\check{\mathbf{P}}^n)^{n+1} \mid \check{h}_0 \cap \cdots \cap \check{h}_l \in U\} \cap \check{\Omega}$$

(resp.:

$$\Gamma = \{(h_0, \ldots, h_n) \in (\mathbf{P}^n)^{n+1} \mid \langle h_{l+1}, \ldots, h_n \rangle \in U\} \cap \Omega.)$$

Proof: Let $(\check{h}_0, \ldots, \check{h}_n)$ be a point of $\check{\Gamma}$ (resp. $(h_0, \ldots, h_n)$) defining a system x of coordinates. For this choice let us consider the subset $E_x(I)$ for the lower (resp. upper) lexicographic filtration. The linear subvariety L defined by $x_0 = \cdots = x_l = 0$ does not intersect $Z(I)$. Hence $linf_x(I)$, being at most l (take the proof of 4.3.4), is equal to l. On the other hand, $Z(I)$ can be projected surjectively on the linear subvariety B generated by the points $h_0, \ldots, h_l$ from the center L (take the proof of 4.3.3), thus $lsup_x(I)$ being at least l (take the proof of 4.3.2) is equal to l.

5.3.3. REMARK

So the competition is open between two approaches: computing the dimension through *linf* involves the construction of a standard basis with respect to the lower lexicographic filtration, which is known to have a better complexity behaviour as the upper lexicographic one. Unfortunately it is difficult to characterize $\check{\Gamma}$, since one needs to recognize successive parameters for a ring . . . , which are precisely defined through dimension!

On the other hand, we can easily find a point in Γ if we know how to find a point outside of a hypersurface. We describe below an algorithm leading to a generic system of $n+1$ points, i.e. such that the first $n-l$ generate a linear subvariety of codimension $l+1$, which does not intersect $Z(I)$.

5.3.4. ALGORITHM TO FIND GENERIC COORDINATES FOR THE UPPER LEXICOGRAPHIC FILTRATION

Input data: a list I of polynomials. Output data: a list of points.

```
PROJ_GENERIC(I) := PROJ_GENERIC2(I NIL)

PROJ_GENERIC2(I 1) :=
          IF I = (0)
          THEN 1
          ELSE (h H): EXTERIOR(I)
                 PROJ_GENERIC2(PROJECT(I (h H))
                        CONS(h 1))
```

PROJECT(I(h H))

performs first the change of basis sending the point h on the point $(0, \ldots, 0, 0, 1)$ and the hyperplane H on the hyperplane $x_n = 0$. Then it gives an ideal defining the projection of $Z(I)$ from $(0, \ldots, 0, 0, 1)$ on the hyperplane $x_n = 0$ using 2.12 and any algorithm of construction mentioned in 5.2

EXTERIOR(I)

gives a point outside of $Z(I)$ and a hyperplane not passing through this point. It is of course enough to pick up among the generators of I an element, say P, for instance of minimal degree, say δ, and to find a point which is not a zero of this polynomial (function OUTSIDE(P)).

Even if we show that the complexity of this algorithm is low (actually polynomial in δ^n as proved in the next section), the result will be uncertain, since the algorithm will be applied to equations of successive projections, whose degrees have to be controlled. This will be done in a forthcoming paper.

5.3.5. PROPOSITION: COMPLEXITY TO FIND A POINT OUTSIDE OF A HYPERSURFACE

Let P be a polynomial of degree δ. The elementary time complexity to find a point on which the polynomial does not vanish is polynomial in δ^n. Furthermore if the input polynomial has integral coefficients of maximal size M, the time complexity is polynomial in $M\delta^n$.

First the polynomial P cannot vanish on the $(\delta + 1)^n$ following points: $x_0 = 1$ and $0 \leqslant x_i \leqslant \delta$ for $i = 1, \ldots, n$, because if so, it is identically zero. This fact is easily obtained by induction on n: it is trivial for $n = 1$, let us assume it is known for homogeneous polynomials of degree at most δ. If P is the polynomial $\sum_{i=0}^{\delta} P_i(x_0, \ldots, x_{n-1})x_n^i$, which vanishes on the $(\delta + 1)^n$ given points, this means that for every point of $\mathbf{P}^{n-1}$ among the $(\delta + 1)^{n-1}$ following ones: $x_0 = 1, 0 \leqslant x_i \leqslant \delta$ for $i = 1, \ldots, n - 1$, the univariate polynomial P in x_n is identically zero, hence the P_i's vanish, and are identically zero using the induction hypothesis.

Second there are at most $\binom{\delta + n}{n}$ monomials in P, and to evaluate a monomial at a given point one needs at most δ multiplications in k.

Thus if we denote by ϵ, α and μ upper bounds for the cost of respectively comparison to zero, addition and multiplication, the time complexity is at most:

$$(\delta + 1)^n\left(\epsilon + \binom{\delta + n}{n}(\alpha + d\mu)\right).$$

Then the number of elementary operations in the base field is obtained for the uniform cost measure $(\epsilon = \alpha = \mu = 1)$ and is bounded above by some polynomial in δ whose leading term is $\frac{1}{n!}\delta^{2n+1}$, anyway a rough upper bound will be $\frac{1}{n!}(\delta + n)^{2n+1}$.

Eventually if I is defined over the integers, the value of P at any previous testing point is at most $2^M\binom{\delta + n}{n}\delta^\delta$, hence its size is at most $M + \delta\, log(\delta) + log(\delta + n) + \cdots + log(\delta + 1)$, itself less than say $M + (\delta + n)^2$.

Hence the cost ϵ of comparison to zero is at most $C_\epsilon M(\delta + n)^2$ where C_ϵ is some constant. The cost α of adding two monomials (resp. μ to obtain a monomial by successive multiplications) is at most $C_\alpha M\delta^2$ (resp. $C_\mu M\delta^2$) where C_α and C_μ are some constants. So finally the time complexity, whatever it is, will be at most

$$CM\frac{1}{n!}(\delta+n)^{3n+3},$$

where C is some constant.

In conclusion, for fixed n, if δ is going to infinity the asymptotic behaviour of the time complexity is polynomial in δ^n.

We will now try to catch the dimension through *linf*.

5.4. QUEEN'S SIDE OPENINGS

The fundamental observation is now that degrees of elements, when constructing a standard basis with respect to the lower lexicographic filtration, grow slowly along some coordinate axes.

5.4.1. PROPOSITION

Consider the set $E(I)$ relative to the lower lexicographic filtration; by definition, there are points $A_{linf_x-1},\ldots,A_n$ of $E(I)$ lying on the last $n-linf_x$ axes. Then:

$$|A_i| \leqslant (n-i+1)(d-1)+1$$

for $linf_x+1 \leqslant i \leqslant n$.

Proof: Let i be some integer between $linf_x+1$ and n. The ideal $I+(x_0,\ldots,x_{i-1})$ defines the empty subvariety and is still generated by polynomials of degree at most d. By proposition 2.11, we are reduced to the particular case of $linf_x=-1$ and $i=0$. For such an ideal, the degrees of elements of a standard basis relative to the lower lexicographic filtration are at most $1+(n+1)(d-1)$. To show this it is easy to adapt the proof of Theorem 3.9 (Giusti, 1984) to the case of non generic coordinates and arbitrary characteristic in the following lemma:

LEMMA: *Let I be an ideal generated by polynomials $f_1,\ldots,f_t$ of degree $d_1 \geqslant \cdots \geqslant d_t$ defining the empty subvariety. Then $D(E(I))$ is at most $1+\sum_{i=1}^{n+1}(d_i-1)$.*

Proof: The first step consists to use 3.6 to replace $D(E(I))$ by the regularity $H(I)$, which is an upper bound. Then by generic combinations, we can extract a regular sequence of maximal length $n+1$ from I, whose elements are of degree $d_1 \geqslant \cdots \geqslant d_{n+1}$, and generate an ideal J contained in I. From the canonical exact sequence $R/J \rightarrow R/I \rightarrow 0$ we deduce that the regularity of I is at most the regularity of J. The resolution of R/J by the Koszul complex shows that the Hilbert function of R/J, hence its regularity, depends only on the degrees $d_1,\ldots,d_{n+1}$ of the generators of J. This regularity can now be easily computed in the special case $I=(x_0^{d_1},\ldots,x_n^{d_{n+1}})$ (cf. 3.5).

5.5. A FIRST ROUGH ESTIMATE TO COMPUTE THE LOWER LEXICOGRAPHIC DIMENSION

What is the complexity to catch lower lexicographic dimension? Due to lack of knowledge about genericity, we have to compute by brute force: we can perform a generic change of coordinates, by introducing $(n+1)^2$ new indeterminates a_{ij}, generic entries of a change of basis matrix. The new base field is then $K = k(a_{ij})$, and the first coordinate hyperplanes form a system of parameters.

Now it is enough by 5.4.1 to construct $E(I)$ in degree less than $(n + 1)d$. So we remain inside the vector space of monomials of degree at most $(n + 1)d$, whose dimension $\delta(d, n)$ is anyway roughly bounded by some polynomial in $(nd)^n$. To compute the standard basis we have at most to triangulate by column operations a $\delta \times t\delta$ matrix (Lazard, 1983; Giusti, 1984), with entries polynomials in the a_{ij}'s of degree at most d. The cost in basic operations in the new field K is anyway at most polynomial in $(ntd)^n$.

And during this triangulation, the polynomial entries appearing being minors of the initial matrix, have a degree most $d\delta$. At any step of the triangulation, the multiplication or addition of two entries needs a number of operations in the base field k which is polynomial in $(ntd)^{n^3}$. Hence the elementary time complexity is of same nature.

Space questions, if the input polynomials have integral coefficients of size at most M: the size of any minor does not exceed a polynomial in $(Mtnd)^n$, so eventually the time complexity is still polynomial in $(Mtnd)^{n^3}$.

Finally remark that this procedure is essentially Lazard's one (Lazard, 1977; algorithm 7.2), translated in the language of standard bases.

5.5.1. REMARK: AN INTERESTING SPECIAL CASE

The fact that the dimension is negative can be decided in a time which is polynomial in $(Mtnd)^n$, since in this case the complement of $E(I)$ is finite and inside a cube of side less than $(n + 1)d$.

5.6. A LAST ALGORITHM, COMPUTING SIMULTANEOUSLY GENERIC COORDINATES AND DIMENSION

5.6.1. THE ALGORITHM

To describe it more comfortably, the algorithm is written with functions having as input and output data 4-uples of the same following type:

—An integer i between 0 and $n + 1$,
—A field K which is either k or an extension $k(\lambda_0, \ldots, \lambda_{n-i-1})$ over $n - i$ indeterminates.
—A system x of coordinates $x = (x_0, \ldots, x_n)$ of the projective space,
—A list J of polynomials in the variables $x_0, \ldots, x_n$ with coefficients in the field K. When K is an extension of k, the coefficients of the polynomials of J are actually in $k[\lambda_0, \ldots, \lambda_{n-i-1}]$.

```
GENERIC(i, K, x, J) :=
    IF i = n + 1
        THEN (i, K, x, J)
        ELSE (i, K, x, J) : STANDARD(i, K, x, J)
            IF AXIS (i, K, x, J)
                THEN GENERIC (i + 1, K, x, J)
                ELSE (i, K, x, J) : STANDARD(CHANGE(i, K, x, J))
                    A: AXIS(i, K, x, J)
                    (i, K, x, J) : SPECIALIZE(A, (i, K, x, J))
                    IF A
                        THEN GENERIC(i + 1, K, x, J)
                        ELSE (i, K, x, J)
```

STANDARD(i, K, x, J) does not change the 3 first components, but replaces the elements of the last one, of degree at most $(i+1)(d-1)+1$, by the list of elements of a standard basis of the ideal generated by J, with respect to the lower lexicographic filtration, up to the same degree. Moreover if the field K is a non trivial extension of the ground field, the triangulation leading to the standard basis is done in integral way, without introducing any denominators in the coefficients elements of K.

AXIS(i, K, x, J) does not change the 3 first components, but replaces the last one by the element of J (if any) with leading monomial depending only on the variable $n-i$.

CHANGE(i, K, x, J) will only act on 4-uples with a field equal to the ground field k. This function:

—does not change the first component,
—replaces the ground field by the extension $k(\lambda_0, \ldots, \lambda_{n-i-1})$,
—changes the variables in the following way:

$$x_j: x_j + \lambda_j x_{n-i} \quad \text{for} \quad 0 \leqslant j \leqslant n-i-1$$

$$x_j: x_j \quad \text{if} \quad j \geqslant n-i,$$

—transforms every polynomial of J according to the previous change of variables.

SPECIALIZE((i, K, x, J), (i, K, x, U)) acts only in the case where K is an extension of k over $n-i$ indeterminates, and J contains at most one element. This function acts only on its second argument:

—does not change the first component,
—replaces the extension field K by the ground field k,
—does not change the coordinates,
—specializes every element of U by setting each λ_j's to some constant value a_j element of k. If J is empty, each a_j is chosen to be zero, so that SPECIALIZE is the inverse map of CHANGE and we go back to the previous coordinates. If not, let us consider the leading coefficient of the unique element of J, which is a polynomial in $\lambda_0, \ldots, \lambda_{n-i-1}$. Applying the function OUTSIDE introduced in 5.3.4, we can find a point $a_0, \ldots, a_{n-i-1}$ on which this leading coefficient does not vanish.

5.6.2. THEOREM: THE RESULT

Starting with the following 4-*uple as input data*:

—$i = 0$,
—*the ground field* k,
—x *the given system of coordinates*,
—*a list* J *of homogeneous polynomials in* $k[x_0, \ldots, x_n]$, *of degree less than* d, *generating an ideal* I,

the previous algorithm stops on the following 4-*uple*:

—*the codimension* $c = n - l$ *of* $Z(I)$,
—*the ground field* k,
—*a generic system of coordinates* y *for* $Z(I)$ *as defined in* 5.3.1 (*denote by* ϕ *the change of coordinates such that* $x = \phi(y)$),

—*a list of homogeneous polynomials in* $k[y]$, *forming a standard basis of* $\phi^*(I)$ *truncated in degree less than* $(c+1)(d-1)+1$.

In other words, if we denote by $E(J)$ *the stable subset generated by the leading monomials of the elements of a list* J *of polynomials in* $k[y_0, \ldots, y_n]$, *and define similarly the integer* $linf_y(J)$ (4.2), *we claim that* $E(J)$ *is the truncated part of* $E(I)$ *in degree less than* $(c+1)(d-1)+1$, *and that it is enough to read the dimension since* $linf_y(I) = linf_y(J) = l$.

Proof: Each time we enter the function GENERIC, the 4-uple argument (i, K, y, J) has the following property: for every of the i axes $y_{n-i+1}, \ldots, y_n$, there is among the elements of J a polynomial whose leading monomial lies on it.

We will prove this assertion by induction on i. If i is zero, there is nothing to prove. Let us now follow the history of a 4-uple argument (i, K, y, J). By induction hypothesis, $E(J)$ cuts the last i axes. To check if an element of a standard basis has a leading monomial on the $(n-i)$th axis, it is enough by 5.4.1 to construct it in degree less than $(i+1)(d-1)+1$. If it is the case, we enter again with an argument satisfying the assertion at level $i+1$.

If not, let us consider the ideal *Igen* of the new polynomial algebra over an extension field, in the new variables introduced by the function CHANGE; $Z(I)$ and $Z(Igen)$ have the same dimension. Now let us look at the stable sets $E_{lowlex}(I)$ (resp. $E_{lowlex}(Igen)$) and $E_{mixlex}(I)$ (resp. $E_{mixlex}(Igen)$) with respect to the lower and mixed (of order $n-i$) lexicographic filtrations. If there is a point in the truncated part of $E_{lowlex}(Igen)$ on the axis y_{n-i}, the same is true for the suitable specialisation, and we enter again at level $i+1$ satisfied.

If not, there is no point of $E_{mixlex}(I)$ in the plane of the first $n-i+1$ coordinates, because if so, this point would correspond to a non zero polynomial depending only this first variables, which would be pushed by the change of coordinates on the axis y_{n-i}; but the leading monomial would not change for the other filtration, hence $E_{lowlex}(Igen)$ would have a point on the same axis, and furthermore another one in the truncated part. But in this case, looking at $E_{\tilde{mixlex}}(I)$, we see that the linear subvariety defined by $x_0 = \cdots = x_{n-i} = 0$ does not cut $Z(I)$, hence the dimension is at most $n-i$; and $Z(I)$ can be projected surjectively on the linear subvariety defined by $x_{n-i+1} = \cdots = x_n = 0$, hence the dimension at least $n-i$ is equal to $n-i$. And the algorithm stops with the wanted property.

Remains the case where we arrive until i is $n+1$; but this is the other way to get out, and the conclusion is also fulfilled.

5.6.3. THEOREM: COMPLEXITY OF THE PREVIOUS ALGORITHM

The elementary time complexity of the previous algorithm is polynomial in $(ntd)^{n^2}$. *Furthermore if the input data are defined over the integers, the time complexity is polynomial in* $(Mtnd)^{n^2}$.

Once we remarked that the matrices to be triangulated have entries polynomials of at most n variables, of degree at most d, and coefficients of maximal size M, the proof involves the same technics as 5.5, with the additional study of OUTSIDE in 5.3.5.

References

Bayer, D., Stillman, M. (1985). A theorem on refining division orders by the reverse lexicographic order, Lecture notes for the meeting on Algebraic Geometry and Computing, Trento, Italy.

Buchberger, B. (1965). *Ein Algorithmus zum Auffinden der Basiselemente des Restklassenringes nach einem nulldimensionalen Polynomideal.* Ph.D. Dissertation, University of Innsbruck, Austria.

Buchberger, B. (1970). Ein algorithmisches Kriterium für die Lösbarkeit eines algebraisches Gleichungssystems. *Aequationes Math.* **4**, 374–383.

Buchberger, B. (1976). A theoretical basis for the reduction of polynomials to canonical form. *ACM SIGSAM Bull.* **39**, 19–29.

Carrà Ferro, G. (1986). Some upper bounds for the multiplicity of an autoreduced subset of $\mathbf{N}^m$ and their applications, Proceedings of the third International Conference on Algebraic Algorithms and Error-Correcting Codes, AAECC-3, Grenoble (France) 1985, *Lecture Notes in Computer Science* **229**, 306–315.

Demazure, M. (1985). *Notes informelles de calcul formel, prépublications du Centre de Mathématiques de l'Ecole Polytechnique.* 3. Le monoïde de Mayr–Meyer. 4. Le théorème de complexité de Mayr–Meyer.

Galligo, A. (1983). Algorithmes de construction de bases standard, prépublication de l'Université de Nice.

Giusti, M. (1984). Some effectivity problems in polynomial ideal theory, in Eurosam 84. *Lecture Notes in Computer Science*, **174**, 159–171.

Giusti, M., Lazard, D. (1986). Complexity of standard basis computations, related algebraic problems and their common double exponential behaviour. *Lecture Notes for the Conference "Computer and Commutative Algebra" (Genova, Italy) (1986)*, Prépublication du Centre de Mathématiques de l'Ecole Polytechnique.

Grauert, H. (1972). Über die Deformationen isolierter Singularitäten analytischer Mengen. *Inventiones Math.* **15**, 3.

Hironaka, H. (1964). Resolution of singularities of an algebraic variety over a field of characteristic zero. *Ann. Math.* **79**, 109–326.

Kandri-Rody, A. (1985). *Proceedings Combinatorial Algorithms in Algebraic Structures, Otzenhaus 1985.* J. Avenhaus and K. Madlener eds., Fachbereich Informatik, Universität Kaiserslautern.

Kredel, H., Weispfennig, V. (1988). Computing dimension and independent sets for polynomial ideals, these Proceedings.

Lazard, D. (1977). Algèbre linéaire sur $K[x_1, \ldots, x_n]$ et élimination. *Bull. Soc. Math.* **105**, 165–190.

Lazard, D. (1981). Résolution des systèmes d'équations algébriques. *Theoretical Computer Science*, **15**, 77–110.

Lazard D. (1982). Commutative algebra and computer algebra. *Lecture Notes in Comp. Sciences*, **144**, 40–48.

Lazard, D. (1983). Gröbner bases, Gaussian elimination and resolution of systems of algebraic equations, in Eurocal 83. *Lecture Notes in Computer Science*, **162**,

Lejeune-Jalabert, M. (1984–1985). Effectivité des calculs polynomiaux, Cours de DEA 1984–1985, Institut Fourier, Université de Grenoble I.

Macaulay, F. S. (1927). Some properties of enumeration in the theory of modular systems, *Proc. London Math. Soc.* **26**, 531–555.

Mayr, E., Meyer, A. (1982). The complexity of the word problem for commutative semigroups and polynomial ideals. *Adv. in Maths.* **46**, 305–329.

Using Gröbner Bases to Determine Algebra Membership, Split Surjective Algebra Homomorphisms Determine Birational Equivalence

DAVID SHANNON† AND MOSS SWEEDLER‡

† *Department of Mathematics, Transylvania University, Lexington, KY 40508, U.S.A.*

‡ *Department of Mathematics, Cornell University, Ithaca, NY 14853, U.S.A.*

(*Received* 22 *November* 1986; *Revised* 28 *May* 1987)

k is a field, $X_1, \ldots, X_n$ are indeterminates over k and $f_1, \ldots, f_m \in k[X_1, \ldots, X_n]$. This note presents a simple algorithm, based on Gröbner bases, to test if a given polynomial g of $k[X_1, \ldots, X_n]$ lies in $k[f_1, \ldots, f_m]$. If so, the algorithm produces a polynomial P of m variables where $g = P(f_1, \ldots, f_m)$. Say $\omega: B \to k[X_1, \ldots, X_n]$ is a homomorphism where $\omega(b_i) = f_i$, for algebra generators $\{b_i\} \subset B$. If ω is onto, our algorithm gives a homomorphism $\lambda: k[X_1, \ldots, X_n] \to B$, where the composite $\omega\lambda$ is the identity map. In particular, the algorithm computes the inverse of algebra automorphisms of the polynomial ring. A variation of our test if $k[f_1, \ldots, f_m] = k[X_1, \ldots, X_n]$, tells if $k(f_1, \ldots, f_m) = k(X_1, \ldots, X_n)$. Existing computer algebra systems, such as IBM's SCRATCHPAD II, have Gröbner basis packages which allow the user to specify a term ordering sufficient to carry out the algorithm.

Continuing with the notation of the abstract, the algorithm uses "tag" variables in a fashion related to Spear's method§ for finding the relations among the f_i's. We introduce tag variables, indeterminates $T_1, \ldots, T_m$, one for each f_i. Let $k[X_1, \ldots, X_n, T_1, \ldots, T_m]$ have a term ordering where:

$$X_i > k[T_1, \ldots, T_m] \quad \text{for } i = 1, \ldots, n. \tag{1}$$

By (1) we mean: each X_i is larger than any monomial which lies in $k[T_1, \ldots, T_m]$. For example, the pure lexicographic order with $X_1 > \ldots > X_n > T_1 > \ldots > T_m$ satisfies (1). The ideal in $k[X_1, \ldots, X_n, T_1, \ldots, T_m]$ generated by $\{f_1 - T_1, \ldots, f_m - T_m\}$ is denoted J. Let G be a Gröbner basis for J.

2. Membership test algorithm. Let g lie in $k[X_1, \ldots, X_n]$ and reduce g over G until reduction terminates with, say,

$$h \in k[X_1, \ldots, X_n, T_1, \ldots, T_m].$$

$g \in k[f_1, \ldots, f_m]$ if and only if $h \in k[T_1, \ldots, T_m]$. In this case: $g = h(f_1, \ldots, f_m)$.

§ The technique of using tag variables which are smaller than the "main" variables is described in Gianni (1986, sect. 3, p. 4). The technique is attributed to Spear.

PROOF. J is the kernel of the homomorphism

$$\gamma: k[X_1, \ldots, X_n, T_1, \ldots, T_m] \to k[X_1, \ldots, X_n]$$
$$e(X_1, \ldots, X_n, T_1, \ldots, T_m) \to e(X_1, \ldots, X_n, f_1, \ldots, f_m).$$

Since g reduces to h, $g-h \in J$. If $h \in k[T_1, \ldots, T_m]$:

$$g = \gamma(g) = \gamma(h) = h(f_1, \ldots, f_m) \in k[f_1, \ldots, f_m].$$

Conversely, say, $g \in k[f_1, \ldots, f_m]$ and P is a polynomial of m variables where $g = P(f_1, \ldots, f_m)$. Then, $g - P(T_1, \ldots, T_m)$ lies in J. Hence, $h - P(T_1, \ldots, T_m) \in J$ and so $h - P(T_1, \ldots, T_m)$ reduces to zero over G. If h does not lie in $k[T_1, \ldots, T_m]$, then some monomial of h—with non-zero coefficient—involves X_i's. In view of (1), the lead monomial of h involves X_i's (or powers of X_i's). Let LM denote the lead monomial of h. No monomial of $P(T_1, \ldots, T_m)$ involves X_i's. Thus, LM is the lead monomial of $h - P(T_1, \ldots, T_m)$. Since h cannot be reduced any further, LM is not divisible by the lead monomial of any element of G. Thus, $h - P(T_1, \ldots, T_m)$ cannot be reduced any further over G, which contradicts the fact that $h - P(T_1, \ldots, T_m)$ reduces to zero. ∎

Note that if in the process of reducing g one reaches a reductum h—which need not be a final reductum—where h lies in $k[T_1, \ldots, T_m]$, then:

$$g = h(f_1, \ldots, f_m) \in k[f_1, \ldots, f_m].$$

In other words, reduction need not always be carried to the end.

3. EXAMPLE. Consider $k[X]$. Let $f_1 = X^3 - X$ and $f_2 = X^2$. Then

$$J = \langle X^3 - X - U, X^2 - V \rangle.$$

With the pure lexicographic order where $X > U > V$, J has a Gröbner basis consisting of:

$$X^2 - V, XU - V^2 + V, XV - X - U, U^2 - V^3 + 2V^2 - V.$$

One reduction of X^5 leads to $X + UV + U$, which cannot be reduced any further. Since $X + UV + U$ does not lie in $k[U, V]$, X^5 does not lie in $k[f_1, f_2]$. This example is a variation on (2.7) in Sweedler (1986).

Assume that G is a Gröbner basis for J and define subsets:

$$G_T = G \cap k[T_1, \ldots, T_m]$$
$$G_M = \{g \in G \mid \text{the lead monomial of } g \text{ is not divisible by the lead monomial of any element of } G_T\}. \quad (4)$$

The subscript T is for Tag and M is for Main. G_T gives the relations among the f_i's. This is Spear's algorithm for determining the relations. A special case of Spear is:

$$f_1, \ldots, f_m \text{ are algebraically independent if and only if } G_T \text{ is empty.} \quad (5)$$

By contrast, our theorems (6) and (10) are primarily concerned with G_M. If G is a reduced Gröbner basis for J, the question of whether

$$k[f_1, \ldots, f_m] = k[X_1, \ldots, X_n]$$

takes particularly nice form. (*Reduced*, meaning that the lead monomial of each $g \in G$ has coefficient one and that the lead and internal monomials of g are not divisible by the lead monomial of other elements of G.) When G is reduced, G_M is the set theoretic complement to G_T in G.

6. THEOREM. *Let* $k[X_1, \ldots, X_n, T_1, \ldots, T_m]$ *have a term ordering satisfying* (1) *and let* G *be a reduced Gröbner basis for* J.

$$k[f_1, \ldots, f_m] = k[X_1, \ldots, X_n]$$

if and only if $G_M = \{g_1, \ldots, g_n\}$ *where* $g_i = X_i - \varphi_i$ *with* $\varphi_i \in k[T_1, \ldots, T_m]$. *If*

$$k[f_1, \ldots, f_m] = k[X_1, \ldots, X_n]$$

and $m = n$, *then* G_T *is empty.*

PROOF. If G_M has the indicated form, then each X_i reduces to $\varphi_i \in k[T_1, \ldots, T_m]$ and by algorithm (2) each X_i lies in $k[f_1, \ldots, f_m]$.

Conversely, assume that

$$k[f_1, \ldots, f_m] = k[X_1, \ldots, X_n].$$

Further, assume that the term ordering is such that $X_1 > \ldots > X_n$. By (2), each X_i must reduce over G_M to a polynomial in $k[T_1, \ldots, T_m]$. Considering X_n, there must be $g_n \in G_M$ with leading monomial X_n. In view of the ordering, $g_n = X_n - \varphi_n$ with $\varphi_n \in k[T_1, \ldots, T_m]$. Next consider X_{n-1}. There must be $g_{n-1} \in G_M$ of the form $X_{n-1} - \varphi_{n-1}$ with $\varphi_{n-1} \in k[X_n, T_1, \ldots, T_m]$. Since G is a reduced Gröbner basis and $g_{n-1} \in G$, any monomials of φ_{n-1} which involve X_n could be reduced by g_n. Hence, $\varphi_{n-1} \in k[T_1, \ldots, T_m]$. In similar fashion, the remaining X_i's give rise to $X_i - \varphi_i$, where $\varphi_i \in k[T_1, \ldots, T_m]$ and a scalar multiple of $X_i - \varphi_i$ lies in G_M. Again, since G is reduced, G_M can contain no other elements.

The last part of the theorem follows from the Spear technique. If $m = n$ and

$$k[f_1, \ldots, f_n] = k[X_1, \ldots, X_n],$$

the f_i's are algebraically independent and so G_T is empty, (5). ■

Let

$$\sigma : k[T_1, \ldots, T_m] \to k[X_1, \ldots, X_n]$$

be the homomorphism determined by

$$\sigma(\varphi(T_1, \ldots, T_m)) = \varphi(f_1, \ldots, f_m).$$

$k[f_1, \ldots, f_m]$ is the image of σ and $\sigma = \gamma | k[T_1, \ldots, T_m]$, where γ is defined at the beginning of the proof of (2).

7. COROLLARY. (*a*) σ *is one-to-one if and only if* G_T *is empty.*

(*b*) σ *is onto if and only if* $G_M = \{g_1, \ldots, g_n\}$ *where* $g_i = X_i - \varphi_i$ *with* $\varphi_i \in k[T_1, \ldots, T_m]$.

(*c*) *Suppose that* σ *is onto,* G_M *is as described in* (*b*) *and let*

$$\mu : k[X_1, \ldots, X_n] \to k[T_1, \ldots, T_m]$$

be the homomorphism determined by $X_i \to \varphi_i$. *The composite*

$$\sigma\mu : k[X_1, \ldots, X_n] \to k[X_1, \ldots, X_n]$$

is the identity homomorphism.

PROOF. (a) Follows from Spear (1977), (5).

(b, c) If σ is onto, then

$$k[f_1, \ldots, f_m] = k[X_1, \ldots, X_n]$$

and G_M has the specified form by (6). Conversely, suppose that G_M has the specified form. By (2), $X_i = \varphi_i(f_i, \ldots, f_n)$. Since $\varphi_i(f_1, \ldots, f_n) = \sigma(\varphi_i)$, and $\varphi_i = \mu(X_i)$: $X_i = \sigma\mu(X_i)$. Thus, $\sigma\mu$ is the identity homomorphism. ■

8. COMMENTS. (a) When (c) holds, μ must be one-to-one and the φ_i's must be algebraically independent in $k[T_1, \ldots, T_m]$.

(b) Suppose that B is a k algebra generated by $b_1, \ldots, b_m$, and the b_i's are not assumed to be algebraically independent over k. Furthermore, assume that $\omega: B \to k[X_1, \ldots, X_n]$ is an algebra homomorphism which is onto and $\omega(b_i) = f_i$. Since $k[T_1, \ldots, T_m]$ is a polynomial ring, there is a unique algebra homomorphism $\rho: k[T_1, \ldots, T_m] \to B$ determined by $\rho(T_i) = b_i$. Then $\omega\rho = \sigma$ since they agree on the T_i's. Using μ from (c), define $\lambda: k[X_1, \ldots, X_n] \to B$ as the composite $\rho\mu$. Then $\omega\lambda = \omega\rho\mu = \sigma\mu$ which is the identity homomorphism by (c). Explicitly, λ is given by:

$$\lambda(g(X_1, \ldots, X_n)) = g(\varphi_1(b_1, \ldots, b_m), \ldots, \varphi_n(b_1, \ldots, b_m)).$$

9. EXAMPLES. (a) Let $f, g \in k[X]$, let $k[X, U, V]$ have the pure lexicographic term ordering with $X > U > V$ and let $J = \langle f - U, g - V \rangle$. $k[f, g] = k[X]$ if and only if J has a reduced Gröbner basis of the form $\{X - \varphi_1, \varphi_2\}$ with $\varphi_1, \varphi_2 \in k[U, V]$.

(b) Let $f, g \in k[X, Y]$, let $k[X, Y, U, V]$ have the pure lexicographic term ordering with $X > Y > U > V$ and let $J = \langle f - U, g - V \rangle$. $k[f, g] = k[X, Y]$ if and only if J has a reduced Gröbner basis of the form $\{X - \varphi_1, Y - \varphi_2\}$ with $\varphi_1, \varphi_2 \in k[U, V]$.

Theorem (6) is the model for a test of birational equivalence, i.e. a test for when

$$k(f_1, \ldots, f_m) = k(X_1, \ldots, X_n).$$

That test is below in theorem (10), which does not assume that G is a reduced Gröbner basis. We shall have more to say about using Gröbner bases to study the field of rational fractions in Shannon & Sweedler (1987). In that paper we generalise theorem (10) to obtain information about the field extension $k(X_1, \ldots, X_n)$ over $k(f_1, \ldots, f_m)$. The generalisation gives a constructive procedure for determining whether $K(X_1, \ldots, X_n)$ is algebraic or transcendental over $k(f_1, \ldots, f_m)$. When $k(X_1, \ldots, X_n)$ is transcendental over $k(f_1, \ldots, f_m)$, the procedure gives the transcendence degree. When $k(X_1, \ldots, X_n)$ is algebraic over $k(f_1, \ldots, f_m)$, the procedure gives the index.

10. THEOREM. *Let* $k[X_1, \ldots, X_n, T_1, \ldots, T_m]$ *have a term ordering satisfying* (1) *and*

$$X_i > k[T_1, \ldots, T_m]X_{i+1} \quad \text{for } i = 1, \ldots, n-1. \tag{11}$$

By (11) we mean: each X_i is larger than any monomial which is the product of X_{i+1} by a monomial which lies in $k[T_1, \ldots, T_m]$. The pure lexicographic order with

$$X_1 > \ldots > X_n > T_1 > \ldots T_m$$

satisfies (1) and (11).

$$k(f_1, \ldots, f_m) = k(X_1, \ldots, X_n)$$

if and only if G_M contains a subset $\{g_1, \ldots, g_n\}$, where $g_i = \delta_i X_i - \varphi_i$ with

$$\varphi_i \in k[X_{i+1}, \ldots, X_n, T_1, \ldots, T_m]$$

and $\delta_i \in k[T_1, \ldots, T_m]$. If $k(f_1, \ldots, f_m) = k(X_1, \ldots, X_n)$ and $m = n$, then G_T is empty.

PROOF. The statement about G_T being empty follows from Spear, exactly as in theorem (6).

If G_M contains the indicated subset, then

$$g_n = \delta_n X_n - \varphi_n \in J.$$

Apply γ—defined at the beginning of the proof of (2)—to $\delta_n X_n - \varphi_n$. Since J is the kernel of γ:

$$\delta_n(f_1, \ldots, f_m) X_n = \gamma(\delta_n X_n) = \varphi_n(f_1, \ldots, f_m).$$

If $\delta_n(f_1, \ldots, f_m) = 0$, then $\delta_n(T_1, \ldots, T_m) \in J$ and $\delta_n(T_1, \ldots, T_m)$ reduces to zero over G_T. In this case, g_n could be non-trivially reduced over G_T, which contradicts the fact that lead monomials of elements of G_M are not divisible by lead monomials of elements of G_T. Hence, $\delta_n(f_1, \ldots, f_m) \neq 0$ and X_n lies in $k(f_1, \ldots, f_m)$. Next, apply γ to

$$g_{n-1} = \delta_{n-1} X_{n-1} - \varphi_{n-1} \in J.$$

As above, this gives

$$\delta_{n-1}(f_1, \ldots, f_m) X_{n-1} = \varphi_{n-1}(X_n, f_1, \ldots, f_m). \tag{12}$$

$\delta_{n-1}(f_1, \ldots, f_m) \neq 0$ by the same argument that $\delta_n(f_1, \ldots, f_m) \neq 0$. Since X_n lies in $k(f_1, \ldots, f_m)$, (12) implies that X_{n-1} lies in $k(f_1, \ldots, f_m)$. Continuing in this fashion shows that all the X_i's lie in $k(f_1, \ldots, f_m)$.

Conversely, assume that

$$k(f_1, \ldots, f_m) = k(X_1, \ldots, X_n).$$

Since $X_i \in k(f_1, \ldots, f_m)$, there must be polynomials α_i and β_i in $k[T_1, \ldots, T_m]$ with

$$\alpha_i(f_1, \ldots, f_m) X_i = \beta_i(f_1, \ldots, f_m) \neq 0. \tag{13}$$

Among all such pairs of polynomials α_i and β_i, pick a pair where the lead monomial of α_i is as small as possible in the term ordering. Such a minimal pair exists because the set of monomials is well ordered. Define P_i as: $\alpha_i X_i - \beta_i$. P_i reduces to zero over G because P_i lies in the kernel of γ which is J. The lead monomial of P_i is $\eta_i X_i$, where η_i is the lead monomial of α_i. Since P_i reduces to zero over G, $\eta_i X_i$ must be divisible by the lead monomial of an element in G. If it is divisible by the lead monomial, LM, of an element of G_T, then LM is a monomial only involving $T_1, \ldots, T_m$ and LM divides η_i. In this case, α_i reduces over G_T. Let $\alpha_i' \in k[T_1, \ldots, T_m]$ be the result of reducing α_i over G_T. Since $\alpha_i - \alpha_i' \in J$,

$$\alpha_i'(f_1, \ldots, f_m) = \gamma(\alpha_i') = \gamma(\alpha_i) = \alpha_i(f_1, \ldots, f_m) \neq 0.$$

Thus, (13) holds with α_i replaced by α_i'. Since α_i' is obtained from α_i by reduction, the lead monomial of α_i' is strictly less than the lead monomial of α_i. The minimality of the lead monomial of α_i is contradicted! This shows that $\eta_i X_i$, the lead monomial of P_i, is not divisible by the lead monomial of any element of G_T. Hence, $\eta_i X_i$ must be divisible by the lead monomial of an element of $G \backslash G_T$, where "\" denotes set theoretic complement. This element, g_i, must have lead monomial of the form $\tau_i X_i$ with τ_i a monomial in $T_1, \ldots, T_m$ (and τ_i divides η_i).

Suppose IM is an internal monomial of g_i, meaning a monomial of g_i with non-zero coefficient and not the lead monomial. We wish to show that IM is of the form:

$$\begin{aligned} &\pi X_i \quad \text{with} \quad \pi \in k[T_1, \ldots, T_m] \\ \text{or} \quad & \\ &IM \in k[X_{i+1}, \ldots, X_n, T_1, \ldots, T_m]. \end{aligned} \tag{14}$$

If X_i divides IM, then†

$$IM/X_i < \eta_i X_i/X_i = \eta_i \in k[T_1, \ldots, T_m].$$

By (1), this implies that $IM/X_i \in k[T_1, \ldots, T_m]$ and IM is of the first form at (14). If X_j divides IM with $j < i$, then IM is greater than or equal to X_j. By (11), $X_j \geqslant X_{i-1}$ and $X_{i-1} > \tau_i X_i$. This contradicts the fact that IM is an internal monomial of g_i. Hence, if IM is not of the first form at (14) it must be of the second. We have shown that by appropriately grouping the terms of g_i, it is of the form:

$$\begin{aligned} &\delta_i X_i - \varphi_i \quad \text{with} \quad \varphi_i \in k[X_{i+1}, \ldots, X_n, T_1, \ldots, T_m] \\ \text{and} \quad & \\ &\delta_i \in k[T_1, \ldots, T_m]. \end{aligned}$$

g_i must lie in G_M, because if the lead monomial of g_i is divisible by the lead monomial, LM, of an element of G_T, then LM divides τ_i and so LM divides η_i. This would imply that α_i can be reduced over G_T which, as above, contradicts the minimality of the lead monomial of α_i. Thus, G_M contains the desired $\{g_1, \ldots, g_n\}$. ■

It is natural to wonder whether G_M must equal $\{g_1, \ldots, g_n\}$, as in theorem (6). The answer is NO as shown by:

1.5. Continuation of example 3. The Gröbner basis in example (3) is a reduced Gröbner basis and splits up:

$$G_M = \{X^2 - V, XU - V^2 + V, XV - X - U\}$$
$$G_T = \{U^2 - V^3 + 2V^2 - V\}.$$

In view of the element $XU - V^2 + V$, it follows that $k(f_1, f_2) = k(X)$. Notice that G_M consists of more than just the element $XU - V^2 + V$.

References

Buchberger, B. (1965). An algorithm for finding a basis for the residue class ring of a zero-dimensional polynomial ideal. Dissertation, Universität Innsbruck, Institut für Mathematik.

Buchberger, B. (1970). An algorithmic criterion for the solvability of algebraic systems of equations. *Aequationes Mathematicae* **4/3,** 374–383.

Buchberger, B. (1976). A theoretical basis for the reduction of polynomials to canonical forms. *ACM Sigsam Bull.* **10/3,** 19–29; *ACM Sigsam Bull.* **10/4,** 19–24.

Buchberger, B. (1979). A criterion for detecting unnecessary reductions in the construction of Gröbner bases. *Proc. EUROSAM 79, Springer Lec. Notes Comp. Sci.* **72,** 3–21.

Buchberger, B. (1984). A critical-pair/completion algorithm for finitely generated ideals in rings. Decision problems and complexity. *Proc. Symp. Rekursive Kombinatorik, Münster, 1983,* Boerger, E., Hasenjaeger, G., Roedding, D., eds). *Springer Lec. Notes Comp. Sci.* **171,** 137.

Buchberger, B. (1985). Gröbner bases: an algorithmic method in polynomial ideal theory. In: (Boese, N. K., ed.) *Multidimensional Systems Theory*, pp. 184–232. Dordrecht: Reidel.

Gianni, P., Trager, B., Zacharias, G. (1986). Gröbner bases and primary decomposition of polynomial ideals. Preprint.

† Using the fact that $IM < \tau_i X_i$ since IM is an internal monomial.

Shannon, D., Sweedler, M. (1987). Using Gröbner bases to determine the algebraic or transcendental nature of field extensions within the field of rational functions. In preparation.

Spear, D. (1977). A constructive approach to commutative ring theory. *Proc. 1977 MACSYMA User's Conference*, pp. 369–376.

Sweedler, M. (1986). Ideal bases and valuation rings. Preprint.

On an Installation of Buchberger's Algorithm

RÜDIGER GEBAUER AND H. MICHAEL MÖLLER†

Springer-Verlag, New York, 175 Fifth Avenue, New York, NY 10010, USA

†FB Mathematic und Informatik, Fernuniversität, D-5800 Hagen 1, FRG

(*Received* 26 *January* 1987, *and in revised form* 23 *October* 1987)

Buchberger's algorithm calculates Groebner bases of polynomial ideals. Its efficiency depends strongly on practical criteria for detecting superfluous reductions. Buchberger recommends two criteria. The more important one is interpreted in this paper as a criterion for detecting redundant elements in a basis of a module of syzygies. We present a method for obtaining a reduced, nearly minimal basis of that module. The simple procedure for detecting (redundant syzygies and) superfluous reductions is incorporated now in our installation of Buchberger's algorithm in SCRATCHPAD II and REDUCE 3.3. The paper concludes with statistics stressing the good computational properties of these installations.

1. Introduction

The concept of Groebner bases for polynomial ideals, introduced first for performing algorithmic computations in residue classes of polynomial rings by Buchberger (1965), now permits the algorithmic solution of a series of problems in polynomial rings and modules and especially the problem of finding all solutions of systems of algebraic equations; for a survey see Buchberger (1985). Buchberger's algorithm for computing Groebner bases is fitted for automatic computation and is installed in nearly all Computer Algebra Systems.

This algorithm is roughly described as follows. Given a finite set F of polynomials, calculate for each pair of polynomials in F a so called S-polynomial and reduce it relatively to F to a polynomial. If this reduced polynomial is not 0, insert it into F. At termination of the algorithm all S-polynomials reduce to 0 and F is a Groebner basis.

The reduction of the S-polynomials is the most time consuming part of the algorithm. Therefore Buchberger developed criteria for predicting reductions to 0, i.e. criteria for avoiding superfluous reductions. There are two types of criteria. The second one depends only on the two polynomials in question (their head terms have to be without common divisors). However, when we apply the first criterion, only very few instances remain, where the second criterion can be used successfully. The first type depends on the pairs considered before. This type is studied in detail by Buchberger (1979); and the most effective criterion of this type together with the second one (and a strategy, which cancels superfluous elements in F) is presented in Buchberger (1985).

The starting point for this paper was the observation, that Groebner bases can be characterized using a basis of its module of syzygies, as already remarked by some authors, for example Bayer (1982), and that reduction strategies to obtain reduced bases from a special basis, the so called Taylor basis, give simpler Groebner basis tests, as

This work was supported by IBM Germany.

already stated by Möller (1985). In this paper, we present a reduction strategy, which constructs a reduced basis of the module of syzygies (Proposition 3.5). Bayer (1986) assumed that this reduced basis is already irreducible, but this is not true as example 3.6 shows. However our experiences, and the discussion in 3.8, indicate that in only a few instances are all redundant syzygies not detected.

Using the one-to-one correspondence of Taylor basis elements and the pairs for the S-polynomial computation, redundant Taylor basis elements correspond to pairs satisfying a criterion of the first type. This is used to develop a variant of Buchberger's algorithm based on our reduction strategy. The detecting of redundant elements, i.e. the detecting of superfluous reductions, requires only the comparison of exponent vectors of some power products and is for technical reasons splitted into three criteria. These three criteria are very similar to criterion 1 of Buchberger (1985). But they are, in contrast to Buchberger's, independent of the succession of pairs considered before, and each pair detected once as superfluous or already used as a pair for the S-polynomial computation and reduction is no longer needed for subsequent tests of a criterion. This allows more flexibility and leads to a speeding up of the tests of the criteria. The flexibility is also used to implement Buchberger's criterion 2 in an optimal way. Also very important for a fast variant of Buchberger's algorithm is to keep the set F of polynomials as small as possible. Therefore, as in Buchberger's algorithm (1985), whenever a new element is inserted into F, redundant elements of F are cancelled. This is taken into account by a slight modification of the criteria.

The resulting algorithm is already installed in the Computer Algebra Systems SCRATCHPAD II (see Gebauer & Möller, 1987), and REDUCE (release 3.3), (see Hearn, 1987). We illustrate it in detail by an example and compare its complexity in 14 examples with an existing installation of Buchberger's algorithm.

2. Groebner Bases

2.1. Let K be a field and $P = K[x_1, \ldots, x_n]$ the ring of polynomials in $x_1, \ldots, x_n$ over K. T denotes the set of terms (power products) $x_1^{i_1} \ldots x_n^{i_n}$, $i_1, \ldots, i_n$ nonnegative integers. We assume T to be totally ordered by $<_T$, such that

$$(1:=)x_1^0 \ldots x_n^0 <_T \varphi \quad \text{for all} \quad \varphi \in T \setminus \{1\}$$

$$\varphi_i <_T \varphi_j \Rightarrow \varphi\varphi_i <_T \varphi\varphi_j \quad \text{for all} \quad \varphi, \varphi_i, \varphi_j \in T.$$

For $f = \sum_{i=1}^{m} c(f, \varphi_i)\varphi_i$ with $\varphi_1 <_T \varphi_2 <_T \cdots <_T \varphi_m$ and $c(f, \varphi_i) \in K \setminus \{0\}$, we define as in Möller & Mora (1986)

$$Hcoeff(f) := c(f, \varphi_m), \qquad Hterm(f) := \varphi_m$$

$$M_T(f) := c(f, \varphi_m)\varphi_m.$$

2.2. In the following, F will always be a finite set of polynomials, $F = \{f_1, \ldots, f_r\}$, $0 \notin F$, and *w.l.o.g.* $Hcoeff(f_i) = 1$, $i = 1, \ldots, r$. Mainly for avoiding tedious notations, we define

$$T(i) := Hterm(f_i),$$

$$T(i, j) := lcm\{T(i), T(j)\},$$

$$T(i, j, k) := lcm\{T(i), T(j), T(k)\}.$$

2.3. For polynomials $f \in P \backslash \{0\}$ being represented as in 2.1 Buchberger (1965) introduced the reduction

$$f \xrightarrow[F]{} g \quad (f \text{ reduces to } g \text{ modulo } F)$$

which means

$$g = f - c(f, \varphi_k) \frac{\varphi_k}{T(i)} f_i$$

for appropriate $f_i \in F$ and $k \in \{1, \ldots, m\}$, such that $T(i)$ divides φ_k.

f is called irreducible modulo F, if $f = 0$ or if $f \xrightarrow[F]{} g$ holds for no $g \in P$. Denoting by $\xrightarrow[F]{}^{+}$ the transitive reflexive closure of $\xrightarrow[F]{}$, Buchberger showed, that $\xrightarrow[F]{}^{+}$ is Noetherian, i.e. any reduction

$$f \xrightarrow[F]{} g_1 \xrightarrow[F]{} g_2 \xrightarrow[F]{} \cdots$$

is finite: $f \xrightarrow[F]{} g_1 \xrightarrow[F]{} \cdots \xrightarrow[F]{} g_s$, g_s irreducible modulo F.

2.4. Definition. $F = \{f_1, \ldots, f_r\} \subset P \backslash \{0\}$ is called a Groebner basis of $\text{Ideal}(F) := \{\sum_{i=1}^{r} g_i f_i \mid g_i \in P\}$, if the so called S-polynomials

$$S(f_i, f_j) := \frac{T(i,j)}{T(i)} f_i - \frac{T(i,j)}{T(j)} f_j$$

satisfy $S(f_i, f_j) \xrightarrow[F]{}^{+} 0$, $1 \leq i < j \leq r$.

(Buchberger gives in his publications a different definition for Groebner bases, but he showed already in his thesis (Buchberger, 1965), that the definition given here is equivalent to his one.) There are many equivalent definitions for Groebner bases of ideals (and even for submodules). For instance eleven definitions for ideals and submodules are given in Möller & Mora (1986). In the following, we need only three equivalent characterizations:

2.5. Theorem. *Let* $F = \{f_1, \ldots, f_r\} \subset P \backslash \{0\}$ *and* $I = \text{Ideal}(F)$. *Then the following conditions are equivalent.*

($C1$) *F is a Groebner basis of I.*

($C2$) $M_T(F) = \{T(1), \ldots, T(r)\}$ *generates* $M_T(I)$, *the least ideal containing* $M_T(f)$ *for all* $0 \neq f \in I$.

($C3$) *Let L be a basis of the module of syzygies*

$$S^{(1)} := \left\{(h_1, \ldots, h_r) \in P^r \,\middle|\, \sum_{i=1}^{r} h_i T(i) = 0\right\}.$$

Then for each $(g_1, \ldots, g_r) \in L$

$$\sum_{i=1}^{r} g_i f_i \xrightarrow[F]{}^{+} 0.$$

The proof of $C1 \Leftrightarrow C2$ can be found in Möller & Mora (1986). $C1 \Leftrightarrow C3$ is shown for instance by Möller (1985).

2.6. An element f_i of a Groebner basis F is called redundant, if $F' := F \backslash \{f_i\}$ is also a Groebner basis, and if $\text{Ideal}(F) = \text{Ideal}(F')$. If $\text{Ideal}(F) = \text{Ideal}(F')$ and F is a Groebner basis, then by $C2$, F' is a Groebner basis too, if and only if a $j \neq i$ exists, such that $T(j)$ divides $T(i)$, i.e. $T(i,j) = T(i)$.

For testing Ideal(F) = Ideal(F') in the case $T(i,j) = T(i)$ for a $j \neq i$, it is sufficient to test

$$(S(f_i, f_j) =) f_i - \frac{T(i,j)}{T(j)} f_j \in \text{Ideal}(F').$$

Using the reduction procedure, this holds in the case

$$S(f_i, f_j) \underset{F}{\longrightarrow}{}^{+} 0,$$

or equivalently, since here $Hterm(S(f_i, f_j)) <_T Hterm(f_i)$,

$$S(f_i, f_j) \underset{F'}{\longrightarrow}{}^{+} 0.$$

Therefore, if F is a Groebner basis and $T(i,j) = T(i)$ holds for a $j \neq i$, then by the definition of Groebner bases $F' = F \backslash \{f_i\}$ is a Groebner basis of the same ideal. This explains why in Groebner bases redundant elements are cancelled without additional modifications as for instance in Buchberger (1985).

3. A Reduced Basis for the Module Syzygies

3.1. The main tools in this section are the resolution of Taylor (1966) and methods for reducing the bases contained in this resolution. In Möller & Mora (1986), the Taylor resolution and reduction strategies are presented. Since we are dealing here only with the first modules of this resolution, we will not explain the complete technical details and refer the interested reader to the mentioned paper.

3.2. Given terms $T(1), \ldots, T(r)$, we call $(g_1, \ldots, g_r) \in P^r$ homogeneous of degree $\varphi \in T$, if for every $i \in \{1, \ldots, r\}$ a $c_i \in K$ exists, such that $g_i T(i) = c_i \varphi$. Then

$$S^{(1)} := \left\{ (g_1, \ldots, g_r) \in P^r \,\middle|\, \sum_{i=1}^{r} g_i T(i) = 0 \right\}$$

has the Taylor basis $L^{(1)} := \{S_{ij} \mid 1 \leqslant i < j \leqslant r\}$ with

$$S_{ij} := \frac{T(i,j)}{T(i)} e_i - \frac{T(i,j)}{T(j)} e_j$$

homogeneous of degree $T(i,j)$, where e_k is the kth canonical unit vector of P^r. Using this specific basis, $C3 \Rightarrow C1$ of theorem 2.5 is obvious.

3.3. For finding a reduced basis of $S^{(1)}$, we introduce the module of syzygies for $L^{(1)}$. We order the $r(r-1)/2$ syzygies S_{ij} by $<_1$,

$$S_{ij} <_1 S_{kl} :\Leftrightarrow T(i,j) <_T T(k,l) \quad \text{or} \quad (T(i,j) = T(k,l), j \leqslant l, j = l \Rightarrow i < k).$$

Using this order, we denote the canonical kth unit vector in $P^{r(r-1)/2}$ no longer by e_k but by e_{ij}, if S_{ij} is the kth syzygy in this order. For instance let $S_{12} <_1 S_{35} <_1 S_{23}$ be the three first of the S_{ij}, then $e_{12} = (1, 0, \ldots, 0)$, $e_{35} = (0, 1, 0, \ldots, 0)$, $e_{23} = (0, 0, 1, 0, \ldots, 0)$.

The module of syzygies

$$S^{(2)} = \left\{ \sum_{\substack{i,j=1 \\ i<j}}^{r} g_{ij} e_{ij} \in P^{r(r-1)/2} \,\middle|\, \sum_{\substack{i,j=1 \\ i<j}}^{r} g_{ij} S_{ij} = 0 \right\}$$

has the Taylor basis $L^{(2)} := \{S_{ijk} \mid 1 \leqslant i < j < k \leqslant r\}$ with

$$S_{ijk} = \frac{T(i,j,k)}{T(i,j)} e_{ij} - \frac{T(i,j,k)}{T(i,k)} e_{ik} + \frac{T(i,j,k)}{T(j,k)} e_{jk}.$$

S_{ijk} is homogeneous of degree $T(i,j,k)$ if we call now $\Sigma\, g_{ij}e_{ij}$ homogeneous of degree $\varphi \in T$, if for all $1 \leqslant i < j \leqslant r$ a $c_{ij} \in K$ exists, such that $g_{ij}T(i,j) = c_{ij}\varphi$.

Let us denote the maximal syzygy being involved in S_{ijk} by $MS(i,j,k)$, i.e.

$$MS(i,j,k) := \max_{<_1}\{S_{ij}, S_{ik}, S_{jk}\}.$$

If and only if $MS(i,j,k)$ and S_{ijk} are homogeneous of the same degree, i.e. $T(i,j,k) = \max\{T(i,j), T(i,k), T(j,k)\}$, then one of the three nonvanishing components of S_{ijk} is a constant. In that case, $MS(i,j,k)$ can be expressed in terms of lesser syzygies *w.r.t.* $<_1$. For instance let $MS(i,j,k) = S_{ik}$ and $T(i,j,k) = T(i,k)$. Then using $S_{ijk} \in S^{(2)}$

$$(*) \quad S_{ik} = \frac{T(i,j,k)}{T(i,j)} S_{ij} + \frac{T(i,j,k)}{T(j,k)} S_{jk}.$$

This allows to remove S_{ik} from $L^{(1)}$. The set $L^{(1)} \backslash \{S_{ik}\}$ still generates $S^{(1)}$, because in every basis representation of an $S \in S^{(1)}$, S_{ik} can be replaced by the lesser syzygies S_{ij} and S_{jk} using (*).

3.4. The procedure for removing elements from $L^{(1)}$ can be applied iteratively. Whenever an S_{ijk} and its corresponding $MS(i,j,k)$ have the same degree, then $MS(i,j,k)$ can be expressed by lesser syzygies and hence it can be removed from the (eventually already reduced) generating set for $S^{(1)}$, as an elementary inductive argument shows.

The tests for detecting reducible elements of $L^{(1)}$ can be done by inspecting the elements of $L^{(2)}$. These tests do not require the explicit representations of the elements of $L^{(2)}$ but only divisibility tests of terms.

We say criterion M holds for (i,k) briefly $M(i,k)$, if a $j < k$ exists, such that $T(j,k)$ divides properly $T(i,k)$. (M stands for Multiple.)

We say criterion F holds for (i,k) briefly $F(i,k)$, if a $j < i$ exists, such that $T(j,k) = T(i,k)$. (F stands for the fact that in the set $\{S_{lk} \mid \text{degree } S_{lk}, 1 \leqslant l < k\}$ the First w.r.t. $<_1$ is different from S_{ik}.)

We say criterion B_k holds for (i,j), briefly $B_k(i,j)$, if $j < k$ and $T(k)$ divides $T(i,j)$ and $T(i,k) \neq T(i,j) \neq T(j,k)$. ($B$ stands for the fact that when we are considering already elements of type S_{ik} for reduction, we have to go Backwards w.r.t. $<_1$ for reducing S_{ij}.)

3.5. PROPOSITION. *The module of syzygies* $S^{(1)}$ *us generated by*

$$L^* := \{S_{ij} \mid 1 \leqslant i < j \leqslant r, \neg M(i,j), \neg F(i,j), \neg B_k(i,j) \text{ for all } k > j\}.$$

PROOF. If $M(i,k)$ holds true, then a $j < k$ exists, such that $T(j,k)$ divides properly $T(i,k)$, especially $T(i,k) = T(i,j,k)$ and $j \neq i$. This means, a syzygy S_{ijk} (in case $i < j$) or S_{jik} (in case $j < i$) exists which is homogeneous of degree $T(i,k)$ and has S_{ik} for its maximal syzygy:

$$S_{jk} <_1 S_{ik} \quad \text{because} \quad T(j,k) <_T T(i,k)$$

$$S_{ij} \;\text{ or }\; S_{ji} <_1 S_{ik} \quad \text{because} \quad j < k, i < k \quad \text{and} \quad T(i,j) \leqslant_T T(i,k).$$

If $F(j,k)$ holds, then $T(j,k) = T(i,k)$ for a $j < i$. This means that S_{jik} is homogeneous of degree $T(j, i, k) = T(i, k)$ and by similar arguments as before $MS(j, i, k) = S_{ik}$.

If $B_k(i, j)$ holds, then analogously $MS(i, j, k) = S_{ij}$.

The arguments in 3.4 give the assertion.

3.6. The reduced basis L^* is not always a completely reduced basis as the following example shows.

EXAMPLE. Let $r = 4$ and $T(1) = x^2y^2$, $T(2) = y^2z$, $T(3) = x^2z$, $T(4) = xyz$. Then

$$T(1, 2) = T(1, 3) = T(2, 3) = T(1, 4) = x^2y^2z,$$

$$T(2, 4) = xy^2z, \qquad T(3, 4) = x^2yz.$$

$M(1, 4)$, $F(2, 3)$, and $B_4(2, 3)$ hold but no other criterion of 3.5. Therefore the reduced basis by 3.5 is

$$L^* = \{S_{12}, S_{13}, S_{24}, S_{34}\}.$$

But

$$S_{13} = S_{14} - yS_{34} \quad \text{by} \quad S_{134} \in S^{(2)},$$

$$S_{14} = S_{12} + xS_{24} \quad \text{by} \quad S_{124} \in S^{(2)}.$$

This implies

$$S_{13} = S_{12} + xS_{24} - yS_{34}.$$

Hence S_{13} is redundant in L^* and

$$\{S_{12}, S_{24}, S_{34}\}$$

also generates $S^{(1)}$.

3.7. A modification of criterion F would have given the minimal basis in the example. Criterion $F(2, 3)$ was based on the syzygy S_{123} yielding

$$0 = S_{12} - S_{13} + S_{23}$$

and used to cancel S_{23} because S_{12} and S_{13} are kept in L^*. But S_{13} is redundant, if L^* contains S_{12} and S_{23}. Hence $\{S_{12}, S_{23}, S_{14}, S_{24}, S_{34}\}$ is a basis of $S^{(1)}$, too. By $M(1, 4)$ and $B_4(2, 3)$ as before, S_{23} and S_{14} may be omitted. This gives the irreducible basis $\{S_{12}, S_{24}, S_{34}\}$.

A consequent application of this data is, when already redundant syzygies of degree $<_T \tau$ and syzygies S_{ij} of degree τ with $j < k$ are cancelled by the criteria, then to delay the decision, what element S_{ik} of the set

$$S_{\tau,k} = \{S_{jk} \mid 1 \leqslant j < k, T(j, k) = \tau\}$$

not to cancel. We may take an arbitrary $S_{ik} \in S_{\tau,k}$, because $i \neq j$, $S_{jk} \in S_{\tau,k}$ we have (*w.l.o.g.* $i < j$)

$$0 = \frac{\tau}{T(i, j)} S_{ij} - S_{ik} + S_{jk},$$

and S_{ij} can be expressed by lower order syzygies of the basis.

3.8. Condition $C3$ of theorem 2.5 allows an easier Groebner basis test, when the basis of $S^{(1)}$ is a minimal basis. Since $S^{(1)}$ is homogeneous, it is reasonable to restrict the considerations to homogeneous bases, such that the notions minimality and irreducibility coincide. A (homogeneous) basis L of $S^{(1)}$ is irreducible if and only if no syzygy for L exists, which has a $c \in K \backslash \{0\} =: K^x$ for a component (allowing the cancellation of an additional element of L). If L is constructed by a successive cancellation of reducible elements from $L^{(1)}$, then each syzygy for L with a component $c \in K^x$ originates from an S_{ijk}, which also has a component from K^x, allowing the cancellation of the same element, cf. Möller & Mora (1986).

The strategy for finding the reduced basis L^* works in a similar way. We take each S_{ijk}, which has a component from K^x, but we always decide to take $MS(i, j, k)$ for redundant. If S_{ijk} has more than one component from K^x, then our choice of the redundant element may cause that we cancel the "wrong" basis element as seen in 3.7.

Only when many S_{ijk} exist, such that at least two of the three non-zero components are constants, and when we often select the "wrong" basis element for cancellation in such situation, then we have still many reducible elements in L^*. Fortunately, we found only more or less artificial examples like 3.6, where this occurs.

4. Buchberger's Algorithm

4.1. Buchberger's algorithm deals with the problem of finding a Groebner basis of a polynomial ideal, when a finite basis of the ideal is given. This algorithm was originally introduced by Buchberger (1965) and refined in subsequent papers. For a survey see Buchberger (1985).

4.2. We will present briefly a version of the algorithm recommended by Buchberger (1985). In order to avoid the technical details for reducing Groebner bases, we concentrate on the construction without reduction.

INPUT: $\{f_1, \ldots, f_r\} \subset P \backslash \{0\}$.

INITIALIZATION: $B := \{\{i, j\}/1 \leqslant i < j \leqslant r\}$;
$G := \{f_1, \ldots, f_r\}$; $R := r$.

ITERATION: *while* there exists $\{I, J\} \in B$ *repeat*
  if $\neg$ criterion 1 *and* $T(I)T(J) \neq T(I, J)$ *then*
    $h := S(f_I, f_J)$;
    $h := NF(h, G)$;
    if $h \neq 0$ *then*
      $f_{R+1} := h$; $G := G \cup \{f_{R+1}\}$;
      $B := B \cup \{\{i, R+1\}/1 \leqslant i \leqslant R\}$;
      $R := R + 1$;
  $B := B \backslash \{\{I, J\}\}$.

OUTPUT: G, a Groebner basis of $(f_1, \ldots, f_r)$.

Here, $NF(h, G)$ means a polynomial irreducible modulo G, such that $h \xrightarrow[G]{}^{+} NF(h, G)$. Criterion 1 applied to $\{I, J\}$ means that there is a $K \in \{1, \ldots, R\} \backslash \{I, J\}$ with $T(I, J) = T(I, J, K)$ and $\{I, K\} \notin B$, $\{J, K\} \notin B$. The criterion $T(I)T(J) = T(I, J)$ is criterion 2 of Buchberger (1985).

4.3. The correctness of algorithm 4.2 is shown by Buchberger (1979). Let us prove it by means of $C3$ of theorem 2.5. The syzygies S_{ij} correspond bijectively to all $\{i, j\}$ which are assigned once in the algorithm to B and removed later from B. We order the S_{ij} by $<_B$, such that $S_{ij} <_B S_{kl}$, if $\{i, j\}$ is removed from B earlier than $\{k, l\}$. If criterion 1 holds for $\{I, J\} \in B$, i.e. $\{I, K\} \notin B$, $\{J, K\} \notin B$, $T(I, J, K)$, then let for simplicity of notation $I < J < K$. The syzygy S_{IJK} shows

$$0 = S_{IJ} - \frac{T(I, J, K)}{T(I, K)} S_{IK} + \frac{T(I, J, K)}{T(J, K)} S_{JK}.$$

By the ordering $<_B$, $S_{IK} <_B S_{IJ}$ and $S_{JK} <_B S_{IJ}$. Hence S_{IJ} is expressible in terms of lower order syzygies. Thus, if criterion 1 holds for $\{I, J\}$, then S_{IJ} is redundant. For the remaining syzygies S_{IJ} we have in case $T(I)T(J) = T(I, J)$

$$S(f_I, f_J) \xrightarrow[\{f_I, f_J\}]{}^{+} 0, \qquad \text{i.e. } S(f_I, f_J) \xrightarrow[G]{}^{+} 0,$$

as already shown in Buchberger (1965), and otherwise

$$S(f_I, f_J) \xrightarrow[G]{}^{+} NF(S(f_I, f_J), G) = f_{R+1} \xrightarrow[f_{R+1}]{} 0.$$

Therefore at termination $B = \emptyset$, we have

$$\frac{T(I, J)}{T(I)} f_I - \frac{T(I, J}{T(J)} f_J \xrightarrow[G]{}^{+} 0 \quad \text{for all} \quad S_{IJ},$$

which are not redundant, and hence G is a Groebner basis by $C3$ of theorem 2.5.

4.4. A consequent use of the reduction strategy in 3.5 gives the following modifications of Buchberger's algorithm.

INPUT: $\{f_1, \ldots, f_r\} \subset P \backslash \{0\}$.

INITIALIZATION: $G := \{f_1\}$; $D := \emptyset$;
  for $t := 2$ *to* r
    $D :=$ updatePairs (D, t);
    $G := G \cup \{f_t\}$;
  $R := r$.

ITERATION: *while* there exists $(I, J) \in D$ *repeat*
    $h := S(f_I, f_J)$;
    $h := NF(h, G)$;
    if $h \neq 0$ *then*
      $f_{R+1} := h$;
      $D :=$ updatePairs $(D, R + 1)$;
      $G := G \cup \{f_{R+1}\}$; $R := R + 1$;
    $D := D \backslash \{(I, J)\}$.

OUTPUT: G, a Groebner basis of $\{f_1, \ldots, f_r\}$.

Here the subalgorithm updatePairs works in the following way, when applied to a set of pairs D and a positive integer t. Cancel in D all pairs (i, j), which satisfy $T(i, j) = T(i, j, t)$, $T(i, t) \neq T(i, j) \neq T(j, t)$, i.e. all pairs (i, j) with $B_t(i, j)$. Denote the set of remaining pairs by D'. Let $D1 := \{(i, t) \mid 1 \leqslant i < t\}$. Cancel in $D1$ each (i, t) for which a $(j, t) \in D1$ exists, s.t. $T(i, t)$ is a proper multiple of $T(j, t)$, i.e. each (i, t) with $M(i, t)$. The subset of $D1$ containing the remaining pairs (i, t) is denoted by $D1'$. In each nonvoid subset $\{(j, t) \mid T(j, t) = \tau\}$ of $D1'$ with $\tau \in T$ fix an element (i, t) satisfying $T(i)T(t) = T(i, t)$ or if no such (i, t) exists, fix an arbitrary (i, t). Cancel the other elements of $\{(j, t) \mid T(j, t) = \tau\}$ in $D1'$. Finally delete in $D1'$ all (i, t) with $T(i)T(t) = T(i, t)$ and denote again by $D1'$ this finally obtained subset of $D1'$. The set $D1' \cup D'$ is returned by the subalgorithm.

By construction of the subalgorithm, we have after the call of $D := \text{updatePairs}\ (D, t)$, that $\{S_{ij} \mid (i, j) \in D\}$ together with some S_{ij} with $1 \leqslant i < j \leqslant t$, $T(i)T(j) = T(i, j)$, constitute a basis of the module of syzygies

$$\left\{(g_1, \ldots, g_t) \in P^t \,\middle|\, \sum_{i=1}^{t} g_i T(i) = 0\right\}.$$

This follows from proposition 3.5 and the modification in 3.7.

4.5. The correctness of algorithm 4.4 is shown in analogy to 4.3. Its termination results from the same arguments as the termination of algorithm 4.2. By construction, each new f_{R+1} is irreducible with respect to $f_1, \ldots, f_R$. Therefore especially

$$(T(1), \ldots, T(R)) \subset (T(1), \ldots, T(R+1)).$$

This gives for (strictly) increasing R a strictly increasing chain of ideals. By Noetherianity, this chain is finite. Thus the iteration is repeated only a finite number of times.

4.6. Buchberger (1985) presented algorithm 4.2 in a version, which already cancels redundant basis elements in G. In a similar way, algorithm 4.4 can be modified. This modification for reducing redundant basis elements is already installed by the authors in SCRATCHPAD II and with minor changements also in REDUCE 3.3. The modification of algorithm 4.4 is based on the following idea.

If the input elements $f_1, \ldots, f_r$ are ordered, such that $T(1) \geqslant_T \cdots \geqslant_T T(r)$, then an f_i is redundant in the final Groebner basis, if and only if for a $j > i$ $T(i, j) = T(i)$ holds, see 2.6. ($j < i$ is excluded by the order of the input elements for $i \leqslant r$ and for $i > r$ it is impossible because then f_i is a f_{R+1} and $T(R+1)$ has no divisor $T(j), j < R+1$.) Then $T(j, t)$ divides $T(i, t)$ for all $t > j$. Hence $M(i, t)$ holds or $T(i, t) = T(j, t)$. Therefore S_{it} is redundant or equivalent to S_{jt} by 3.7.

Thus, when $T(i, j) = T(i)$, then f_i is removed from the actual G and in the subsequent calls of updatePairs (D, t), $t > j$, the pair (i, t) is ignored.

4.7. The cancelling of redundant basis elements in the actual set G leads in both algorithms to space savings and to faster tests of criterion 1 in algorithm 4.2 or faster applications of updatePairs in algorithm 4.4 respectively. However, for several reasons it is to be expected that algorithm 4.4 is faster than algorithm 4.2, as the statistics in section 5 will confirm,

(1) B contains usually more elements than D, because pairs $\{I, J\}$ are assigned to B before being tested by criterion 1 or criterion 2, whereas in updatePairs all possible tests are already done, before pairs (i, j) are assigned to D.
(2) If in the iteration of algorithm 4.2 the pair $\{I, J\}$ is in one loop $\{I, J_1\}$ and in a later loop $\{I, J_2\}$ with the same I, then the test of criterion 1 includes in both cases the testing of the same $\{I, K\}$ for some K. Such surplus tests do not occur in updatePairs.

(3) Following a recommendation of Buchberger, in algorithm 4.2 the pairs $\{I, J\} \in B$ is always selected, such that

$$T(I, J) = \min\{T(K, L)/\{K, L\} \in B\},$$

but it is left to chance, what pair $\{I, J\} \in B$ with minimal $T(I, J)$ is selected. updatePairs selects among all (i, t) with $1 \leqslant i < t$ and same $T(i, t)$ one element which satisfies criterion 2 and omits the other (i, t). This chance of omitting some pairs if one satisfies criterion 2 is sometimes lost in algorithm 4.2 as the careful analysis of some involved examples showed and it causes, that in some examples more reductions $S(f_I, f_J) \underset{G}{\rightarrow}{}^{+} 0$ are detected by the criteria in algorithm 4.4 than in algorithm 4.2.

5. Examples

5.1. In 4.6 we described how algorithm 4.4 has to be modified in order to obtain a Groebner basis without redundant elements. The following example illustrates this version of algorithm 4.4.

Let $P := Q[x, y, z]$, Q the field of rationals, and $<_T$ be the lexicographical term ordering with $x <_T y <_T z$. We want to calculate a Groebner basis of (f_1, f_2, f_3) with

$$f_1 := zy^2 + 2x + \frac{1}{2},$$

$$f_2 := zx^2 - y^2 - \frac{1}{2}x,$$

$$f_3 := -z + y^2x + 4x^2 + \frac{1}{4},$$

see example 6.15 of Buchberger (1985). In the iteration of the algorithm, we always select $(I, J) \in D$, such that

$$T(I, J) = \min\{T(K, L) \mid (K, L) \in D\}.$$

f_1 and f_2 are redundant because of $T(1, 3) = T(1)$ and $T(2, 3) = T(2)$. Therefore the initialization gives first $(t = 2) G = \{f_1, f_2\}$ and $D = \{(1, 2)\}$ and then $(t = 3)$

$$G = \{f_3\}, D = \{(1, 3), (2, 3)\}.$$

Because of $B_3(1, 2)$ the pair $(1, 2)$ was removed from D.

The first pair (I, J) is $(2, 3)$. Then

$$f_4 := NF(S(f_2, f_3), G) = -y^2x^3 + y^2 - 4x^4 - \frac{1}{4}x^2 + \frac{1}{2}x$$

gives $D = \{(1, 3)\}$, because f_1 and f_2 are redundant and $T(3)T(4) = T(3, 4)$, such that neither $(1, 4)$ nor $(2, 4)$ nor $(3, 4)$ is inserted into D. f_3 is not redundant. Therefore $G = \{f_3, f_4\}$.

The only choice for the next (I, J) is $(1, 3)$. Then

$$f_5 := NF(S(f_1, f_3), G) = y^4x + 4x^2y^2 + \frac{1}{4}y^2 + 2x + \frac{1}{2}$$

gives $D = \{(4, 5)\}$, because again f_1 and f_2 are redundant and $T(3)T(5) = T(3, 5)$, such that neither (1, 5) nor (2, 5) nor (3, 5) but (4, 5) is inserted into D. We also get $G = \{f_3, f_4, f_5\}$.

The only choice for the next (I, J) is (4, 5). Then

$$f_6 := NF(S(f_4, f_5), G) = y^4 + \frac{1}{2}y^2x + 2x^3 + \frac{1}{2}x^2$$

gives $D = \{(5, 6)\}$ by the same arguments as before and in addition by $M(4, 6)$. We also get $G\{f_3, f_4, f_6\}$ because of $T(5, 6) = T(5)$.

The only choice for the next (I, J) is (5, 6). Then

$$f_7 := NF(S(f_5, f_6), G) = x^2y^2 + \frac{1}{14}y^2 - \frac{4}{7}x_4 - \frac{1}{7}x^3 + \frac{4}{7}x + \frac{1}{7}$$

gives $D = \{(4, 7), (6, 7)\}$ by similar arguments as before and $G\{f_3, f_6, f_7\}$ because of $T(4, 7) = T(4)$.

The next (I, J) is (4, 7). Then

$$f_i := NF(S(f_4, f_7), G) = y^2x + 14y^2 - 8x^5 - 58x^4 + \frac{9}{2}x^2 + 9x$$

gives $D = \{(6, 8), (7, 8)\}$ because of $T(3)T(8) = T(3, 8)$ and $B_8(6, 7)$ and $G = \{f_3, f_6, f_8\}$ because of $T(7, 8) = T(7)$. Then

$$f_9 := NF(S(f_7, f_8), G) = y^2 + \frac{112}{2745}x^6 - \frac{84}{305}x^5 - \frac{1264}{305}x^4 - \frac{13}{549}x^3 + \frac{84}{305}x^2 + \frac{1772}{2745}x + \frac{2}{2745}$$

gives $D = \{(6, 9), (8, 9)\}$ because of $B_9(6, 8)$ and $G = \{f_3, f_9\}$ because of $T(6, 9) = T(6)$ and $T(8, 9) = T(8)$. Then

$$f_{10} := NF(S(f_8, f_9), G) = x^7 + \frac{29}{4}x^6 - \frac{17}{16}x^4 - \frac{11}{8}x^3 + \frac{1}{32}x^2 + \frac{15}{16}x + \frac{1}{4}$$

gives $D = \{(6, 9)\}$ because of $T(9)T(10) = T(9, 10)$ and $T(3)T(10) = T(3, 10)$ and $G = \{f_3, f_9, f_{10}\}$. Then

$$NF(S(f_6, f_9), G) = 0.$$

Now $D = \emptyset$ and the algorithm terminates giving the Groebner basis

$$G = \{f_3, f_9, f_{10}\}.$$

5.2. The following statistics compare the algorithms 4.2 and 4.4. All examples can be found in Gebauer (1985).

								Algorithm 4.2				Algorithm 4.4			
Example	*a*	*b*	*c*	*d*	*e*	*f*	*g*	*h*	*k*	*l*	*m*	*h*	*k*	*l*	*m*
Ex1	7	6	*l*	*RN*	3	2	6	8/3	59	15	1·68	8/3	2	7	0·96
Ex5	6	6	*l*	*RN*	3	10	6	16/7	151	22	16·32	15/2	3	6	8.51
Ex27	7	7	*g*	*RN*	3	2	6	12/15	139	19	5·58	12/12	11	12	2·66
Ex12	6	6	*g*	*RN*	3	3	13	10/19	62	16	28·18	10/17	11	13	11·03
Ex2	3	3	*l*	*RN*	3	7	3	7/3	38	10	0·60	7/1	2	3	0·56
Ex8	3	3	*g*	*RN*	3	4	6	3/5	12	6	0·51	3/5	5	6	0·56
Ex3	4	4	*l*	*RN*	2	7	5	13/16	66	17	6·98	13/9	6	6	3·19
Ex10	4	4	*g*	*RN*	2	4	7	6/8	31	10	5·34	6/5	5	7	2·10
Ex4	5	5	*l*	*RN*	2	16	5	106/126	2392	111	5749·13	106/100	29	17	542·38
Ex11	5	5	*g*	*RN*	2	5	13	10/21	75	15	52·69	10/20	16	13	22·27
Ex14	6	6	*g*	*RFI*	3	5	13	13/12	120	19	203·13	13/9	7	13	60·04
Ex28	6	5	*g*	*RFI*	7	0	1	38/66	746	44	167·99	38/65	33	25	51·88
Ex9	3	3	*g*	*RN*	9	10	19	18/21	178	21	41·11	18/21	13	19	27·73
Ex29	6	6	*g*	*RN*	2	6	22	18/53	191	24	904·41	18/50	34	22	237·21

a Number of input polynomials.
b Number of variables.
c Lexicographical (*l*) or graduated (*g*) term ordering.
d Coefficient field of rational numbers (*RN*) or of rational functions over the integers (*RF I*).
e Maximal degree of input polynomials.
f Maximal degree of output polynomials.
g Length of Groebner basis.
h Number of *NF* computations: number of non-vanishing/vanishing *NF*'s.
k Maximal cardinality of set *B* or *D* respectively.
l Maximal cardinality of *G*.
m Computing time in seconds on an IBM 3090 mainframe.

References

Bayer, D. A. (1982). *The Division Algorithm and the Hilbert Scheme.* Ph.D. Thesis, Harvard University.

Bayer, D. A. (1986). Private communication.

Buchberger, B. (1965). *Ein Algorithmus zum Auffinden der Basis-elemente des Restklassenrings nach einem nulldimensionalen Polynomideal.* Ph.D. Thesis, Universität Innsbruck.

Buchberger, B. (1979). A criterion for detecting unnecessary reductions in the construction of Groebner bases. *Proc. EUROSAM 79, Springer L.N. in Comp. Sci.* **72,** 3–21.

Buchberger, B. (1985). Groebner Bases: An Algorithmic Method in Polynomial Ideal Theory. In: (ed. N. K. Bose) *Multidimensional Systems Theory.* D. Reidel Publ. Comp. Pp. 184–232.

Gebauer, R. (1985). *A collection of examples for Groebner calculations.* IBM Thomas J. Watson Research Center, Yorktown Heights, NY 10598.

Gebauer, R. & Möller, H. M. (1987). Groebner Bases, to appear in *SCRATCHPAD II Newsletter*, Vol. 2, No. 1.

Hearn, A. C. (1987). *REDUCE User's Manual*: Version 3.3. The Rand Corporation, Santa Monica, CA 90406.

Möller, H. M. (1985). A reduction strategy for the Taylor resolution. *Proc. EUROCAL 85, Springer L.N. in Comp. Sci.* **162,** 526–534.

Möller, H. M. & Mora, F. (1986). New constructive methods in classical ideal theory. *J. of Algebra,* **100,** 138–178.

Taylor, D. K. (1966). *Ideals Generated by Monomials in an R-sequence.* Ph.D. Thesis. University of Chicago.

A p-adic Approach to the Computation of Gröbner Bases

FRANZ WINKLER

Institut für Mathematik and Research Institute for Symbolic Computation, Johannes Kepler Universität, A-4040 Linz, Austria

(*Received* 5 *March* 1987, *and in revised form* 27 *November* 1987)

A method for the p-adic lifting of a Gröbner basis is presented. If F is a finite vector of polynomials in $\mathbb{Q}[x_1, \ldots, x_v]$ and p is a lucky prime for F (it turns out that there are only finitely many unlucky primes) then in a first step the normalized reduced Gröbner basis $G^{(0)}$ for F modulo p is computed, together with matrices $Y^{(0)}$ and $R^{(0)}$ such that $Y^{(0)} . G^{(0)} \equiv F$ (mod p) and $R^{(0)} . G^{(0)} \equiv 0$ (mod p), where the rows of $R^{(0)}$ are the syzygies of $G^{(0)}$ derived from the reduction of the S-polynomials of $G^{(0)}$ to 0. These congruences can be lifted to congruences modulo p^i, for any natural number i, finally leading to the normalized reduced Gröbner basis for F in $\mathbb{Q}[x_1, \ldots, x_v]$.

Introduction

p-adic methods have been successfully applied to a variety of problems in computer algebra, such as the computation of greatest common divisors of multivariate polynomials over the integers (Moses & Yun, 1973; Miola & Yun, 1974), and factorization of multivariate integral polynomials (Zassenhaus, 1969; Musser, 1975; Wang & Rothschild, 1975; Wang, 1978). For an introduction to p-adic lifting and an overview of applications we refer to (Lauer, 1983).

In all these applications the p-adic methods help to control the otherwise enormous growth of coefficients. A similar problem with coefficient growth is encountered in the computation of Gröbner bases of polynomial ideals over the rational number field $\mathbb{Q}$. Trinks remarks in (Trinks, 1984): "Dealing with $K = \mathbb{Q}$, we usually start with a system $f_1, \ldots, f_m$ having coefficients of modest size, but during the algorithm the size of coefficients tends to increase and often makes results unattainable due to space and time limits." So it is just natural to develop a p-adic method for the computation of Gröbner bases over $\mathbb{Q}$. However, to our knowledge, this problem has not received much attention yet. A method for a special case (solution of a system of algebraic equations with finitely many, simple solutions) is treated in (Trinks, 1984; Malle & Trinks, 1984) and used for the solution of an example in (Matzat & Zeh-Marschke, 1986). In this paper we report some new results on p-adic lifting of Gröbner bases, extending (Winkler, 1987a).

For future reference we review some standard definitions and facts. Throughout this paper we assume that K is a field, $K[x_1, \ldots, x_v]$ the ring of polynomials in $x_1, \ldots, x_v$ over K, and $\lhd$ is an admissible ordering of the power products in the variables $x_1, \ldots, x_v$ (as defined in Buchberger, 1985). If $f \in K[x_1, \ldots, x_v]$, then $lpp(f)$ denotes the leading power

Work reported herein has been supported by the *Österreichische Forschungsgemeinschaft*.

product of f w.r.t. $\lhd$ and $lc(f)$ the leading coefficient of f, i.e. the coefficient of $lpp(f)$ in f. If $F \subset K[x_1, \ldots, x_\nu]$, then $lpp(F) = \{lpp(f) \mid f \in F\}$. $pp(f)$ denotes the set of power products occurring in f and $pp(F) = \{pp(f) \mid f \in F\}$.

Every subset F of $K[x_1, \ldots, x_\nu]$ generates an ideal in $K[x_1, \ldots, x_\nu]$, namely

$$ideal(F) = \left\{ \sum_{i=1}^{n} h_i \cdot f_i \mid 0 \leqslant n, h_i \in K[x_1, \ldots, x_\nu], f_i \in F \right\}.$$

F is called a *basis* of the ideal generated by F. In fact, from Hilbert's basis theorem (Hilbert, 1980; van der Waerden, 1967) we know that every ideal in $K[x_1, \ldots, x_\nu]$ has a finite basis. For convenience we will often write a basis F for a polynomial ideal as a vector $(F_1, \ldots, F_m)^T$ rather than a set $\{F_1, \ldots, F_m\}$.

Every set $F \subseteq R[x_1, \ldots, x_\nu]$, R a commutative ring, induces a *reduction relation* $\rightarrow_F$ on $R[x_1, \ldots, x_\nu]$ in the following way:

$$g_1 \rightarrow_F g_2 \quad \text{iff } g_2 = g_1 - \frac{a}{lc(f)} \cdot u \cdot f$$

for some $f \in F$, u a power product in $x_1, \ldots, x_\nu$, such that $u \cdot lpp(f)$ occurs in g_1 with coefficient a and $lc(f)$ is invertible in R. In words: g_1 *is reducible to* g_2 *w.r.t.* F. If no such u and f exist, then g_1 is *irreducible w.r.t.* F. This reduction relation $\rightarrow_F$ is Noetherian. A polynomial f is reducible w.r.t. F if and only if f contains a term which is in *ideal*($lpp(F)$). If $g \rightarrow_F^* h$ (g is reducible to h w.r.t. F in finitely many steps) and h is irreducible w.r.t. F then h is called a *normal form of* g *w.r.t.* F. Whenever $g_1 \rightarrow_F^* g_2$, $F = (F_1, \ldots, F_m)^T$, then $g_1 - g_2 = (h_1, \ldots, h_m) \cdot F$ for some polynomials $h_1, \ldots, h_m$. But not every vector $(h_1, \ldots, h_m)$ corresponds to a reduction w.r.t. F. A vector $(h_1, \ldots, h_m)$ such that $(h_1, \ldots, h_m) \cdot F = 0$ is called a *syzygy* of F. The set of syzygies of F form a module over $R[x_1, \ldots, x_\nu]$, the module of syzygies of F.

For $f, g \in R[x_1, \ldots, x_\nu]$ such that $lc(f)$ and $lc(g)$ are invertible, the *S-polynomial* of f and g is defined as follows:

$$Spol(f, g) = \frac{1}{lc(f)} \cdot \frac{lcm(lpp(f), lpp(g))}{lpp(f)} \cdot f - \frac{1}{lc(g)} \cdot \frac{lcm(lpp(f), lpp(g))}{lpp(g)} \cdot g.$$

A finite basis F of an ideal I in $K[x_1, \ldots, x_\nu]$ is a *Gröbner basis* for I iff every polynomial f in I is reducible to 0 (in finitely many steps) w.r.t. F. If $ideal(F') = I$, then in abuse of notation we also say "F is a Gröbner basis for F'" instead of "F is a Gröbner basis for I". The Gröbner basis F is called *reduced* iff, for every polynomial $f \in R$, f is irreducible w.r.t. $F \backslash \{f\}$. F is *normalized* iff every polynomial f in F is *monic*, i.e. $lc(f) = 1$. The normalized reduced Gröbner basis for an ideal I is uniquely determined (Buchberger, 1976).

There are many characterizations of Gröbner bases; we only list some of them as far as they are relevant for this paper. Let $F = (F_1, \ldots, F_m)^T$ be a finite basis of the ideal I.

— F is a Gröbner basis for I if and only if
— every S-polynomial $Spol(f, g)$ of elements $f, g \in F$ is reducible to 0 w.r.t. F if and only if
— every polynomial has a unique normal form w.r.t. F.

These and other characterizations can be found in Buchberger (1985) and Möller & Mora (1986). For an introduction to the theory of Gröbner bases we refer to Buchberger (1985). If F is a Gröbner basis, then the set $S(F)$, consisting of the vectors $(h_1, \ldots, h_m)$

derived from a reduction of the S-polynomials of F to 0 w.r.t. F, is a basis for the module of syzygies of F, see (Winkler, 1986). Whenever F is a sequence of polynomials in $K[x_1, \ldots, x_v]$ and G is a Gröbner basis for F, then there are matrices X, Y, R over $K[x_1, \ldots, x_v]$ such that $G = X \,.\, F$, $F = Y \,.\, G$, $R \,.\, G = 0$, where the rows of R are the elements of $S(G)$. We call the matrices X and Y transformation matrices and we call the matrix R a syzygy matrix of F and G.

The notion of a Gröbner basis can be extended to modules over the polynomial ring (Galligo, 1979). All the properties of a Gröbner basis mentioned above carry over to this generalization.

In the following we will be dealing primarily with Gröbner bases in the ring $\mathbb{Q}[x_1, \ldots, x_v]$, which we will denote $\mathscr{A}$. By $\mathbb{Z}_m$ we mean the ring of integers modulo m. The ring $\mathbb{Z}_m[x_1, \ldots, x_v]$, $m \in \mathbb{N}$, will be denoted $\mathscr{A}_m$. The aim of this paper is to approximate the Gröbner basis G of a polynomial ideal in $\mathscr{A}$ by a basis G' in $\mathscr{A}_{p^n}$, p a prime. If p^n is sufficiently large, then we will be able to recover the "true" coefficients of G from their approximations in G'. The coefficients which we want to approximate will be *Farey rationals*. The *N-Farey rationals*, $N \in \mathbb{N}$, are defined as

$$\mathscr{F}_N = \left\{ \frac{a}{b} \,\middle|\, a, b \in \mathbb{Z}, \ -N \leqslant a \leqslant N, \ 1 \leqslant b \leqslant N, \ gcd(a, b) = 1 \right\}.$$

Furthermore, for a prime p, $\mathscr{F}_{p,N} = \mathscr{F}_N \cap \{a/b \in \mathbb{Q} \mid gcd(b, p) = 1\}$. The elements of $\mathscr{F}_{p,N}$ can be encoded uniquely in the integers modulo m for a suitable m. More specifically, if p is a prime, $m = p^k$, N such that $N \leqslant \sqrt{(m-1)/2}$, then for any $n \in \mathbb{Z}_m$ there exists at most one $a/b \in \mathscr{F}_{p,N}$ such that $a \equiv b \,.\, n \bmod m$. A proof of this fact is given, for instance, in (Trinks, 1984). The usual canonical mapping is used for mapping $\mathscr{F}_{p,N}$ into $\mathbb{Z}_m$. For the inverse mapping from $\mathbb{Z}_m$ to $\mathscr{F}_{p,N}$ one can use a suitably extended Euclidean algorithm, as described in (Kornerup & Gregory, 1983). Whenever we say that p does not divide q for a prime p and a rational number q, we mean that p divides neither the numerator nor the denominator of q, i.e. the p-adic norm of q is one, $|q|_p = 1$, and when we say that q' is the image of $q = a/b \in \mathscr{F}_{p,N}$ modulo $m = p^k$, we mean that $a \equiv b \,.\, q' \bmod m$.

The structure of the paper is as follows. Section 1 contains a short discussion of lucky primes, i.e. those primes with respect to which the p-adic approximation of a Gröbner basis is possible. In Section 2 existence and uniqueness of such an approximation are investigated, giving rise to the lifting algorithm of Section 3. In Section 4 we draw some conclusions and list open problems.

1. Lucky Primes

W. S. Brown (1971) calls a prime p lucky for the computation of the greatest common divisor of two integral polynomials f and g, if p does not divide the leading coefficient of f and g and the degree of the gcd of f and g modulo p equals the degree of the gcd of f and g over the integers (the gcd of f and g modulo p cannot be less than the degree of the gcd over the integers, as long as p does not divide any of the leading coefficients of f and g). Only such lucky primes can be used in the modular computation of the polynomial greatest common divisor. We need a similar condition on the prime p.

EXAMPLE 1. (a) (from Ebert, 1983) Let $F = \{xy^2 - 2y, x^2y + 3x\} \subset \mathbb{Q}[x, y]$. Let the power products be ordered according to the graduated lexicographic ordering with $x < y$.

The normalized reduced Gröbner basis for F in $\mathbb{Q}[x, y]$ is $G = \{x, y\}$. $G \bmod 5 = \{x, y\}$, but that is not a Gröbner basis for *ideal*(F) in $\mathbb{Z}_5[x, y]$. Actually, the normalized reduced Gröbner basis for *ideal*(F) in $\mathbb{Z}_5[x, y]$ is $\{xy^2 - 2y, x^2y - 2x\}$. So 5 is not "lucky" for F.

(b) Let $F = \{7xy + y + 4x, y + 2\} \subset \mathbb{Q}[x, y]$. Let the power products be ordered according to the graduated lexicographic ordering with $x < y$. The normalized reduced Gröbner basis for F in $\mathbb{Q}[x, y]$ is $G = \{x + \frac{1}{5}, y + 2\}$. 5 divides the second coefficient of the first polynomial in G, so 5 is not "lucky" for F. The normalized reduced Gröbner basis for F in $\mathbb{Z}_5[x, y]$ is $\{1\}$.

(c) Let $F = \{16x^2 + 4xy^2 - 4z + 1, 4x + 2y^2z + 1, -2x^2z + x + 2y^2\} \subset \mathbb{Q}[x, y, z]$. Let the power products be ordered according to the lexicographic ordering with $z < y < x$. The normalized reduced Gröbner basis for F in $\mathbb{Q}[x, y, z]$ has the leading power products $\{z^7, y^2, x\}$ (see Winkler *et al.*, 1985). The normalized reduced Gröbner basis for F in $\mathbb{Z}_7[x, y, z]$ has the leading power products $\{z^6, zy^2, y^4, x\}$. So 7 is not "lucky" for F.

(d) Let $F = \{x^2y + 9x^2 - y, xy + 4x^2 + 3x\} \subset \mathbb{Q}[x, y]$. Let the power products be ordered according to the graduated lexicographic ordering with $x < y$. The normalized reduced Gröbner basis for F in $\mathbb{Q}[x, y]$ is $\{xy + 4x^2 + 3x, y^2 - 16x^2 + 3y - 12x, x^3 - \frac{3}{2}x^2 + \frac{1}{4}y\}$. The normalized reduced Gröbner basis for F in $\mathbb{Z}_5[x, y]$ is $\{xy - x^2 - 2x, y^2 - x^2 - 2y - 2x, x^3 + x^2 - y\}$. So 5 is "lucky" for F.

Ebert (1983) observes that in general the number of polynomials in the normalized reduced Gröbner basis G for some $F \subset \mathbb{Q}[x_1, \ldots, x_v]$ can be greater than, equal to, or less than the number of polynomials in the normalized reduced Gröbner basis G_p for F in $\mathbb{Z}_p[x_1, \ldots, x_v]$, for some prime p. In Example 1 we see that also the leading power products of G can be greater than, equal to, less than, or incomparable to the leading power products of G_p. From Example 1(a) we also see, that it is not sufficient to just require that p not provide any coefficient of F and G. Fortunately, there are only finitely many such "unlucky" primes.

THEOREM 1. *Let* $F = (F_1, \ldots, F_m)^T$ *be a finite sequence of polynomials in* $\mathcal{A}$, $G = (G_1, \ldots, G_m)^T$ *the normalized reduced Gröbner basis for* F *in* $\mathcal{A}$. *For almost all primes* p *the images* $\bar{F} = F \bmod p$, $\bar{G} = G \bmod p$ *exist and* $\bar{G}$ *is the normalized reduced Gröbner basis for* $\bar{F}$ *in* $\mathcal{A}_p$.

PROOF. Let X, Y, R be matrices over $\mathcal{A}$ such that

$$G = X \,.\, F \quad \text{and} \quad F = Y \,.\, G,$$

and the rows of R are the elements of $S(G)$, i.e. the syzygies of G derived from reductions of the S-polynomials of G to 0 w.r.t. G.

$$R \,.\, G = 0.$$

Let p be such that it does not divide any coefficient of F, G, X, Y and R. Then $\bar{F} = F \bmod p$, $\bar{G} = G \bmod p$, $\bar{X} = X \bmod p$, $\bar{Y} = Y \bmod p$ exist and

$$\bar{G} \equiv \bar{X} \,.\, \bar{F}, \qquad \bar{F} \equiv \bar{Y} \,.\, \bar{G} \pmod{p}.$$

So $\bar{G}$ and $\bar{F}$ generate the same ideal in $\mathcal{A}_p$.

Furthermore, $\bar{R} = R \bmod p$ exists.

$$\bar{R} \,.\, \bar{G} \equiv 0 \pmod{p}$$

and the rows of the matrix $\bar{R}$ correspond to reductions of the S-polynomials of $\bar{G}$ to 0 w.r.t. $\bar{G}$ in $\mathscr{A}_p$. So $\bar{G}$ is the normalized reduced Gröbner basis for $\bar{F}$ in $\mathscr{A}_p$.

For a given F there are only a finite number of primes that divide some coefficient of F, G, X, Y or R. □

After Example 1(a) we have remarked that it is not sufficient to require that the prime p not divide any coefficient of F and G. In this case the transformation matrix X is

$$\begin{pmatrix} \frac{1}{15}x^2 & \frac{1}{3}-\frac{1}{15}xy \\ -\frac{1}{2}-\frac{1}{10}xy & \frac{1}{10}y^2 \end{pmatrix}.$$

X has no representation in $\mathscr{A}_5$.

Definition 1. Let F be a finite sequence of polynomials in $\mathscr{A}$, G the normalized reduced Gröbner basis for F. Let p be a rational prime. p is *lucky* for F iff there are transformation matrices X, Y and a syzygy matrix R for F and G such that p does not divide any coefficient in F, G, X, Y and R. □

2. The Lifting of a Gröbner Basis

From now on we assume that p is a rational prime. Let F be a finite sequence of polynomials in $\mathscr{A}$. Let $G^{(0)}$ be a Gröbner basis for F in $\mathscr{A}_p$, and $X^{(0)}$, $Y^{(0)}$ transformation matrices and $R^{(0)}$ a syzygy matrix for F and $G^{(0)}$ over $\mathscr{A}_p$.

$$\begin{aligned} X^{(0)} \,.\, F &\equiv G^{(0)} \\ Y^{(0)} \,.\, G^{(0)} &\equiv F \pmod p \qquad (2.1) \\ R^{(0)} \,.\, G^{(0)} &\equiv 0. \end{aligned}$$

The first two congruences in (2.1) guarantee that F and $G^{(0)}$ generate the same ideal in $\mathscr{A}_p$ and the third congruence guarantees that $G^{(0)}$ is a Gröbner basis. In fact various criteria can be used for eliminating unnecessary S-polynomials or rows of $R^{(0)}$, respectively (see Buchberger, 1979; Winkler, 1984). In addition to the usual requirement for such a criterion, namely that checking only the necessary S-polynomials for reducibility to 0 suffices to ensure the reducibility to 0 of all S-polynomials, we demand that it should work uniformly for coefficient domains $\mathbb{Z}_{p^i}$, $i \geqslant 1$. I.e. if $G^{(0)}$ is a sequence of polynomials over $\mathbb{Z}_p$ and $G^{(i)}$ a sequence of polynomials over $\mathbb{Z}_{p^i}$ such that $G^{(0)} \equiv G^{(i)} \pmod p$ and $pp(G_j^{(0)}) = pp(G_j^{(i)})$ for $1 \leqslant j \leqslant length(G^{(0)})$, then $Spol(G_k^{(0)}, G_l^{(0)})$ is necessary if and only if $Spol(G_k^{(i)}, G_l^{(i)})$ is necessary, $1 \leqslant k, l \leqslant length(G^{(0)})$. From now on we will assume such a criterion (which could, of course, be trivial in the sense that all S-polynomials are deemed necessary). For instance the criteria given in (Buchberger, 1979; Winkler, 1984) satisfy our requirements. If $R^{(0)}$ is such that it contains only rows corresponding to necessary S-polynomials of $G^{(0)}$, then it is also called a syzygy matrix of $G^{(0)}$.

Now one could try to lift the congruences (2.1) to congruences modulo p^i for large enough i. The work required in the lifting process depends, of course, on the number of congruences that have to be lifted. The matrix $X^{(0)}$ can contain very high power products, since F is not a Gröbner basis and therefore the representation of $G^{(0)}$ in terms of F is not just a reduction. This problem does not occur with $Y^{(0)}$ and $R^{(0)}$, since $G^{(0)}$ is a Gröbner basis. The following theorem shows that actually we don't really have to consider the equivalence $X^{(0)} . F \equiv G^{(0)} \pmod{p}$.

THEOREM 2. *Let G and G′ be Gröbner bases in* $K[x_1, \ldots, x_v]$, $lpp(G) = lpp(G')$ and $G \subseteq ideal(G')$. *Then* $ideal(G) = ideal(G')$.

PROOF. Every nonzero polynomial $f \in ideal(G')$ is reducible w.r.t. G' and, since $lpp(G') = lpp(G)$, it is also reducible w.r.t. G and the reduction result is again in $ideal(G')$. Since the reduction w.r.t. G is Noetherian, we get that every nonzero $f \in ideal(G')$ can be reduced to 0 w.r.t. G. Therefore, $ideal(G') \subseteq ideal(G)$. □

Suppose that the prime p is lucky for F in $\mathscr{A}$. So the normalized reduced Gröbner basis G for F contains the same leading power products as the normalized reduced Gröbner basis $G^{(0)}$ for F in $\mathscr{A}_p$. Then in the lifting process it suffices to lift the congruences $Y^{(0)} . G^{(0)} \equiv F$ and $R^{(0)} . G^{(0)} \equiv 0$. By Theorem 2, if one finally gets a Gröbner basis G' in $\mathscr{A}$, then G' will automatically be a Gröbner basis for F. This provided that in the lifting process the leading power products of the approximations to the basis G' remain unchanged. So (2.1) is reduced to

$$\left.\begin{aligned} Y^{(0)} . G^{(0)} &\equiv F \\ R^{(0)} . G^{(0)} &\equiv 0 \end{aligned}\right\} \pmod{p}. \tag{2.2}$$

The following technicality will be used in subsequent proofs.

LEMMA 1. *Let* $\mathscr{A}'$ *be the set of polynomials in* $\mathscr{A}$ *no coefficient of which has a denominator divisible by p. Let I be an ideal in* $\mathscr{A}$, $I' = I \cap \mathscr{A}'$ *and* I_p *the set of polynomials f in* $\mathscr{A}_p$ *such that* $f \equiv g \pmod{p}$ *for some* $g \in I'$. (I_p *is an ideal in* $\mathscr{A}_p$.) *Let* $h \in I'$, $h' \in \mathscr{A}_p$ *be such that* $p^{i-1} . h' \equiv h \pmod{p^i}$. *Then* $h' \in I_p$.

PROOF. h is a multiple of p^{i-1}, so for some $g \in \mathscr{A}'$ we have $h = p^{i-1} . g$ and $h' = g \bmod p$. Since $h \in I'$, also $g \in I'$, so $h' \in I_p$. □

DEFINITION 2. Let $F \in \mathscr{A}^m$, p lucky for F, $G^{(0)}$ the normalized reduced Gröbner basis for F in $\mathscr{A}_p$, $R^{(0)}$ a syzygy matrix for $G^{(0)}$ over $\mathscr{A}_p$; specifically, let the jth row of $R^{(0)}$ be the syzygy of $G^{(0)}$ derived from the reduction of $Spol(G^{(0)}_{l(j)}, G^{(0)}_{r(j)})$ to 0 w.r.t. $G^{(0)}$. Let t be a positive integer or ∞, and for $0 \leqslant i < t$ let $G^{(i)}$, $Y^{(i)}$, $R^{(i)}$ be matrices over $\mathscr{A}_{p^{i+1}}$ such that

$$\left.\begin{aligned} Y^{(i)} . G^{(i)} &\equiv F \\ R^{(i)} . G^{(i)} &\equiv 0 \end{aligned}\right\} \pmod{p^{i+1}},$$

$$G^{(i)} \equiv G^{(i-1)}, \quad Y^{(i)} \equiv Y^{(i-1)}, \quad R^{(i)} \equiv R^{(i-1)} \pmod{p^i} \quad \text{if } i \geqslant 1,$$

every element of $G^{(i)}$ is monic, and $pp(G^{(i)}_k) = pp(G^{(i-1)}_k)$ for $0 < i < t$, $1 \leqslant k \leqslant n$. Then $L = ((G^{(i)}, Y^{(i)}, R^{(i)}))_{0 \leqslant i < t}$ is a *lifting sequence for F* modulo p.

For $0 \leqslant j < t$ the lifting sequence L is called *reducing up to j* iff for all $0 \leqslant i \leqslant j$ the rows of $R^{(i)}$ are the syzygies of $G^{(i)}$ corresponding to reductions of the necessary S-polynomials of $G^{(i)}$ to 0 in $\mathscr{A}_{p^{i+1}}$; specifically, the jth row of $R^{(i)}$ is the syzygy of $G^{(i)}$ corresponding to the reduction of $Spol(G^{(i)}_{l(j)}, G^{(i)}_{r(j)})$ to 0 w.r.t. $G^{(i)}$. □

THEOREM 3. (existence of a lifting sequence): *Let m, n, l be natural numbers, $F \in \mathscr{A}^m$, p a lucky prime for F, $G \in \mathscr{A}^n$ the normalized reduced Gröbner basis for F, and $G^{(0)} \in \mathscr{A}_p^n$ the normalized reduced Gröbner basis for F in $\mathscr{A}_p$. If*

$$\left.\begin{aligned} Y^{(0)} . G^{(0)} &\equiv F \\ R^{(0)} . G^{(0)} &\equiv 0 \end{aligned}\right. \pmod p$$

holds for some (m, n)-matrix $Y^{(0)}$ over $\mathscr{A}_p$, and some (l, n)-matrix $R^{(0)}$ over $\mathscr{A}_p$, then for every $i \in \mathbb{N}$ there exist $G^{(i-1)}$, $Y^{(i-1)}$, $R^{(i-1)}$ over $\mathscr{A}_{p^i}$ such that

$$\left.\begin{aligned} Y^{(i-1)} . G^{(i-1)} &\equiv F \\ R^{(i-1)} . G^{(i-1)} &\equiv 0 \end{aligned}\right. \pmod{p^i} \tag{2.3}$$

and

$$Y^{(i-1)} \equiv Y^{(i-2)}, \qquad R^{(i-1)} \equiv R^{(i-2)} \pmod{p^{i-1}}, \quad \textit{if } i > 1 \tag{2.4}$$

and

$$G^{(i-1)} = G \bmod p^i. \tag{2.5}$$

The increments

$$\frac{1}{p^{i-1}}(G^{(i-1)} - G^{(i-2)}), \qquad \frac{1}{p^{i-1}}(Y^{(i-1)} - Y^{(i-2)}), \qquad \frac{1}{p^{i-1}}(R^{i-1)} - R^{(i-2)})$$

can be chosen as a solution of a system of linear equations in $\mathscr{A}_p$, the polynomials in $G^{(i)}$ are monic, and $pp(G_k^{(i-1)}) = pp(G_k^{(i-2)})$ for $1 \leqslant k \leqslant n$.

PROOF. By induction on i. For $i = 1$ the statements (2.3) and (2.5) obviously hold. (2.4) is void. Now let $i > 1$. By the induction hypothesis there exist $G^{(i-2)}$, $Y^{(i-2)}$, $R^{(i-2)}$ over $\mathscr{A}_{p^{i-1}}$ such that (2.3), (2.4) and (2.5) hold for $i - 1$. Let $G' \in A_p^n$, $G^{(i-1)} \in \mathscr{A}_{p^i}^n$ be such that

$$G^{(i-1)} = G^{(i-2)} + p^{i-1} . G' = G \bmod p^i.$$

So (2.5) obviously holds. We need to construct matrices Y', R' over $\mathscr{A}_p$ such that for

$$Y^{(i-1)} = Y^{(i-2)} + p^{i-1} . Y',$$
$$R^{(i-1)} = R^{(i-2)} + p^{i-1} . R'$$

the congruence (2.3) hold (the definition of $Y^{(i-1)}$ and $R^{(i-1)}$ directly implies that (2.4) holds). So we have to solve

$$\begin{aligned} F &\equiv Y^{(i-1)} . G^{(i-1)} \\ &\equiv Y^{(i-2)} . G^{(i-2)} + p^{i-1} . Y' . G^{(i-2)} + p^{i-1} . Y^{(i-2)} . G' \pmod{p^i}. \end{aligned} \tag{2.6}$$

Rewriting (2.6) as

$$F - Y^{(i-2)} . G^{(i-2)} - p^{i-1} . Y^{(0)} . G' \equiv p^{i-1} . Y' . G^{(0)} \pmod{p^i}, \tag{2.7}$$

we see that the image modulo p^i of the left hand side of (2.7) is a vector.

$$\begin{pmatrix} h_1^{(i-1)} \\ \vdots \\ h_m^{(i-1)} \end{pmatrix}$$

where each component $h_k^{(i-1)}$ is the image modulo p^i of a polynomial h_k in *ideal*(F) = *ideal*(G), no coefficient of which has a denominator divisible by p. Because of (2.3) for $i-1$, the left hand side of (2.7) is divisible by p^{i-1}. So we have to solve

$$\frac{1}{p^{i-1}} \cdot \begin{pmatrix} h_1^{(i-1)} \\ \vdots \\ h_m^{(i-1)} \end{pmatrix} \equiv Y' \,.\, G^{(0)} \quad (\text{mod}\, p).$$

For every k, $1 \leqslant k \leqslant m$, $h_k \in ideal(G)$, so by Lemma 1 $h_k^{(i-1)}/p^{i-1} \in ideal(G^{(0)})$ and therefore we can take the elements of Y' as the multiplicands used in the reductions of $h_1^{(i-1)}/p^{i-1}, \ldots, h_m^{(i-1)}/p^{i-1}$ to 0 w.r.t. $G^{(0)}$.

We still have to find a matrix R' such that

$$(R^{(i-2)} + p^{i-1} \,.\, R') \,.\, (G^{(i-2)} + p^{i-1} \,.\, G') \equiv 0 \quad (\text{mod}\, p^i).$$

So we have to solve

$$R^{(i-2)} \,.\, G^{(i-2)} + p^{i-1} \,.\, R^{(i-2)} \,.\, G' \equiv -p^{i-1} \,.\, R' \,.\, G^{(i-2)} \quad (\text{mod}\, p^i). \tag{2.8}$$

The image modulo p^i of the left hand side of (2.8) is a vector of images of polynomials in *ideal*(G) = *ideal*(F). Because of (2.3) for $i-1$, the left hand side of (2.8) is divisible by p^{i-1}. So we have to solve

$$\frac{1}{p^{i-1}} \cdot (R^{(i-2)} \,.\, G^{(i-2)} + p^{i-1} \,.\, R^{(i-2)} \,.\, G') \equiv -R' \,.\, G^{(0)} \quad (\text{mod}\, p).$$

All the images modulo p of the polynomials on the left hand side are in *ideal*$(G^{(0)})$ (by Lemma 1), so the reduction to 0 modulo $G^{(0)}$ yields the matrix R'.

Observe that $G' = (G'_1, \ldots, G'_n)^T$, $Y' = (Y'_{ij})_{1 \leqslant i \leqslant m, 1 \leqslant j \leqslant n}$ and $R' = (R'_{ij})_{1 \leqslant i \leqslant l, 1 \leqslant j \leqslant n}$ are a solution to the system of linear equations

$$G^{(0)} \,.\, Y' + Y^{(0)} \,.\, G' \equiv \frac{1}{p^{i-1}} \cdot (F - Y^{(i-2)} \,.\, G^{(i-2)}) \quad (\text{mod}\, p)$$

$$G^{(0)} \,.\, R' + R^{(0)} \,.\, G' \equiv \frac{1}{p^{i-1}} \cdot (-R^{(i-2)} \,.\, G^{(i-2)}) \quad (\text{mod}\, p),$$

the polynomials in $G^{(i)}$ are monic, and $pp(G'_k) \subseteq pp(G_k^{(i-2)})$ for $1 \leqslant k \leqslant n$. □

From Theorem 3 we know that for every lucky prime p for F a lifting sequence for F modulo p exists. The next theorem deals with the uniqueness of such a lifting sequence. It turns out that the components $G^{(i)}$ are indeed uniquely determined, whereas the components $Y^{(i)}$, $R^{(i)}$ are usually not. Starting from the normalized reduced Gröbner basis $G^{(0)}$ modulo p, after i lifting steps we get $G^{(i)}$, a sequence of polynomials in $\mathscr{A}_{p^{i+1}} \,.\, \mathbb{Z}_{p^{i+1}}$ is not a field, and since we have not introduced the notion of a Gröbner basis over a ring, we have to prove (a), (b), (c) in Theorem 4.

THEOREM 4 (uniqueness of lifting sequences): *Let $m, n \in \mathbb{N}$, $F \in \mathscr{A}^m$, $G \in \mathscr{A}^n$ the normalized reduced Gröbner basis for F, p lucky for F, and $i \geqslant 0$. Let $L = ((G^{(j)}, Y^{(j)}, R^{(j)}))_{0 \leqslant j < i+1}$ be a lifting sequence for F modulo p. Then*

(*a*) *all the S-polynomials of $G^{(i)}$ can be reduced to* 0 *w.r.t. $G^{(i)}$,*
(*b*) *every polynomial h in ideal$(G^{(i)})$ ($\subseteq \mathscr{A}_{p^{i+1}}$) is reducible w.r.t. $G^{(i)}$,*

(c) *ideal*(F) = *ideal*$(G^{(i)})$ *as ideals in* $\mathscr{A}_{p^{i+1}}$,
(d) $G^{(i)} = G \bmod p^{i+1}$.

PROOF. (a) Since $G^{(0)}$ is a Gröbner basis, we know that all the S-polynomials of $G^{(0)}$ can be reduced to 0 w.r.t. $G^{(0)}$, and actually the rows of $R^{(0)}$ are just the syzygies derived from the reductions of the necessary S-polynomials of $G^{(0)}$.

If L is reducing up to i, then obviously the S-polynomials of $G^{(i)}$ are reducible to 0 w.r.t. $G^{(i)}$.

Otherwise let j, $1 \leqslant j \leqslant i$, be the smallest index such that the rows of $R^{(j)}$ do not correspond to the syzygies derived from reductions of the necessary S-polynomials of $G^{(j)}$ to 0 w.r.t. $G^{(j)}$. Let R' be the matrix over $\mathscr{A}_p$ such that

$$R^{(j)} = R^{(j-1)} + p^j . R'.$$

The elements of $R^{(j-1)}$ can be considered as the multiplicands used in an "incomplete" reduction of the necessary S-polynomials of $G^{(j)}$ w.r.t. $G^{(j)}$. A step in this incomplete reduction consists of partially reducing an occurring term, possibly leaving a coefficient which is a multiple of p^j. The result of this incomplete reduction is a vector of the form

$$p^j . h = p^j . \begin{pmatrix} h_1 \\ \vdots \\ h_l \end{pmatrix}$$

for some $h_1, \ldots, h_l \in \mathscr{A}_p$. Because of $R^{(j)} . G^{(j)} \equiv 0 \pmod{p^{j+1}}$ we have

$$p^j . h \equiv -p^j . R' . G^{(0)} \quad (\bmod\, p^{j+1}),$$

or equivalently,

$$h \equiv -R' . G^{(0)} \pmod{p}.$$

Obviously $h_k \in ideal(G^{(0)})$ for $1 \leqslant k \leqslant l$, so h_k can be reduced to 0 w.r.t. $G^{(0)}$. Actually, whenever a term has to be reduced which is also reduced in the incomplete reduction above, then corresponding basis polynomials can be used in corresponding reduction steps. Let the elements of the matrix $\bar{R}'$ be the multiplicands used in this reduction of h to 0, so that $h \equiv -\bar{R}' . G^{(0)} \pmod{p}$. For the matrix

$$S^{(j)} := R' - \bar{R}'$$

we have

$$S^{(j)} . G^{(0)} \equiv 0 \quad (\bmod\, p).$$

If we now let

$$\bar{R}^{(j)} := R^{(j)} - p^j . S^{(j)},$$

then the rows of $\bar{R}^{(j)}$ correspond to the syzygies derived from reductions of the necessary S-polynomials of $G^{(j)}$ to 0 w.r.t. $G^{(j)}$.

In the sequel we construct matrices $\bar{R}^{(j+1)}, \ldots, \bar{R}^{(i)}$ such that

$$\bar{L} = ((G^{(0)}, Y^{(0)}, R^{(0)}), \ldots, (G^{(j-1)}, Y^{(j-1)}, R^{(j-1)}),$$
$$(G^{(j)}, Y^{(j)}, \bar{R}^{(j)}), \ldots, (G^{(i)}, Y^{(i)}, \bar{R}^{(i)}))$$

is a lifting sequence for F modulo p. Actually we prove the following: for every k with

$j \leqslant k \leqslant i$ there exists a matrix $S^{(k)}$ over $\mathscr{A}_p$ such that for

$$\bar{R}^{(k)} := R^{(k)} - \sum_{r=j}^{k} p^r \,.\, S^{(r)},$$

$$\bar{R}^{(j-1)} := R^{(j-1)}$$

we have

$$\left\{\begin{array}{c} \bar{R}^{(k)} \,.\, G^{(k)} \equiv 0 \pmod{p^{k+1}}, \\ \bar{R}^{(k)} \equiv \bar{R}^{(k-1)} \pmod{p^k}, \quad \text{and} \\ \left(\sum_{r=j}^{k} p^r \,.\, S^{(r)}\right) . \, G^{(k-j)} \equiv 0 \pmod{p^{k+1}} \end{array}\right\}. \tag{2.9}$$

Induction on k.
For $k = j$, we have

$$\bar{R}^{(j)} \,.\, G^{(j)} = R^{(j)} \,.\, G^{(j)} - p^j \,.\, S^{(j)} \,.\, G^{(j)} \equiv 0 \pmod{p^{j+1}},$$

$$\bar{R}^{(j)} = R^{(j)} - p^j \,.\, S^{(j)} \equiv R^{(j-1)} = \bar{R}^{(j-1)} \pmod{p^j},$$

$$p^j \,.\, S^{(j)} \,.\, G^{(0)} \equiv 0 \pmod{p^{j+1}}.$$

Now consider k such that $j < k \leqslant i$. Let the matrix R over $\mathscr{A}_p$ be such that

$$R^{(k)} = R^{(k-1)} + p^k \,.\, R.$$

By the induction hypothesis we have

$$(\bar{R}^{(k-1)} + p^k \,.\, R) \,.\, G^{(k)} \equiv (R^{(k-1)} + p^k \,.\, R) \,.\, G^{(k)} - \left(\sum_{r=j}^{k-1} p^r \,.\, S^{(r)}\right) . \, G^{(k)} \equiv 0 \pmod{p^k}.$$

So

$$-\left(\sum_{r=j}^{k-1} p^r \,.\, S^{(r)}\right) . \, G^{(k-j)} \equiv p^k \,.\begin{pmatrix} g_1 \\ \vdots \\ g_l \end{pmatrix} =: p^k \,.\, g \pmod{p^{k+1}}$$

for some $g_1, \ldots, g_l \in \mathscr{A}_p$. Observe that $k - j < i$. So by the assumption of the theorem and Lemma 1, every g_t, $1 \leqslant t \leqslant l$, is the homomorphic image of a polynomial in *ideal*(G) modulo p, so it can be reduced to 0 w.r.t. $G^{(0)}$. Collecting the multiplicands used in these reductions in the matrix $S^{(k)}$, we get

$$g \equiv S^{(k)} \,.\, G^{(0)} \pmod{p}.$$

For this definition of $S^{(k)}$ the condition (2.9) holds.

$$\bar{R}^{(k)} \,.\, G^{(k)} \equiv \left(R^{(k)} - \sum_{r=j}^{k} p^r \,.\, S^{(r)}\right) . \, G^{(k)} \equiv R^{(k)} \,.\, G^{(k)} - \left(\sum_{r=j}^{k-1} p^r \,.\, S^{(r)}\right) . \, G^{(k)} - p^k \,.\, S^{(k)} \,.\, G^{(k)}$$

$$\equiv R^{(k)} \,.\, G^{(k)} + p^k \,.\, g - p^k \,.\, g \equiv 0 \pmod{p^{k+1}},$$

$$\bar{R}^{(k)} = R^{(k)} - \sum_{r=j}^{k} p^r \,.\, S^{(r)} \equiv R^{(k-1)} - \sum_{r=j}^{k-1} p^r \,.\, S^{(r)} = \bar{R}^{(k-1)} \pmod{p^k},$$

$$\left(\sum_{r=j}^{k} p^r \,.\, S^{(r)}\right) . \, G^{(k-j)} = \left(\sum_{r=j}^{k-1} p^r \,.\, S^{(r)}\right) . \, G^{(k-j)} + p^k \,.\, S^{(k)} \,.\, G^{(k-j)}$$

$$\equiv -p^k \,.\, g + p^k \,.\, g = 0 \pmod{p^{k+1}}.$$

So we have constructed a lifting sequence $\bar{L}$ which is reducing up to j. Repeating this process, we finally get a lifting sequence which is reducing up to i. Thus, all the necessary S-polynomials of $G^{(i)}$ are reducible to 0 w.r.t. $G^{(i)}$, and therefore all S-polynomials are reducible to 0 w.r.t. $G^{(i)}$. This completes the proof of (a).

(b) A polynomial h in $ideal(G^{(i)})$ can be written as

$$h \equiv \sum_{k=1}^{n} h_k \, . \, G_k^{(i)} \pmod{p^{i+1}}, \tag{2.10}$$

for some $h_k \in \mathscr{A}_{p^{i+1}}$. Let u be the highest power product w.r.t. $\lhd$ occurring in some summand on the right hand side of (2.10). If $lpp(h) \lhd u$, then the coefficients of u on the right hand side of (2.10) cancel. Subtracting proper multiples of syzygies of $G^{(i)}$, derived from reducing S-polynomials of $G^{(i)}$ to 0, we can decrease u by a process analogous to that described in (Winkler, 1986), proof of Theorem 5. So, finally, the leading power products on both sides of (2.10) will be the same, and therefore h is reducible w.r.t. $G^{(i)}$.

(c) Since p is lucky, $\bar{G}^{(i)} = G \bmod p^{i+1}$ exists and $ideal(\bar{G}^{(i)}) = ideal(F)$ as ideals in $\mathscr{A}_{p^{i+1}}$. Certainly $ideal(F) \subseteq ideal(G^{(i)})$, since $Y^{(i)} \, . \, G^{(i)} \equiv F \pmod{p^{i+1}}$. Assume that $ideal(G^{(i)}) \nsubseteq ideal(F)$. Let h be a polynomial in $ideal(G^{(i)}) \backslash ideal(F)$. By (b), h is reducible w.r.t. $G^{(i)}$. But every polynomial which is reducible w.r.t. $G^{(i)}$ is also reducible w.r.t. $\bar{G}^{(i)}$, since the leading power products in both bases are the same and the basis elements are monic. Say $h \to_{\bar{G}^{(i)}} g$. Then $g \in ideal(G^{(i)})$, since $ideal(\bar{G}^{(i)}) = ideal(F) \subseteq ideal(G^{(i)})$. But $g \notin ideal(F)$, for otherwise $h \in ideal(F)$. So $g \in ideal(G^{(i)}) \backslash ideal(F)$, and by the same reasoning as for h we get that g can be reduced w.r.t. $\bar{G}^{(i)}$. That process can be repeated indefinitely, leading to an infinite chain of reductions. This, however, is impossible.

(d) Assume that there exists a k, $1 \leqslant k \leqslant n$, such that $G_k^{(i)} \neq \bar{G}_k^{(i)}$. $\bar{G}_k^{(i)}$ can be reduced by $G_k^{(i)}$ to some nonzero polynomial h in $\mathscr{A}_{p^{i+1}}$, and h is irreducible w.r.t. $G^{(i)}$ (because the same power products occur in $G_k^{(i)}$ and $\bar{G}_k^{(i)}$ and $G^{(i)}$ is reduced). So, by (b), $h \notin ideal(G^{(i)})$, and therefore $ideal(G^{(i)}) \neq ideal(\bar{G}^{(i)}) = ideal(F)$ in $\mathscr{A}_{p^{i+1}}$. This, however, is a contradiction to (c). □

If p is a lucky prime for F, then by Theorem 3 the congruence (2.2) can be extended to a lifting sequence of arbitrary length and by Theorem 4 such a lifting sequence guarantees that we get the correct approximation of the normalized reduced Gröbner basis G for F. In order to compute $(G^{(i)}, Y^{(i)}, R^{(i)})$ from $(G^{(i-1)}, Y^{(i-1)}, R^{(i-1)})$ we have to solve the system

$$U \, . \, c = v^{(i)} \tag{2.11}$$

over $\mathscr{A}_p$, where

$$v^{(i)} = \frac{1}{p^i} \cdot \begin{pmatrix} F - Y^{(i-1)} \, . \, G^{(i-1)} \\ \cdots\cdots\cdots\cdots \\ -R^{(i-1)} \, . \, G^{(i-1)} \end{pmatrix}$$

and U is the matrix

$$\begin{array}{l} \\ \\ \text{row } m \\ \\ \\ \text{row } m+l \end{array}
\left(\begin{array}{cccccc:cccccc:c}
G_1^{(0)} \ \ldots \ G_n^{(0)} & & & & & & & & & & & & \\
 & 0 & \ddots & 0 & & & & & 0 & & & & Y^{(0)} \\
 & & & G_1^{(0)} \ \ldots \ G_n^{(0)} & & & & & & & & & \\
\hdashline
 & & & & & & G_1^{(0)} \ \ldots \ G_n^{(0)} & & & & & & \\
 & & 0 & & & & & 0 & \ddots & 0 & & & R^{(0)} \\
 & & & & & & & & & G_1^{(0)} \ \ldots \ G_n^{(0)} & & &
\end{array}\right)$$

(column $n \, . \, m$ marks the last column of the left block; column $n \, . \, (m+1)$ marks the last column before the final column.)

If

$$(y'_{11}, \ldots, y'_{1n}, \ldots, y'_{m1}, \ldots, y'_{mn}, r'_{11}, \ldots, r'_{1n}, \ldots, r'_{l1}, \ldots, r'_{ln}, g'_1, \ldots, g'_n)^T$$

is a solution such that $pp(g'_j) \subseteq pp(G_j^{(0)}) \setminus \{lpp(G_j^{(0)})\}$ for $1 \leqslant j \leqslant n$, then

$$G^{(i)} = G^{(i-1)} + p^i \cdot \begin{pmatrix} g'_1 \\ \vdots \\ g'_n \end{pmatrix},$$

$$Y^{(i)} = Y^{(i-1)} + p^i \cdot \begin{pmatrix} y'_{11} & \ldots & y'_{1n} \\ \vdots & & \vdots \\ y'_{m1} & \ldots & y'_{mn} \end{pmatrix},$$

$$R^{(i)} = R^{(i-1)} + p^i \cdot \begin{pmatrix} r'_{11} & \ldots & r'_{1n} \\ \vdots & & \vdots \\ r'_{l1} & \ldots & r'_{ln} \end{pmatrix}.$$

For the computation of a basis

$$C' = \left\{ \begin{pmatrix} \vdots \\ g_1'^{(1)} \\ \vdots \\ g_n'^{(1)} \end{pmatrix}, \ldots, \begin{pmatrix} \vdots \\ g_1'^{(q)} \\ \vdots \\ g_n'^{(q)} \end{pmatrix} \right\}$$

for the module of solutions of the homogeneous system

$$U \,.\, c = 0 \tag{2.12}$$

and a solution to the inhomogeneous system (2.11) we refer to (Möller & Mora, 1986; Furukawa *et al.*, 1986; Winkler, 1986). Observe that for every $1 \leqslant j, k \leqslant n$ there exists a solution $c' = (\ldots, c'_{n(m+l)+1}, \ldots, c'_{n(m+l)+n})^T$ of (2.12) such that $c'_{n(m+l)+j} = G_k^{(0)}$ and $c'_{n(m+1)+t} = 0$ for $1 \leqslant t \leqslant n, t \neq j$.

The only remaining complication is that during the lifting process no new power products are allowed to be introduced in the basis and the basis polynomials should stay monic. In order to satisfy this requirement, the basis C' of the solution module of the homogeneous system is transformed to a basis

$$C = \left\{ \begin{pmatrix} \vdots \\ g_1^{(1)} \\ \vdots \\ g_n^{(1)} \end{pmatrix}, \ldots, \begin{pmatrix} \vdots \\ g_1^{(r)} \\ \vdots \\ g_n^{(r)} \end{pmatrix} \right\}$$

in which the last n components constitute a Gröbner basis for the module generated by these components (see Möller & Mora, 1986; Winkler, 1987b).

LEMMA 2. *Let* $i \in \mathbb{N}$, $G^{(0)} = (G_1^{(0)}, \ldots, G_n^{(0)})^T$ *the normalized reduced Gröbner basis for* $F = (F_1, \ldots, F_m)^T$ *in* $\mathcal{A}_p$, p *lucky for* F. *Let* C *be a basis for the module of solutions of* (2.12) *such that the last* n *components of the vectors in* C *constitute a Gröbner basis for the module generated by these components. Let* $L = ((G^{(j)}, Y^{(j)}, R^{(j)}))_{0 \leqslant j < i}$ *be a lifting sequence for* F *modulo* p.

(*a*) *There exists a vector* $g' = (g'_1, \ldots, g'_n)^T$ *over* $\mathscr{A}_p$ *such that the last n components of every solution*

$$\bar{c} = (\bar{c}_1, \ldots, \bar{c}_{n(m+l)}, \bar{c}_{n(m+l)+1}, \ldots, \bar{c}_{n(m+l)+n})^T$$

of (2.11) satisfying the additional requirement

$$pp(\bar{c}_{n(m+l)+k}) \subseteq pp(G_k^{(0)}) \setminus \{lpp(G_k^{(0)})\} \quad \text{for } 1 \leqslant k \leqslant n \tag{2.13}$$

are equal to g'.

(*b*) *From an arbitrary solution* $\bar{\bar{c}}$ *of* (2.11) *a solution* $\bar{c}$ *of* (2.11) *which satisfies the additional requirement* (2.13) *can be computed.*

PROOF: (a) By Theorems 3 and 4.

(b) Let $\bar{\bar{c}} = (\ldots, h_1, \ldots, h_n)^T$. Then $h - g' = (h_1 - g'_1, \ldots, h_n - g'_n)^T$ are the last n components of a solution of (2.12). So they can be reduced to 0 w.r.t. the last n components of

$$C = \Bigg\{ \underbrace{\begin{pmatrix} \vdots \\ g_1^{(1)} \\ \vdots \\ g_n^{(1)} \end{pmatrix}}_{C_1}, \ldots, \underbrace{\begin{pmatrix} \vdots \\ g_1^{(r)} \\ \vdots \\ g_n^{(r)} \end{pmatrix}}_{C_r} \Bigg\}.$$

If we let the possible coefficients in g' be parameters and reduce $h - g'$ w.r.t. the last n components of C, we get linear equations for these parameters that have a unique solution (by (a)), and we get a representation of $h - g'$ as a linear combination of the vectors consisting of the last n components of C,

$$g' = h + \sum_{j=1}^{r} a_j \cdot \begin{pmatrix} g_1^{(j)} \\ \vdots \\ g_n^{(j)} \end{pmatrix}.$$

Then

$$\bar{c} = \bar{\bar{c}} + \sum_{j=1}^{r} a_j \cdot C_j$$

is a solution of (2.11) which satisfies (2.13). □

The left hand side U of (2.11) remains the same throughout the lifting process. The right hand side varies and it can be efficiently computed as

$$v^{(i+1)} = \frac{1}{p} \cdot \left(v^{(i)} - \begin{pmatrix} Y'G^{(i-1)} + Y^{(i-1)}G' + p^i Y'G' \\ \cdots\cdots\cdots\cdots \\ R'G^{(i-1)} + R^{(i-1)}G' + p^i R'G' \end{pmatrix} \right),$$

where G', Y', R' are the matrices over $\mathscr{A}_p$ such that $G^{(i)} = G^{(i-1)} + p^i G'$, $Y^{(i)} = Y^{(i-1)} + p^i Y'$, $R^{(i)} = R^{(i-1)} + p^i R'$.

3. The Lifting Algorithm

Suppose we knew how to determine a reasonable bound B for the coefficients of the normalized reduced Gröbner basis G for a given F and how to select a lucky prime p.

Then the following algorithm could be used to compute G by lifting the corresponding Gröbner basis modulo p just high enough so that every coefficient in G has a unique representation. If such a bound B is not known, then the algorithm *lift* can be considered as p-adically approximating G.

Algorithm *lift* (**in:** $F=(F_1,\ldots,F_m)^T$ in $\mathscr{A}^m$,
p, a lucky prime for F,
B, a bound on the coefficients in the normalized reduced Gröbner basis G for F, i.e. every coefficient of G is in $\mathscr{F}_{p,B}$;
out: G, the normalized reduced Gröbner basis for F);

(1) [length of lifting sequence] Compute K such that $2B^2+1\leqslant p^K$.
(2) [initialization] Set $i\leftarrow 1$. Compute the normalized reduced Gröbner basis $G^{(0)}=(G_1^{(0)},\ldots,G_n^{(0)})^T$ for F modulo p and matrices $Y^{(0)}$, $R^{(0)}$ over $\mathscr{A}_p$ such that

$$Y^{(0)}\,.\,G^{(0)}\equiv F \quad\text{and}\quad R^{(0)}\,.\,G^{(0)}\equiv 0 \pmod p,$$

and the rows of $R^{(0)}$ are the syzygies of $G^{(0)}$ derived from the reductions of the necessary S-polynomials of $G^{(0)}$ to 0 w.r.t. $G^{(0)}$.
(3) [solution of homogeneous system] Let the matrix U be as in (2.11). Compute a basis

$$C=\begin{pmatrix}\vdots\\ g_1^{(1)}\\ \vdots\\ g_n^{(1)}\end{pmatrix},\ldots,\begin{pmatrix}\vdots\\ g_1^{(k)}\\ \vdots\\ g_n^{(k)}\end{pmatrix}$$

for the module of solutions of $U\,.\,c=0$ in $\mathscr{A}_p$, such that the last n components of the basis vectors in C constitute a Gröbner basis for the module generated by these components.
(4) [finished?] If $i=K$ then go to (6).
(5) [lift to congruence modulo p^{i+1}] Compute a particular solution

$$(y'_{11},\ldots,y'_{1n},\ldots,y'_{m1},\ldots,y'_{mn},r'_{11},\ldots,r'_{1n},\ldots,r'_{l1},\ldots,r'_{ln},g'_1,\ldots,g'_n)^T$$

of

$$U\,.\,c=\frac{1}{p^i}\cdot\begin{pmatrix}F-Y^{(i-1)}\,.\,G^{i-1)}\\ \cdots\cdots\cdots\cdots\\ -R^{(i-1)}\,.\,G^{(i-1)}\end{pmatrix}$$

over $\mathscr{A}_p$, such that $pp(g'_j)\subseteq pp(G_j^{(0)})\backslash\{lpp(G_j^{(0)})\}$ for $1\leqslant j\leqslant n$. Set

$$G^{(i)}\leftarrow G^{(i-1)}+p^i\,.\,(g'_1,\ldots,g'_n)^T,$$

$$Y^{(i)}\leftarrow Y^{(i-1)}+p^i\,.\,(y'_{kj})_{\substack{1\leqslant k\leqslant m,\\ 1\leqslant j\leqslant n}}\qquad R^{(i)}\leftarrow R^{(i-1)}+p^i\,.\,(r'_{kj})_{\substack{1\leqslant k\leqslant i\\ 1\leqslant j\leqslant n}}.$$

Set $i:=i+1$. Go to (4).
(6) [convert the coefficients back to $\mathbb{Q}$] Compute the unique coefficients in $\mathscr{F}_{p,B}$ corresponding to the coefficients in $G^{(K)}$, getting the normalized reduced Gröbner basis G for F over $\mathbb{Q}$. □

EXAMPLE 2. We carry out the algorithm *lift* for computing a p-adic approximation of a Gröbner basis in $\mathbb{Q}[x,y]$. As the ordering $<$ of the power products we choose the lexicographic ordering with $x < y$. Let the input basis F be

$$F = \begin{pmatrix} x^2y^2 - \frac{14}{3}x^3y \\ xy^3 - \frac{12}{7}x^2y^2 + 8x \\ y^4 + 8y - 24x^3 \end{pmatrix}.$$

The normalized reduced Gröbner basis for F over $\mathbb{Q}$ is

$$G = \begin{pmatrix} x^2 \\ xy^3 + 8x \\ y^4 + 8y \end{pmatrix}.$$

So the coefficients of both F and G are relatively small, whereas the highest coefficient appearing in the computation of G is 1098247/1190896.

As the prime p we choose 5, which is indeed a lucky prime for F. The Gröbner basis for F modulo 5 is

$$G^{(0)} = \begin{pmatrix} x^2 \\ xy^3 - 2x \\ y^4 - 2y \end{pmatrix},$$

the transformation matrix from $G^{(0)}$ to F modulo 5 is $Y^{(0)}$ and the rows of $R^{(0)}$ are the syzygies derived from the necessary S-polynomials $Spol(G_1^{(0)}, G_2^{(0)})$, $Spol(G_2^{(0)}, G_3^{(0)})$ of $G^{(0)}$.

$$Y^{(0)} = \begin{pmatrix} y^2 + 2xy & 0 & 0 \\ -y^2 & 1 & 0 \\ x & 0 & 1 \end{pmatrix}, \qquad R^{(0)} = \begin{pmatrix} y^3 - 2 & -x & 0 \\ 0 & y & -x \end{pmatrix}.$$

In lifting the Gröbner basis $G^{(0)}$ modulo 5 to a basis modulo some higher power of 5 we have to solve a system of inhomogeneous linear equations. The left hand side of this system is the matrix U, whose transposed is

$$\begin{pmatrix}
x^2 & 0 & 0 & 0 & 0 \\
xy^3 - 2x & 0 & 0 & 0 & 0 \\
y^4 - 2y & 0 & 0 & 0 & 0 \\
0 & x^2 & 0 & 0 & 0 \\
0 & xy^3 - 2x & 0 & 0 & 0 \\
0 & y^4 - 2y & 0 & 0 & 0 \\
0 & 0 & x^2 & 0 & 0 \\
0 & 0 & xy^3 - 2x & 0 & 0 \\
0 & 0 & y^4 - 2y & 0 & 0 \\
0 & 0 & 0 & x^2 & 0 \\
0 & 0 & 0 & xy^3 - 2x & 0 \\
0 & 0 & 0 & y^4 - 2y & 0 \\
0 & 0 & 0 & 0 & x^2 \\
0 & 0 & 0 & 0 & xy^3 - 2x \\
0 & 0 & 0 & 0 & y^4 - 2y \\
y^2 + 2xy & -y^2 & x & y^3 - 2 & 0 \\
0 & 1 & 0 & -x & y \\
0 & 0 & 1 & 0 & -x
\end{pmatrix}$$

We compute a basis C for the module of solutions of $U \,.\, c = 0$ in $\mathbb{Z}_5[x, y]$, such that the last 3 components of the basis vectors in C constitute a Gröbner basis for the module generated by these components. C includes the vectors

$$\begin{pmatrix} \vdots \\ x^2 \\ 0 \\ 0 \end{pmatrix}, \begin{pmatrix} \vdots \\ xy^3 - 2x \\ 0 \\ 0 \end{pmatrix}, \begin{pmatrix} \vdots \\ y^4 - 2y \\ 0 \\ 0 \end{pmatrix},$$

$$\begin{pmatrix} \vdots \\ 0 \\ x^2 \\ 0 \end{pmatrix}, \begin{pmatrix} \vdots \\ 0 \\ xy^3 - 2x \\ 0 \end{pmatrix}, \begin{pmatrix} \vdots \\ 0 \\ y^4 - 2y \\ 0 \end{pmatrix},$$

$$\begin{pmatrix} \vdots \\ 0 \\ 0 \\ x^2 \end{pmatrix}, \begin{pmatrix} \vdots \\ 0 \\ 0 \\ xy^3 - 2x \end{pmatrix}, \begin{pmatrix} \vdots \\ 0 \\ 0 \\ y^4 - 2y \end{pmatrix}.$$

In the first lifting step we compute a particular solution $\bar{\bar{c}}$ to the system

$$U \,.\, c = \frac{1}{5} \cdot \begin{pmatrix} F - Y^{(0)} \,.\, G^{(0)} \\ \cdots\cdots\cdots\cdots \\ -R^{(0)} \,.\, G^{(0)} \end{pmatrix} \tag{3.1}$$

over $\mathbb{Z}_5[x, y]$. The basis vectors in C are used to reduce $\bar{\bar{c}}$ to a solution $\bar{c}$ of (3.1) satisfying the requirement that none of the last 3 components of $\bar{c}$ contains any power product that does not already appear in the corresponding element of $G^{(0)}$ and, moreover, the coefficient of $lpp(G_j^{(0)})$ in $\bar{c}_{15+j}$ is 0, for $1 \leqslant j \leqslant 3$. As the vector $\bar{c}$ we get

$$\bar{c} = \begin{pmatrix} 2xy \\ 0 \\ 0 \\ -xy^4 + 2x^2y^3 + 2y^2 + 2xy + x^2 \\ x^2y - 2x^3 \\ 0 \\ 0 \\ 0 \\ 0 \\ 2 \\ 0 \\ 0 \\ -xy^5 + 2x^2y^4 + 2xy^2 + x^2y \\ 0 \\ x^3y - 2x^4 \\ 0 \\ 2x \\ 2y \end{pmatrix}$$

Now we use the components of $\bar{c}$ to update the approximations to G, Y, and R, getting

$$G^{(1)} = \begin{pmatrix} x^2 \\ xy^3 + 8x \\ y^4 + 8y \end{pmatrix},$$

$$Y^{(1)} = \begin{pmatrix} y^2 + 12xy & 0 & 0 \\ -5xy^4 + 10x^2y^3 + 9y^2 + 10xy + 5x^2 & 5x^2y - 10x^3 + 1 & 0 \\ x & 0 & 1 \end{pmatrix},$$

$$R^{(1)} = \begin{pmatrix} y^3 + 8 & -x & 0 \\ -5xy^5 + 10x^2y^4 + 10xy^2 + 5x^2y & y & 5x^3y - 10x^2 - x \end{pmatrix}.$$

$G^{(1)}$ (with its coefficients mapped back to $\mathbb{Q}$) is already the normalized reduced Gröbner basis for F in $\mathbb{Q}[x, y]$. □

4. Conclusion

As we have shown, it is possible to give a lifting algorithm that computes a p-adic approximation to the normalized reduced Gröbner basis G for the ideal generated by a finite set of polynomials F in $\mathbb{Q}[x_1, \ldots, x_v]$. For the lifting process to be valid, we have to guarantee that the prime p is not one of finitely many unlucky primes. Unfortunately, up to now we do not have an effective criterion for determining luckyness. What we would like to have is a criterion similar to the one for the polynomial factorization problem, where we only have to check that p does not divide the leading coefficient and the resultant of the primitive squarefree polynomial f, and f remains squarefree modulo p. So determining luckyness remains an open problem.

Another open problem is the computation of a reasonable upper bound on the coefficients of the normalized reduced Gröbner basis for the ideal generated by a set of polynomials F. Such a bound is essential for the termination criterion of the algorithm *lift*.

Although we cannot yet present a totally effective procedure for the problem of lifting a Gröbner basis, we hope that this is the starting point for further investigation into the subject.

References

Brown, W. S. (1971). On Euclid's algorithm and the computation of polynomial greatest common divisors. *JACM*, **18/4,** 478–504.

Buchberger, B. (1976). Some properties of Gröbner bases for polynomial ideals. *ACM SIGSAM Bull.* **10**(4), 19–24.

Buchberger, B. (1979). A criterion for detecting unnecessary reductions in the construction of Gröbner-Bases. *Proc. EUROSAM 79*, Marseille, 1979, LNCS **72,** 3–21, E. W. Ng, ed., Springer-Verlag, Heidelberg.

Buchberger, B. (1985). Gröbner bases: An algorithmic method in polynomial ideal theory. In (Bose, N. K., ed.) *Multidimensional Systems Theory*. D. Reidel Publ. Comp. pp. 184–232.

Ebert, G. L. (1983). Some comments on the modular approach to Gröbner-bases. *ACM SIGSAM Bull.* **17**(2), 28–32.

Furukawa, A., Sasaki, T., Kobayashi, H. (1986). Gröbner basis of a module over $K[x_1, \ldots, x_n]$ and polynomial solutions of a system of linear equations. In (Char, B. W., ed.) *Proc. SYMSAC'86.* ACM 1986. pp. 222–224.

Galligo, A. (1979). Théorème de division et stabilité en géométrie analytique locale. *Ann. Inst. Fourier*, **29,** 107–184.

Hilbert, D. (1890). Über die Theorie der algebraischen Formen. *Math. Annalen*, **36,** 473–534.

Kornerup, P., Gregory, R. T. (1983). Mapping integers and Hensel codes onto Farey fractions. *BIT* **23,** 9–20.

Lauer, M. (1983). Computing by homomorphic images. In (Buchberger, B. *et al.*, eds.) *Computer Algebra — Symbolic and Algebraic Computation*, 2nd edition. Springer-Verlag, Heidelberg.

Malle, G., Trinks, W. (1984). *Zur Behandlung algebraischer Gleichungssysteme mit dem Computer*. Mathematisches Institut, Universität Karlsruhe, unpublished manuscript.

Matzat B. H., Zeh-Marschke, A. (1986). Realisierung der Mathieugruppen M_{11} und M_{12} als Galoisgruppen über Q. *J. Number Theory* **23,** 195–202.

Miola, A., Yun, D. Y. Y. (1974). The computational aspects of Hensel-type univariate greatest common divisor algorithms. In (Jenks, R. D., ed.) *Proc. EUROSAM'74. SIGSAM Bulletin* **8/3,** 46–54.

Möller, H. M., Mora, F. (1986). New constructive methods in classical ideal theory. *J. of Algebra*, **100,** 138–178.

Moses, J., Yun, D. Y. Y. (1973). The EZGCD algorithm. *Proc. ACM Annual Conference*, Atlanta, pp. 159–166.

Musser, D. R. (1975). Multivariate polynomial factorization. *JACM* **22,** 291–308.

Trinks, W. (1984). On improving approximate results of Buchberger's algorithm by Newton's method. *SIGSAM Bull.* **18**(3), 7–11.

van der Waerden, B. L. (1967). *Algebra II*. Springer-Verlag, Heidelberg.

Wang, P. S. H. (1978). An improved multivariate polynomial factorization algorithm. *Math. Comp.* **32,** 1215–1231.

Wang, P. S. H., Rothschild, L. P. (1975). Factoring multivariate polynomials over the integers. *Math. Comp.* **29,** 935–950.

Winkler, F. (1984). *The Church-Rosser Property in Computer Algebra and Special Theorem Proving: An Investigation of Critical-Pair/Completion Algorithms*. Dissertation, Institut für Mathematik, J. Kepler Universität, Linz.

Winkler, F. (1986). Solution of Equations I: Polynomial Ideals and Gröbner Bases. Lecture notes, Short Course "Symbolic and Algebraic Computation". *Conf. on "Computers & Mathematics"*. Stanford University, 1986, Series in Computational Mathematics, R. D. Jenks, ed., Springer-Verlag.

Winkler, F. (1987a). *p*-adic methods for the computation of Gröbner bases, extended abstract. *EUROCAL'87*, Leipzig, GDR.

Winkler, F. (1987b). A recursive method for computing a Gröbner basis of a module in $K[x_1, \ldots, x_v]^r$. *A.A.E.C.C.-5*, Menorca, Spain.

Winkler, F., Buchberger, B., Lichtenberger, F., Rolletschek, H. (1985). "Algorithm 628 — An algorithm for constructing canonical bases of polynomial ideals. *ACM Trans. on Math. Software* **11,** 66–78.

Zassenhaus, H. (1969). On Hensel factorization, I. *J. Number Theory* **1,** 291–311.

Constructive Lifting in Graded Structures: A Unified View of Buchberger and Hensel Methods

A. MIOLA AND T. MORA*

Dipartimento di Informatica e Sistemistica dell'Università degli Studi di Roma "La Sapienza", Italy

**Dipartimento di Matematica dell'Università di Genova, Italy*

(*Received* 15 *January* 1987, *and in revised form* 1 *March* 1988)

This paper discusses a general lifting technique for solving polynomial equations in graded structures $\underline{A}$, where the solution is understood to lie in the completion $\underline{A}^\wedge$ of $\underline{A}$. It shows that the classical Hensel lifting and the main constructions related to the Buchberger algorithm for Gröbner bases are both instances of this technique. So, while the setting of it is too general to allow for an effective solution of equations, this technique stresses a theoretical relation between two basic algorithms in computer algebra and could be used as a theoretical model to attack computational problems under the same viewpoint.

1. Introduction and Motivations

Two of the most powerful concepts in Computer Algebra are the *Hensel–Zassenhaus lifting construction* (Hensel, 1913; Zassenhauss, 1969) mainly used to solve different algebraic algorithmic problems, which can be expressed as polynomial equations (Miola & Yun, 1974; Yun, 1974; 1976); and *Gröbner bases* (Buchberger, 1965; 1985), by means of which it has been possible to introduce efficient algorithmic solutions for a variety of problems in commutative algebra.

Recently, Robbiano (1986) introduced the new concept of *graded structures*, mainly in order to clarify, in a unified frame, the strict relations between the concept of Gröbner bases and the one of *standard bases*.

The latter were introduced by Hironaka (1964), independently by Buchberger, and have a strong connection with the classical topological interpretation of Hensel's results. An interpretation of the Buchberger algorithm as a lifting technique was already introduced in (Zacharias, 1978; Möller, 1985). As a matter of fact, the connection with the classical lifting methods in commutative algebra, allowed by this interpretation, is one of the main themes underlying Robbiano's research.

The aim of our paper is to pursue this connection further. Actually we introduce a method for the solution of polynomial equations over a graded structure. This method is based on a general lifting technique in graded structures, which allows to recover (initial segments of) Cauchy sequences approaching the actual solution of the given problem, which, in general, lies, rather than in the original graded structure, in its completion w.r.t. a canonical topology, as it is natural with a topological lifting technique.

It is to be remarked that the method presented doesn't imply the existence of a general

Research partially supported by "Progetto Calcolo Algebrico, M.P.I."

lifting algorithm in graded structures, and it can be assumed only to be an abstract specification of algorithms.

This is because, unless specific properties of the equations to be solved, related to existence and uniqueness, are known, the technique could fail, in the sense that the approximate solutions produced by the algorithm could fail to converge. Moreover, also under the assumption that a unique solution exists in the original graded structure, the technique could fail to produce it in finitely many steps. However, the method shows a general environment suitable to describe, and incorporate as instances, some particular lifting algorithms in specific domains.

In particular, our method has a sufficient generality to recover, as it will be shown in Section 5, both the Hensel-Zassenhaus lifting and Buchberger algorithm as specific instances. Actually, we can show that particular good conditions holding for these two cases, even being different for each case, guarantee the convergence of the general lifting process.

Therefore, we think that the main contribution of this approach is to yield a common and neat algebraic interpretation of both these techniques, giving also a common frame where to look for possible further generalizations and for new applications of lifting techniques in computer algebra.

Also, some of the examples of the failure of the lifting technique (notably in Sections 5.2 and 5.3) point to an inherent limitation of it, and should help to avoid pitfalls consequent to uncautious applications (also in the well-known Gröbner and Hensel cases).

In Section 2 we review, for the unfamiliar reader, the concept and the basic examples of graded structures, as introduced in Robbiano (1986). In Section 3 we present some constructions over graded structures which are relevant to the rest of the paper. Section 4 presents the main result of the paper, namely a general lifting technique for solving polynomial equations over (the completion of) a graded structure. The final section is devoted to the interpretation of Hensel and Gröbner techniques in the proposed new context.

2. Graded Structures: Review and Examples

2.1. DEFINITIONS

As introduced by Robbiano (1986), a *filtered* (or *graded*) *structure* $\underline{A} := (A, \Gamma, \bar{V})$ is a triple consisting of:

A, a commutative ring with 1, which we further assume here to be a noetherian integral domain;

Γ, a totally ordered group, whose ordering is denoted by $<$;

$\bar{V} := \{V_\gamma : \gamma \in \Gamma\}$ a set of additive subgroups of A satisfying the following axioms:

($R1$) V_γ is contained in V_δ if $\gamma < \delta$

($R2$) $V_\gamma V_\delta$ is contained in $V_{\gamma+\delta}$

($R3$) for each $a \in A$, $a \neq 0$, there is $\gamma \in \Gamma$ s.t. $a \in V_\gamma$ and for each $\delta \in \Gamma$, s.t. $\delta < \gamma$, $a \notin V_\delta$.

The following objects can be canonically associated with $\underline{A}$:

$v: A - \{0\} \to \Gamma$, a function which to every $a \in A$, $a \neq 0$, associates the minimum $\gamma \in \Gamma$ s.t. $a \in V_\gamma$. It satisfies the following properties:

($R4$) $v(ab) = v(a) + v(b)$ for every $a, b, ab \neq 0$;

($R5$) $v(a - b) \leqslant \max(v(a), v(b))$ if $a, b, a - b \neq 0$;

($R6$) $v(1) = 0$;

$G_\gamma := V_\gamma / \bigcup_{\delta < \gamma} V_\delta$, for each $\gamma \in \Gamma$, an additive group. $G := \bigoplus_{\gamma \in \Gamma} G_\gamma$, which is a Γ-graded ring, if endowed with the canonical multiplication (if $a \in G_\gamma = V_\gamma / \bigcup_{\gamma' < \gamma} V_{\gamma'}$, and $b \in G_\delta = V_\delta / \bigcup_{\delta' < \delta} V_{\delta'}$, let $a_1 \in V_\gamma$, $b_1 \in V_\delta$ be s.t. a (resp. b) is the residue class of a_1 (resp. b_1) mod $\bigcup_{\gamma' < \gamma} V_{\gamma'}$ (resp. $\bigcup_{\delta' < \delta} V_{\delta'}$); then by ($R2$), $a_1 b_1 \in V_{\gamma + \delta}$ and the residue class c of $a_1 b_1$ doesn't depend, again because of ($R2$), on the choice of a_1 and b_1. So the product of a and b is defined to be c). Non-zero elements of G_γ are said to be *homogeneous* of degree γ.

We will further assume here that G is a noetherian integral domain.

$\Gamma_0 := \{\gamma \in \Gamma: G_\gamma \neq 0\} = \{\gamma \in \Gamma: \gamma \in Im(v)\}$, which, under our assumptions, can be proved to be a finitely generated semigroup (Robbiano, 1986; Proposition 1.3(b)).

We will further assume Γ_0 is *inf-limited*, by which we mean that for each infinite decreasing sequence $\gamma_1 > \gamma_2 > \cdots > \gamma_n > \cdots$ of elements in Γ_0 and for each $\gamma \in \Gamma_0$, there is n s.t. $\gamma > \gamma_n$ ($-\mathbb{N}$ is an example of an inf-limited semigroup). in: $A \to G$, a function s.t. $\text{in}(0) = 0$ and, if $a \neq 0$, in(a) is the residue class of a mod. $\bigcup_{\gamma < V(a)} V_\gamma$.

It satisfies the following properties:

($R7$) $\text{in}(a) \in G_{v(a)}$ for every $a \neq 0$;

($R8$) $\text{in}(a) = 0$ iff $a = 0$;

($R9$) $\text{in}(1) = 1$;

($R10$) for each $g \in G$ homogeneous, there is $a \in A$ s.t. $\text{in}(a) = g$;

($R11$) $\text{in}(a)\text{in}(b) = \text{in}(ab)$, for a, b s.t. $ab \neq 0$;

($R12$) $v(a - v) = v(a) = v(b)$ iff $\text{in}(a - b) = \text{in}(a) - \text{in}(b)$, when $a, b, a - b \neq 0$;

($R14$) if $v(a) < v(b)$, then $v(b - a) = v(b)$ and $\text{in}(b - a) = \text{in}(b)$, when $a, b, a - b \neq 0$.

Because of ($R10$), there are functions $\phi: \bigcup_\Gamma G_\gamma \to A$ s.t. for each homogeneous $g \in G$, $a := \phi(g)$ is s.t. $\text{in}(a) = g$. In the following we will assume that such a function in*: $\bigcup_\Gamma G_\gamma \to A$ is fixed s.t. $\text{in}^*(1) = 1$. in* is obviously s.t. in in* is the identity on $\bigcup_\Gamma G_\gamma$. When we will need, we will use the notation $\underline{A} = (A, \Gamma, G, v, \text{in})$ to refer explicitly to the objects canonically associated to $\underline{A}$.

2.2. EFFECTIVENESS CONDITIONS

In the following, we will present some algorithms. In order to guarantee their applicability, we need also some effectiveness conditions. In particular, we will require that A is an *effective* graded structure (Mora, 1987), i.e.

($E1$) A, Γ, G are effective;
($E2$) v, in are computable functions;
($E3$) in* is computable;
($E4$) if $h_1, \ldots, h_s, h$ are homogeneous elements of G, it is possible to compute

homogeneous $g_1, \ldots, g_s \in G$ s.t. $\Sigma\, g_i h_i = h$, $\deg(g_i) + \deg(h_i) = \deg(h)$, if such elements exist;

(E5) If $h_1, \ldots, h_s$ are homogeneous elements of G, it is possible to compute a homogeneous basis for the G-modulus of syzygies of $(h_1, \ldots, h_s)$, $\{(g_1, \ldots, g_s)\colon \Sigma\, g_i h_i = 0\}$, i.e. a basis consisting of tuples $(g_1, \ldots, g_s)$ s.t. every g_i is homogeneous and there is $\gamma \in \Gamma$ s.t. for each i either $g_i = 0$ or $\deg(g_i) + \deg(h_i) = \gamma$;

and moreover we will require that polynomial equations can be solved in every G_γ.

2.3. EXAMPLES

EXAMPLE 1. $A := \mathbb{Z}$, $G := \mathbb{Z}$, $V_n := \mathbb{Z}$ and $V_{-n} := (p^n)$ for $n \geqslant 0$, where p is a fixed prime number.

Then one has $\Gamma_0 = \{n \in \mathbb{Z}\colon n \leqslant 0\}$, $G = \bigoplus_{n<0} \mathbb{Z}_p \approx \mathbb{Z}_p[T]$ (the isomorphism is obtained by identifying the almost always null sequence $(g_n)_{n \leqslant 0}$, with $g_n \in \mathbb{Z}_p$, to $\Sigma\, g_n T^{-n}$; homogeneous elements of G are then exactly the homogeneous polynomials). If $a \in \mathbb{Z}$ is s.t. $a \in (p^n) - (p^{n+1})$, i.e. $a = bp^n$ with b not divisible by p, one has $v(a) = -n$ and $\text{in}(a) = \text{res}(b)T^n$, $\text{res}(b)$ denoting the residue class of $b \bmod p$.
In this way one gets the usual p-adic valuation in $\mathbb{Z}$. If p is odd, there are at least two natural choices for in*:

(1) $\text{in}^*(gT^n) := ap^n$ with $a \in \mathbb{Z}$, $0 < a < p$, $\text{res}(a) = g$.
(2) $\text{in}^*(gT^n) := ap^n$ with $a \in \mathbb{Z}$, $-p/2 < a < p/2$, $\text{res}(a) = g$

As it is well known the second one has better computational properties.

EXAMPLE 2. $A := \mathbb{Z}[X_1, \ldots, X_n]$, $\Gamma := \mathbb{Z}$, $V_n := A$ and $V_{-n} := (p^n)$ for $n \geqslant 0$, where p is a fixed prime number.

Then one has $\Gamma_0 = \{n \in \mathbb{Z}\colon n \leqslant 0\}$, $G = \bigoplus_{n \leqslant 0} \mathbb{Z}_p[X_1, \ldots, X_n] \approx \mathbb{Z}_p[X_1, \ldots, X_n, T]$ (see example 1; homogeneous elements of G are of the form aT^n, $a \in \mathbb{Z}_p[X_1, \ldots, X_n]$). If $a \in \mathbb{Z}[X_1, \ldots, X_n]$ is s.t. $a \in (p^n) - (p^{n+1})$ i.e. $a = p^n b$, with $b \in \mathbb{Z}[X_1, \ldots, X_n]$ not divisible by p, one has $v(a) = -n$ and $\text{in}(a) = \text{res}(v)T^n$. As above, there are two natural choices for in*.

EXAMPLE 3. Let A be a ring, I an ideal of A s.t. $\cap I^n = (0)$. A is turned into a topological ring if it is endowed with the I-adic topology, i.e. the topology with (I^n) as a set of neighboroughs of 0.

This is a classical notion in commutative algebra, which generalizes the p-adic topology on $\mathbb{Z}$ and whose interpretation in the graded structure theory (Robbiano, 1986; Example 1) makes our example 1 a particular case. One has $\Gamma := \mathbb{Z}$, $V_n := A$ and $V_{-n} := I^n$ for $n \geqslant 0$. $\Gamma_0 = \{n \in \mathbb{Z}\colon n \leqslant 0\}$, $G = \bigoplus_{n \leqslant 0} I^n/I^{n+1}$ (where $I^0 := A$): if $a \in I^n - I^{n+1}$ then $v(a) = -n$, and $\text{in}(a)$ is the residue class of a mod I^{n+1}, which is in I^n/I^{n+1}, i.e. it is a homogeneous element in G of degree $-n$.

EXAMPLE 4. Assuming $A := K[X_1, \ldots, X_n]$ and $I := (X_1, \ldots, X_n)$ in Example 3, we have that $G = A$ (more exactly, G is isomorphic to A), and if $a \in A$ is s.t. $a = a_d + a_{d+1} + \cdots + a_{d+k}$, with a_j a homogeneous form of degree j and $a_d \neq 0$, then $v(a) = -d$ and $\text{in}(a) = a_d$. In this case one can define in* to be the identity.

EXAMPLE 5. If we turn Example 4 "upside down" we get the following graded structure: $A := K[X_1, \ldots, X_n]$, $\Gamma := \mathbb{Z}$, $V_n := \{f \in A: \deg(f) \leqslant n\}$ if $n \geqslant 0$, $V_n := \{0\}$ if $n < 0$, $\Gamma_0 = M$, $G = K[X_1, \ldots, X_n]$ with the usual notion of degree, and if $a = \sum_{i=0,d} a_i$, a_i homogeneous of degree i, $a_d \neq 0$, we have $v(a) = d$ and $\mathrm{in}(a) = a_d$.

EXAMPLE 6. Let $A := K[X_1, \ldots, X_n]$, $\Gamma := \mathbb{Z}^n$ with some ordering $<$ on it. If $m := X_1^{j(1)} \ldots X_n^{j(n)}$, define $\log(m) := (j(1), \ldots, j(n)) \in \mathbb{N}^n$. Then each $a \in A - \{0\}$ can be uniquely written as $a = \sum_{i=1,t} c_i m_i$, $c_i \in K - \{0\}$, m_i terms, $\log(m_1) > \log(m_2) > \cdots > \log(m_t)$. Defining $v(a) := \log(m_1)$, $V_\gamma := \{a \in A: a = 0 \text{ or } v(a) \leqslant \gamma\}$ for each $\gamma \in \mathbb{Z}^n$, one obtains a graded structure where $\Gamma_0 = \mathbb{N}^n$, $G = K[X_1, \ldots, X_n]$, and if $a = \sum_{i=1,t} c_i m_i$, then $\mathrm{in}(a) = c_1 m_1$.

2.4. COMMENTS

Example 3 on one side, and Examples 4–6 on the other, are the main motivations for Robbiano's theory of graded structures.

While the motivation of Example 3 is clear, since one obtains a generalisation of a classical and very important concept in commutative algebra, the motivation provided by the other examples needs to be explained.

If I is an ideal of A, $\{\mathrm{in}(f): f \in I\}$ generates a homogeneous ideal $\mathrm{in}(I)$ of G. One can then choose a finite set F of elements of I s.t. $\{\mathrm{in}(f): f \in F\}$ generates $\mathrm{in}(I)$. In the case of Example 6, if $<$ is s.t. $0 \leqslant \gamma$ for each $\gamma \in \Gamma_0$ (i.e. if $<$ is a term-ordering) such a set is exactly a Gröbner basis of I (Buchberger, 1965; 1985). In the case of Example 5 it is a Macaulay basis or H-basis of I, a concept which played a computational role analogous (albeit not so efficient) to the one of Gröbner bases within polynomial ideal theory, before the introduction of the latter concept (Hermann, 1926; Renschuch, 1976). In the case of Example 4, and of Example 6 if $<$ is not a term-ordering, such a set has a strict relation with Hironaka's (1964) concepts of standard bases; the exact relation is not immediate to describe, so we refer the interested reader to (Robbiano, 1986; Mora, 1987).

3. Auxiliary Constructions on a Graded Structure

3.1. THE COMPLETION OF A GRADED STRUCTURE

PROPOSITION. *Let* $\underline{A} := (A, \Gamma, \bar{V}) = (A, \Gamma, G, v, \mathrm{in})$ *be a graded structure. Then* A *is a topological ring w.r.t. the filtration* $\bar{V}$.

Let us denote $A^\wedge$ *the ring completion of* A; *if* $f \in A^\wedge - \{0\}$, *and* $(a_n: n \in N)$ *is a Cauchy sequence in* A *converging to* f, *then for* n *sufficiently large,* $v(a_n)$ *and* $\mathrm{in}(a_n)$ *are constant and independent of the choice of the Cauchy sequence; so we can define* $v(f) := v(a_n)$ *and* $\mathrm{in}(f) := \mathrm{in}(a_n)$ *for large* n.

If we define $V_\gamma^\wedge := \{f \in A^\wedge : f = 0 \text{ or } v(f) \leqslant \gamma\}$, $\bar{V}^\wedge := \{V_\gamma^\wedge : \gamma \in \Gamma\}$ *then* $\underline{A}^{\hat{A}} := (A^\wedge, \Gamma, V^\wedge) = (A^\wedge, \Gamma, G, v, in)$ *is a graded structure.*

PROOF. (MORA, 1987).

3.2. POLYNOMIALS ON A GRADED STRUCTURE

PROPOSITION. *Let* $\underline{A} := (A, \Gamma, \bar{V}) = (A, \Gamma, G, v, \mathrm{in})$ *be a graded structure and let* $\delta_1, \ldots, \delta_m \in \Gamma_0$.

Let us extend v to a function $A[Z_1, \ldots, Z_m] - \{0\} \to \Gamma$ (which we will again denote by v) defining $v(t) := \Sigma\, n_i\delta_i$, if t is a term and $(n_1, \ldots, n_m) := \log(t)$, and, for $a \in A[Z_1, \ldots, Z_m] - \{0\}$, $a := \Sigma\, c_i t_i$, $c_i \in A - \{0\}$, t_i terms with $t_i \neq t_j$ if $i \neq j$, defining $v(a) := \max(v(c_i) + v(t_i))$.

For each $\gamma \in \Gamma$, define $U_\gamma := \{a \in A[Z_1, \ldots, Z_m]$: $a = 0$ or $v(a) \leqslant \gamma\}$, $\bar{U} := \{U_\gamma : \gamma \in \Gamma\}$.

Then $A[Z_1, \ldots, Z_m] := (A[Z_1, \ldots, Z_m], \Gamma, \bar{U}) = (A[Z_1, \ldots, Z_m], \Gamma, G[Z_1, \ldots, Z_m], v, \mathrm{in})$ is a graded structure on $A[Z_1, \ldots, Z_m]$, where $G[Z_1, \ldots, Z_m]$ is Γ-graded by $\deg(Z_i) := \delta_i$, and if $a = \Sigma\, c_i t_i \in A[Z_1, \ldots, Z_m] - \{0\}$ is represented as above, $\mathrm{in}(a) = \Sigma\, \mathrm{in}(c_i) t_i$, the sum being done on those indexes i s.t. $v(c_i) + v(t_i) = v(a)$.

PROOF. One has just to verify the axioms ($R1$), ($R2$), ($R3$), which are quite straightforward.

Remark that $A[Z_1, \ldots, Z_m]^\wedge$ is a ring which, usually, strictly contains $A^\wedge[Z_1, \ldots, Z_m]$.

3.3. FORMAL TAYLOR EXPANSION

Let $A := (A, \Gamma, \bar{V}) = (A, \Gamma, G, v, \mathrm{in})$ be a graded structure and let $A[Z_1, \ldots, Z_m] := (A[Z_1, \ldots, Z_m], \Gamma, \bar{U}) = (A[Z_1, \ldots, Z_m], \Gamma, G[Z_1, \ldots, Z_m], v, \mathrm{in})$ be the induced graded structure on $A[Z_1, \ldots, Z_m]$, with $v(Z_i) = \delta_i$, as defined above. If $P \in A[Z_1, \ldots, Z_m]$, denote by $D_j(P)$ the formal derivative of P w.r.t. Z_j. If $\mathrm{in}(P) =: \Sigma\, Q_i(Z_1, \ldots, Z_{j-1}, Z_{j+1}, \ldots, Z_m) Z_j^i$ and there is at least one i s.t. $Q_i \neq 0$ and $v(i) = 0$ (remark that for all i, $v(i) \leqslant 0$) then $v(D_j(P)) + v(Z_j) = v(P)$.

If $a_1, \ldots, a_m \in A$, there is $R \in A[Z_1, \ldots, Z_m]$ (obtained through the formal Taylor expansion of P at $(a_1, \ldots, a_m)$, and involving both the values $a_1, \ldots, a_m$ and the higher formal derivatives of P) s.t. for all $(c_1, \ldots, c_m)$, $c_i \in A$:

$$P(a_1 + c_1, \ldots, a_m + c_m) = P(a_1, \ldots, a_m) + \sum c_j D_j(P)(a_1, \ldots, a_m) + R(a_1 + c_1, \ldots, a_m + c_m):$$

e.g. if $P := Z_1^3 + Z_1 Z_2^2$, then

$$\begin{aligned} R &= (Z_1 - a_1)^2 D_{11}(P)(a_1, a_2)/2! + (Z_1 - a_1)(Z_2 - a_2) D_{12}(P)(a_1, a_2) \\ &\quad + (Z_2 - a_2)^2 D_{22}(P)(a_1, a_2)/2! + (Z_1 - a_1)^3 D_{111}(P)(a_1, a_2)/3! \\ &\quad + (Z_1 - a_1)(Z_2 - a_2)^2 D_{122}(P)(a_1, a_2)/2! \\ &= 3a_1(Z_1 - a_1)^2 - 2a_2(Z_2 - a_2) - a_1(Z_2 - a_2)^2 \\ &\quad + (Z_1 - a_1)^3 + (Z_1 - a_1)(Z_2 - a_2)^2 \end{aligned}$$

If $v(c_j) \leqslant v(Z_j)$, $v(a_j) \leqslant v(Z_j)$, $v(D_j(P)) + v(Z_j) = v(P)$, $v(D_j(P)) = v(D_j(P)(a_1, \ldots, a_m))$ for all j, a cumbersome verification shows that:

$$v(R(a_1 + c_1, \ldots, a_m + c_m)) \leqslant v(c_j) + v(D_j(P)) = v(c_j) + v(D_j(P)(a_1, \ldots, a_m))$$

Moreover, if $v(c_j) < v(Z_j)$ for all j,

$$v(R(a_1 + c_1, \ldots, a_m + c_m)) < v(c_j) + v(D_j(P)(a_1, \ldots, a_m))$$

so that if

$$v(P(a_1 + c_1, \ldots, a_m + c_m)) < v(P(a_1, \ldots, a_m)),$$

one has

$$v(P(a_1, \ldots, a_m)) = v(\Sigma\, c_j D_j(P)(a_1, \ldots, a_m))$$

and

$$\text{in}(P(a_1, \ldots, a_m)) + \text{in}(\Sigma\, c_j D_j(P)(a_1, \ldots, a_m)) = 0.$$

4. Constructive Lifting Method

4.1. STATEMENT OF THE PROBLEM

Let $\underline{A} := (A, \Gamma, \bar{V}) = (A, \Gamma, G, v, \text{in})$ be a graded structure and let $\underline{A}[Z_1, \ldots, Z_m] := (A[Z_1, \ldots, Z_m], \Gamma, \bar{U}) = (A[Z_1, \ldots, Z_m], \Gamma, G[Z_1, \ldots, Z_m], v, \text{in})$ be the induced graded structure on $A[Z_1, \ldots, Z_m]$, with $v(Z_i) = \delta_i$. We intend to solve the following problem:

Given $P \in A[Z_1, \ldots, Z_m]$, to find $f_1, \ldots, f_m \in A^\wedge$ s.t.

(i) $P(f_1, \ldots, f_m) = 0$
(ii) $v(f_i) \leqslant v(Z_i)$ for all i
(iii) there is i s.t. $v(f_i) = v(Z_i)$.

Such $f_1, \ldots, f_m$ will be called *a solution over* $A^\wedge$ *of the polynomial equation* $P(z_1, \ldots, z_m) = 0$.

Remark that we are distinguishing between variables Z_i in $A[Z_1, \ldots, Z_m]$ and unknowns z_i in $A^\wedge$.

Conditions (ii) and (iii) mean that we are looking for solutions $f_1, \ldots, f_m$ which are "bounded in valuation". This is not only a technical restriction in order to use the graded structure machinery. In fact, the solution required by the Buchberger algorithm (*cf. infra* 5.3 and 5.5) must satisfy this restrictions.

LEMMA 1. *Let* $f_1, \ldots, f_m \in A^\wedge$ *satisfy* (i), (ii), (iii). *Denote* $H_i := \text{in}(f_i)$ *if* $v(f_i) = v(Z_i)$, $H_i := 0$ *otherwise.*

Assume that:

($RC1$) *for all* j, $v(D_j(P)) + v(Z_j) = v(P)$

($RC2$) *for all* j, $\text{in}(D_j(P))(H_1, \ldots, H_m) \neq 0$.

Then for all i, *there is a Cauchy sequence* $(a_{ni}: n \in N)$ s.t. $f_i = \lim a_{ni}$ *and a decreasing sequence of elements of* Γ $v(P) = \gamma_1 > \cdots > \gamma_n > \gamma_{n+1} > \cdots$, s.t., *for all* n:

(1) *if* $f_j = 0$, *then* $a_{nj} = 0$
(2) $v(P(a_{n-11}, \ldots, a_{n-1m})) \leqslant \gamma_n$ (*where* $a_{oj} := 0$)
(3) *for all* j, $v(a_{nj} - f_j) < \gamma_n - v(D_j(P))$
(4) *if* $a_{nj} \neq a_{n-1j}$, *then* $v(a_{nj} - a_{n-1j}) = \gamma_n - v(D_j(P))$
(5) *there is* i s.t. $v(a_{ni}) = v(Z_i)$

PROOF. For $n = 1$ let $a_{1j} := \text{in}^*(H_j)$, $\gamma_1 := v(P)$. Then (1)–(5) are trivially verified. Assume $a_{n1}, \ldots, a_{nm}, \gamma_n$ have been produced satisfying (1)–(5) and let $g_j := f_j - a_{nj}$, $\gamma_{n+1} := \max\{v(g_j) + v(D_j(P))\}$. Then, for some j:

$$\gamma_{n+1} = v(g_j) + v(D_j(P)) = v(f_j - a_{nj}) + v(D_j(P)) < \gamma_n$$

Define $J := \{j: v(g_j) + v(D_j(P)) = \gamma_{n+1}\}$. Let $h_j := \text{in}(g_j)$ if $j \in J$, $h_j := 0$ otherwise, $c_j := \text{in}^*(h_j)$ and $a_{n+1j} := a_{nj} + c_j$.

(1) again is trivial. One has:

$$0 = P(f_1, \ldots, f_m) = P(a_{n1}, \ldots, a_{nm}) + \sum_{j=1,m} g_j D_j(P)(a_{n1}, \ldots, a_{nm}) + R(a_{n1} + g_1, \ldots, a_{nm} + g_m)$$

Since $v(g_j) < v(Z_j)$ (by (3)), $v(D_j(P)) + v(Z_j) = v(P)$ (by (RC1)) and $v(D_j(P)(a_{n1}, \ldots, a_{nm})) = v(D_j(P))$ (by (RC2)) one has $v(R(a_{n1} + g_1, \ldots, a_{nm} + g_m)) < v(g_j) + v(D_j(P)) \leqslant \gamma_{n+1}$ for all j (*cf.* 3.3); so $v(P(a_{n1}, \ldots, a_{nm})) \leqslant \gamma_{n+1}$ and (2) is satisfied. Either $h_j = 0$, in which case $a_{n+1j} - f_j = a_{nj} - f_j = g_j$ and $v(g_j) \leqslant \gamma_{n+1} - v(D_j(P))$; or $h_i = \text{in}(g_j) = \text{in}(c_j)$ and $a_{n+1j} - f_j = c_j - g_j$, so $v(a_{n+1j} - f_j) < v(g_j) = \gamma_{n+1} - v(D_j(P))$. This proves (3).

Also, if $a_{nj} \neq a_{n+1j}$, i.e. $h_i \neq 0$, then $v(c_i) = \gamma_{n+1} - v(D_j(P))$, which proves (4).

Also if $v(a_{ni}) = v(Z_i)$ then $v(a_{n+1i}) = v(a_{ni})$, proving (5). Finally (3) and (4) imply immediately that (a_{ni}) is a Cauchy sequence whose limit is f_i.

Proposition 1. *Let* $f_1, \ldots, f_m \in A^\wedge$ *satisfy (i), (ii), (iii). Denote* $H_i := \text{in}(f_i)$ *if* $v(f_i) = v(Z_i)$, $H_i := 0$ *otherwise. Assume that*:

(*RC*1) *for all* j, $v(D_j(P)) + v(Z_j) = v(P)$

(*RC*2) *for all* j, $\text{in}(D_j(P))\,(H_1, \ldots, H_m) \neq 0$.

Then for all i, *there is a Cauchy sequence* $(a_{ni}: n \in N)$ s.t. $f_i = \lim a_{ni}$, *and a decreasing sequence of elements of* Γ $v(P) = \gamma_1 > \cdots > \gamma_n > \gamma_{n+1} > \cdots$, s.t. $(a_{ni}: n \in N)$ *satisfy the following properties*:

(*A*) *either* $v(a_{1i}) = v(Z_i)$ *or* $a_{1i} = 0$
(*B*) *for some* i, $a_{1i} \neq 0$
(*C*) $v(P(a_{n1}, \ldots, a_{nm}) < \gamma_n$ *for all* n
(*D*) *if* $a_{nj} \neq a_{n-1j}$, *then* $v(a_{nj} - a_{n-1j}) = \gamma_n - v(D_j(P))$.

Proof. A corollary of Lemma 1.

The following proposition 2 is a sufficient condition in order to guarantee that sequences $(a_{nj}: n \in N)$, $j = 1 \ldots m$ are Cauchy sequences whose limits $f_1, \ldots, f_m$ satisfy (i), (ii), (iii). However it is not a necessary one, since we require $v(P(a_{n-11}, \ldots, a_{n-1m})) = \gamma_n$ instead than $v(P(a_{n-11}, \ldots, a_{n-1m})) \leqslant \gamma_n$.

Lemma 2. *For* $i = 1 \ldots m$, *let* $(a_{ni}: n \in N)$ *be a sequence of elements of* A.
Denoting $\gamma_1 := v(P)$, $\gamma_{n+1} := v(P(a_{n1}, \ldots, a_{nm}))$, *assume that*:

(*A*) *either* $v(a_{1i}) = v(Z_i)$ *or* $a_{1i} = 0$
(*B*) *for some* i, $a_{1i} \neq 0$
(*C*) $v(P(a_{n1}, \ldots, a_{nm})) < \gamma_n$ *for all* n
(*D*) *if* $a_{nj} \neq a_{n-1j}$, *then* $v((a_{nj} - a_{n-1j}) = \gamma_n - v(D_j(P))$.

Then the following conditions hold:

(1) *for all* i, n *if* $a_{ni} \neq 0$, *then* $v(a_{ni}) \gtrsim \gamma_n - v(D_i(P))$ *and* $\text{in}(a_{ni}) = \text{in}(a_{n+1i})$

(2) *for all* i, n $v(a_{ni}) \leqslant v(Z_i)$
(3) *there is* i *s.t. for all* n, $v(a_{ni}) = v(Z_i)$

PROOF. (1) Let l be the least index s.t. $a_{lj} \neq 0$; then $v(a_{lj}) = \gamma_l - v(D_j(P))$ and $v(a_{l+1j} - a_{lj}) = \gamma_{l+1} - v(D_j(P)) < \gamma_l - v(D_j(P)) = v(a_{lj}$, so $\text{in}(a_{lj}) = \text{in}(a_{l+1j})$. So assume we have proven (1) for n, and $a_{nj} \neq a_{n+1j}$ (otherwise there is nothing to prove). Then $v(a_{n+1j} - a_{nj}) = \gamma_{n+1} - v(D_j(P)) < \gamma_n - v(D_j(P)) \leqslant v(a_{nj})$, so $\text{in}(a_{nj}) = \text{in}(a_{n+1j})$.

(2) and (3) are trivial since, if k is the least index s.t. $a_{kj} \neq 0$, then for all $n \geqslant k$:

$$v(a_{nj}) = \gamma_k - v(D_j(P)) \leqslant v(P) - v(D_j(P)) = v(Z_j)$$

PROPOSITION 2. *Under the assumptions of Lemma* 2, *denote* $H_i := \text{in}(a_{1i})$ *if* $v(a_{1i}) = v(Z_i)$, $H_i := 0$ *otherwise. Assume that*:

($RC1$) *for all* j, $v(D_j(P)) + v(Z_j) = v(P)$

($RC2$) *for all* j, $\text{in}(D_j(P))(H_1, \ldots, H_m) \neq 0$.

Then for all i, $(a_{ni}: n \in N)$ *is a Cauchy sequence*; *if* f_i *denotes its limit, then* $f_1, \ldots, f_m$ *satisfy* (i), (ii), (iii).

PROOF. (a_{ni}) is trivially a Cauchy sequence because of assumptions (C) and (D). Let $f_i < A^\wedge$ denote its limit. Remark that, because of conditions (D) and (1) of Lemma 2, for sufficiently large n, $\text{in}(f_i) = \text{in}(a_{ni})$, unless $f_i = 0$ (in which case $a_{ni} = 0$ for all n, because of (1)); also, because of (B) and (2), if n is sufficiently large, $v(f_i - a_{ni}) \leqslant \gamma_{n+1} - v(D_i(P)) < v(a_{ni}) \leqslant v(Z_i)$. Fix a sufficiently large n and let $g_i := f_i - a_{ni}$. One has:

$$P(f_1, \ldots, f_m) = P(a_{n1}, \ldots, a_{nm}) + \sum_{j=1,m} g_j D_j(P)(a_{n1}, \ldots, a_{nm}) + R(a_{n1} + g_1, \ldots, a_{nm} + g_m)$$

Since:

$$v(g_j) < v(Z_j) \text{ (by the remark above)},$$

$$v(D_j(P)) + v(Z_j) = v(P) \quad \text{(by } (RC1))$$

and

$$v(D_j(P))(a_{n1}, \ldots, a_{nm}) = v(D_j(P)) \quad \text{(by } (RC2)),$$

one has

$$v(R(a_{n1} + g_1, \ldots, a_{nm} + g_m)) < v(g_j) + v(D_j(P)) \leqslant \gamma_{n+1} \quad \text{for all } j.$$

Also, $v(P(a_{n1}, \ldots, a_{nm})) < \gamma_n$, and $v(g_j D_j(P)(a_{n1}, \ldots, a_{nm})) \leqslant \gamma_{n+1}$. This implies that, for sufficiently large n, $v(P(f_1, \ldots, f_m)) < \gamma_n$, so that $P(f_1, \ldots, f_m) = 0$, proving (i). (ii) and (iii) are, then, trivial consequences of (2) and (3).

4.2. APPROXIMATE SOLUTIONS OF THE PROBLEM

The results of the previous paragraph provide the motivation for the following:

DEFINITION. Given $P \in A[Z_1, \ldots, Z_m]$ as in 4.1 and $\gamma \in \Gamma_0$, we say that $(a_{n1}: n \leqslant N), \ldots, (a_{nm}: n \leqslant N)$ is a *γ-approximate solution* of the polynomial equation $P(z_1, \ldots, z_m) = 0$ iff, denoting $\gamma_1 := v(P)$, $\gamma_{n+1} := v(P(a_{ni}, \ldots, a_{nm}))$ for all $n \leqslant N$, one has:

(A) either $v(a_{1i}) = v(Z_i)$ or $a_{1i} = 0$

(B) for some i, $a_{1i} \neq 0$
(C) $v(P(a_{n1}, \ldots, a_{nm})) < \gamma_n$ for all $n \leqslant N$
(D) if $a_{nj} \neq a_{n-1j}$, then $v(a_{nj} - a_{n-1j}) = \gamma_n - v(D_j(P))$, for all $n \leqslant N$

and moreover:

(E) $\gamma_{N+1} \leqslant \gamma$.

Given $P \in A[Z_1, \ldots, Z_m]$ and $\gamma \in \Gamma_0$, to find a γ-approximate solution of the polynomial equation $P(z_1, \ldots, z_m) = 0$, it is clearly sufficient:

(1) to find an *initial approximation*: i.e. to find $a_{11}, \ldots, a_{1m} \in A$ s.t.
 (*A.I*) either $v(a_{1i}) = v(Z_i)$ or $a_{1i} = 0$
 (*B.I*) for some i, $a_{1i} \neq 0$
 (*C.I*) $v(P(a_{11}, \ldots, a_{1m}) < v(P)$
(2) given a γ'-approximate solution $(a_{n1}: n \leqslant N), \ldots, (a_n m: n \leqslant N)$, to *refine it*, i.e. to find $a_{N+11}, \ldots, a_{N+1m}$, s.t.
 (*C.R*) $v(P(a_{N+11}, \ldots, a_{N+1m})) < v(P(a_{N1}, \ldots, a_{Nm}))$
 (*D.R.*) *if* $a_{N+1j} \neq a_{Nj}$, then $v(a_{N+1j} - a_{Nj}) = v(P(a_{N1}, \ldots, a_{Nm})) - v(D_j(P))$.

The next paragraphs are devoted to present techniques to solve both these problems, and then to show how to apply them to produce a lifting algorithm to compute an approximate solution of a polynomial equation over A. The aim of such an algorithm is obviously to find the initial segment of a Cauchy sequence that approximates an actual solution of the polynomial equation over $A^\wedge$.

4.3. EFFECTIVENESS OF THE LIFTING TECHNIQUE

Before presenting such an algorithm, it is therefore fair to stress that it falls short of its aim, due to the unavoidable weakness of the given definition of approximate solutions, because of the following reasons:

(1) assume an initial approximation $a_{11}, \ldots, a_{1m}$ has been obtained. Denote $H_i := \text{in}(a_{1i})$ if $a_{1i} \neq 0$ (i.e. $v(a_{1i}) = v(Z_i)$), $H_i := 0$, otherwise. Unless both conditions

 (*RC*1) for all j, $v(D_j(P)) + v(Z_j) = v(P)$

 (*RC*2) for all j, $\text{in}(D_j(P))\ (H_1, \ldots, H_m) \neq 0$

 are satisfied, our technique, which is based on propositions 1 and 2 of section 4.1, cannot be applied.
(2) let $f_1, \ldots, f_m$ be a solution over $A^\wedge$ of the polynomial equation $P(z_1, \ldots, z_m) = 0$, and let (a_{ni}) be the Cauchy sequences, γ_i be the elements in Γ_0, whose existence is implied by Proposition 1 of Section 4.1. Assume that $\gamma_n = v(P(a_{n-11}, \ldots, a_{n-1m}))$ for $n \leqslant N$ but $\gamma_{N+1} > v(P(a_{N1}, \ldots, a_{Nm}))$. Then $(a_{ni}: n \leqslant N)$, $i = 1 \ldots m$, is an approximate solution; however $(a_{ni}: n \leqslant N+1)$, $i = 1 \ldots m$, is not and our algorithm is unable to compute it.
(3) it is possible that, while there are $a_1, \ldots, a_m \in A$ s.t. $v(P(a_1, \ldots, a_m)) < \gamma$, there are however no $f_1, \ldots, f_m \in A^\wedge$ s.t. $P(f_1, \ldots, f_m) = 0$ and $\text{in}(a_i) = \text{in}(f_i)$. This means that a γ-approximate solution could be found which is not the initial segment of a Cauchy sequence approximating a solution over $A^\wedge$.

(4) in particular the lifting algorithm could produce approximate solutions also in case the polynomial equation has no solution over $A^\wedge$.

(5) our technique is based on the solution of one polynomial equation and of several linear systems over some G_γ. Such equations and systems could have non-unique (and possibly infinitely many) solutions, some of which produce approximate solutions which are initial segments of Cauchy sequences approximating a solution over $A^\wedge$, while others produce approximate solutions which are *not* initial segments of Cauchy sequences approximating a solution over $A^\wedge$. In general it is impossible to predict which are the solutions to be chosen.

In order to guarantee that the solutions chosen at each step are actually approximating the desired solutions in $A^\wedge$, theoretical informations related to existence and uniqueness of solutions, and to vincula to their "shape" (such as degree bounds for polynomials) are needed to be plugged into the considered equations.

Also, such informations could help to infer the existence of a solution over $A^\wedge$ from the existence of approximate solutions.

This will clearly appear in the examples we will discuss in the next section.

It is to be remarked also that the choice of in* plays an important role in this technique.

In fact, solutions in A, obtained as approximated by definitely constant Cauchy sequences (i.e. being found in finitely many steps) for a given choice of in*, could as well not satisfy the same condition for a different choice of in*.

In other words, an unfortunate choice of in* could imply that some elements of A cannot be approximated by definitely constant Cauchy sequences and so cannot be found in finitely many steps.

For instance, in example 1 of section 2.4, negative elements in $\mathbb{Z}$ are approximated by definitely constant Cauchy sequences if choice (2) is made for in*, but they are not, if one makes choice (1).

4.4. FINDING AN INITIAL APPROXIMATION

Let $\underline{A}$, $\underline{A}[Z_1, \ldots, Z_m]$, P be as in 4.1.

LEMMA. *If the equation*

$$(*)\ \mathrm{in}(P)(z_1, \ldots, z_m) = 0,$$

$$z_i \in G,\ z_i \textit{ homogeneous},\ \deg(z_i) = v(Z_i)$$

has no non-trivial solution, then there are no $f_1, \ldots, f_m \in A^\wedge$ *satisfying* (*i*), (*ii*), (*iii*).

Conversely if (*) *has a non-trivial solution* $(H_1, \ldots, H_m)$ *define* $a_{1i} := \mathrm{in}^*(H_i)$ if $H_i \neq 0$, $a_{1i} := 0$ *otherwise. Then*:

(*A.I*) *either* $v(a_{1i}) = v(Z_i)$ *or* $a_{1i} = 0$

(*B.I*) *for some* i, $a_{1i} \neq 0$

(*C.I*) $v(P(a_{11}, \ldots, a_{1m})) < v(P)$

PROOF. If $f_1, \ldots, f_m \in A^\wedge$ satisfy (i), (ii), (iii), let $H_i := \mathrm{in}(f_i)$ if $v(f_i) = v(Z_i)$, $H_i := 0$ otherwise. Since $P(f_1, \ldots, f_m) = 0$ then necessarily $\mathrm{in}(P)(H_1, \ldots, H_m) = 0$; so (*) has a non-trivial solution.

The converse is trivial.

4.5. REFINING A GIVEN APPROXIMATION

Assume then $(a_{n1}: n \leqslant N), \ldots, (a_{nm}: n \leqslant N)$ is a *γ-approximate solution* of the polynomial equation $P(z_1, \ldots, z_m) = 0$ where $\gamma := v(P(a_{N1}, \ldots, a_{Nm}))$. Denoting $H_i := \text{in}(a_{1i})$ if $a_{1i} \neq 0$, $H_i := 0$, otherwise, assume also:

$(RC1)$ for all j, $v(D_j(P)) + v(Z_j) = v(P)$

$(RC2)$ for all j, $\text{in}(D_j(P))\,(H_1, \ldots, H_m) \neq 0$

LEMMA 1. *Under the assumptions above, if $h_1, \ldots, h_m$ is a non-trivial solution of the equation*

$$(**) \qquad \sum_{j=1,m} z_j(\text{in}(D_j(P)))(H_1, \ldots, H_m) = -\text{in}(P(a_{N1}, \ldots, a_{Nm})),$$

$$z_j \in G,\ z_j \textit{ homogeneous},\ \deg(z_j) + v(D_j(P)) = \gamma$$

let $c_j := \text{in}^*(h_j)$ if $h_j \neq 0$, $c_j := 0$ otherwise and let $a_{N+1j} := a_{Nj} + c_j$ for all j. Then:

$(C.R)$ $v(P(a_{N+11}, \ldots, a_{N+1m})) < v(P(a_{N1}, \ldots, a_{Nm}))$

$(D.R.)$ *if $a_{N+1j} \neq a_{Nj}$, then* $v(a_{N+1j} - a_{Nj}) = v(P(a_{N1}, \ldots, a_{Nm})) - v(D_j(P))$.

PROOF. Because of the assumptions on h_j, either $c_j = 0$ or $v(c_j) + v(D_j(P)) = \gamma$; therefore $(D.R.)$ is satisfied. One has:

$$P(a_{N+11}, \ldots, a_{N+1m}) = P(a_{N1}, \ldots, a_{Nm}) + \sum_{j=1,m} c_j D_j(P)(a_{N1}, \ldots, a_{Nm}) + R(a_{N1} + c_1, \ldots, a_{Nm} + c_m)$$

Since, because of $(RC2)$, $v(R(a_{N1} + c_1, \ldots, a_{Nm} + c_m)) < v(c_j) + v(D_j(P))$ for all j, and, because of $(RC1)$ and the assumptions on h_j, $v(c_j) + v(D_j(P)) = \gamma$ for at least one j, we have $v(R(a_{N1} + c_1, \ldots, a_{Nm} + c_m)) < \gamma$ and if $J := \{j: h_j \neq 0\}$ also:

$$\text{in}\left(\sum_{j=1,m} c_j D_j(P)(a_{N1}, \ldots, a_{Nm})\right) = \sum_{j \in J} \text{in}(c_j)\ \text{in}(D_j(P)(a_{N1}, \ldots, a_{Nm}))$$

$$= \sum_{j=1,m} h_j(\text{in}(D_j(P)))(H_1, \ldots, H_m)$$

$$= -\text{in}(P(a_{N1}, \ldots, a_{Nm}))$$

So $v(P(a_{N+11}, \ldots, a_{N+1m})) < \gamma$ and $(C.R.)$ is satisfied.

4.6. A LIFTING ALGORITHM

Let $\underline{A}$, $\underline{A}[Z_1, \ldots, Z_m]$, P be as in 4.1. Let $\gamma \in \Gamma_0$, $\gamma \leqslant v(P)$. On the basis of the preceeding results, we sketch here an algorithm which produces a γ-approximate solution of the equation $P(z_1, \ldots, z_m) = 0$, unless it fails because of the reasons discussed in Section 4.3.

If there are no $H_1, \ldots, H_m$, $H_i \in G$, H_i homogeneous, $\deg(H_i) = v(Z_i)$,
not all of them zero, s.t. in $(P)(H_1, \ldots, H_m) = 0$ **then**
return "the problem has no solution"

else
  compute $H_1, \ldots, H_m$, $H_i \in G$, H_i homogeneous, $\deg(H_i) = v(Z_i)$,
    not all of them zero, s.t. $\text{in}(P)(H_1, \ldots, H_m) = 0$
  if $\text{in}(D_j(P))(H_1, \ldots, H_m) = 0$ **or** $v(D_j(P)) + v(Z_j) < v(P)$ **then**
    return "regularity conditions are not satisfied"
  else
    $n := 1$
    for $i = 1 \ldots m$ **do**
      a_{1i}: $\text{in}^*(H_i)$
    while $v(P(a_{n1}, \ldots, a_{nm})) \geqslant \gamma$ **do**
      $\delta := v(P(a_{n1}, \ldots, a_{nm}))$
      if there are no $h_1, \ldots, h_m$, $h_i \in G$, h_i homogeneous, $\deg(h_i) = \delta - v(D_j(P))$, not all of
          them zero, s.t. $\sum_{j=1,m} h_j\,(\text{in}(D_j(P)))(H_1, \ldots, H_m) = -\text{in}(P(a_{n1}, \ldots, a_{nm}))$ **then**
        return "there is no refinement of the approximate solution
          $((a_{i1}: i = 1 \ldots n), \ldots, (a_{im}: i = 1 \ldots n))$"
      else
        compute $h_1, \ldots, h_m$, $h_i \in G$, h_i homogeneous, $\deg(h_i) = \delta - v(D_j(P))$, not all of
            them zero, s.t. $\sum_{j=1,m} h_j\,(\text{in}(D_j(P)))(H_1, \ldots, H_m) = -\text{in}(P(a_{n1}, \ldots, a_{nm}))$
        for $i = 1 \ldots m$ **do**
          $a_{n+1i} := a_{ni} + \text{in}^*(h_i)$
        $n := n + 1$
    return $((a_{i1}: i = 1 \ldots n), \ldots, (a_{im}: i = 1 \ldots n))$

4.7. A LIFTING THEOREM

In contrast with the negative results of Section 4.3, the following result gives sufficient conditions to guarantee the existence of solutions to a polynomial equation over $A^\wedge$ and to assure that the lifting algorithm discussed above will produce initial segments of a Cauchy sequence approximating such a solution.

THEOREM. *Let* $\underline{A} := (A, \Gamma, \bar{V}) = (A, \Gamma, G, v, \text{in})$ *be a graded structure and let* $A[Z_1, \ldots, Z_m] := (A[Z_1, \ldots, Z_m], \Gamma, \bar{U}) = (A[Z_1, \ldots, Z_m], \Gamma, G[Z_1, \ldots, Z_m], v, \text{in})$ *be the induced graded structure on* $A[Z_1, \ldots, Z_m]$, *with* $v(Z_i) = \delta_i$. *Let* $P \in A[Z_1, \ldots, Z_m]$. *Let* $H_1, \ldots, H_m \in G$ *be* s.t. $\deg(H_i) = v(Z_i)$, *not all of them being zero. If*:

(*RC*1) *for all j*, $v(D_j(P)) + v(Z_j) = v(P)$

(*RC*2) *for all j*, $\text{in}(D_j(P))(H_1, \ldots, H_m) \neq 0$

(*RC*3) *for all* $\gamma < v(P)$, *denoting* $\gamma(i) := \gamma - v(D_i(P))$:

$$\text{in}(D_1(P))(H_1, \ldots, H_m)G_{\gamma(1)} \oplus \cdots \oplus \text{in}(D_m(P))(H_1, \ldots, H_m)G_{\gamma(m)} = G_\gamma$$

then:

(1) *if* $(a_{n1}: n \leqslant N), \ldots, (a_{nm}: n \leqslant N)$ *is a* γ *approximate solution of the polynomial equation* $P(z_1, \ldots, z_m) = 0$, s.t. $a_{1i} = 0$ *iff* $H_i = 0$, *and* $\text{in}(a_{1i}) = \text{in}(H_i)$ *otherwise, then there are Cauchy sequences* $(b_{n1}: n \in N), \ldots, (b_{nm}: n \in N)$, s.t. *for all* $i = 1 \ldots m$, *for all* $n \leqslant N$, $b_{ni} = a_{ni}$, *and which satisfy the assumptions of* 4.1. *Lemma* 1, *so that their limits are a solution over* $A^\wedge$ *of the polynomial equation* $P(z_1, \ldots, z_m) = 0$.
(2) *there is a solution* $f_1, \ldots, f_m$ *over* $A^\wedge$ *of the polynomial equation* $P(z_1, \ldots, z_m) = 0$, s.t. $v(f_i) < v(Z_i)$ *iff* $h_i = 0$, *and* $\text{in}(f_i) = H_i$ *otherwise.*

PROOF. $(RC3)$ implies that, for all $\gamma < v(P)$, for all $h \in G_\gamma - \{0\}$, there are $h_1, \ldots, h_m \in G$, homogeneous,

$$\deg(h^j) = \gamma - v(D_j(P)), \text{ s.t. } \sum_{j=1,m} h_j \operatorname{in}(D_j(P))(H_1, \ldots, H_m) = h.$$

This in turn implies that any approximate solution $(a_{n1}: n \leqslant N), \ldots, (a_{nm}: n \leqslant N)$, with either $a_{1i} = 0$ or $\operatorname{in}(a_{1i}) = \operatorname{in}(H_i)$ can be refined.

By induction, then there are sequences $(a_{n1}: n \in N), \ldots, (a_{nm}: n \in N)$ satisfying conditions (A), (B), (C), (D) of 4.1, Proposition 1. Both conclusions then follow immediately from 4.1, Proposition 2.

5. Two Classical Cases: Hensel and Gröbner

5.1. HENSEL–ZASSENHAUS LIFTING OF A POLYNOMIAL FACTORIZATION

Let $\underline{A}$ be the graded structure of example 2, i.e. $\underline{A} := (\mathbb{Z}[X], \mathbb{Z}, \{(p^n)\}) = (\mathbb{Z}[X], \mathbb{Z}, \mathbb{Z}_p[X, T], v, \operatorname{in})$. Let $\underline{A}[Z_1, Z_2]$ be given by $v(Z_1) := V(Z_2) := 0$. Let $a \in \mathbb{Z}[X]$ s.t. $v(a) = 0$. Remark that, denoting by c the leading coefficient of a, $v(c) = 0$, i.e. c is not a multiple of p.

The aim of the Berlekamp–Hensel–Zassenhaus factorization algorithm is to find polynomials f_1, f_2 over the p-adic completion of $\mathbb{Z}$ s.t. $a = f_1 f_2$.

So the polynomial equation to be solved is $P(z_1, z_2) = 0$ where $P := Z_1 Z_2 - a$.

Since we have $\operatorname{in}(P) = Z_1 Z_2 - \operatorname{in}(a) = Z_1 Z_2 - \operatorname{res}(a)$, where $\operatorname{res}(b)$ is the residue class of $b \in \mathbb{Z}[X] \bmod p$, the initial approximation is obtained by solving the homogeneous equation (in $G_0 = \mathbb{Z}_p[X]$) $z_1 z_2 - \operatorname{res}(a) = 0$, i.e. by finding H_1, H_2 in $\mathbb{Z}_p[X]$ s.t. $H_1 H_2 = \operatorname{res}(a)$ (this is classically done by means of the Berlekamp algorithm) and then choosing $a_{11}, a_{12} \in \mathbb{Z}[X]$ s.t. $\operatorname{res}(a_{1i}) = \operatorname{in}(a_{1i}) = H_i$.

Since $D_1(P) = Z_2$, $D_2(P) = Z_1$, regularity conditions $(RC1)$ and $(RC2)$ are obviously satisfied.

Assume moreover that $GCD(H_1, H_2) = 1$; this implies that also condition $(RC3)$ holds.

The theorem in section 4.7 then guarantees that solutions of the polynomial equation $P(z_1, z_2) = 0$ exist over $A^\wedge$ and that the lifting algorithm of section 4.6 allows to compute initial segments of a Cauchy sequence approximating such a solution.

However, this result is weaker than Hensel Lemma and the Hensel–Zassenhaus Algorithm, since $A^\wedge$ is a ring that strictly contains $\mathbb{Z}^\wedge[X]$, the polynomial ring over the p-adic completion of $\mathbb{Z}$ (for instance $\sum_{i=0,\infty} p^i X^i$ is in the former, but not in the latter, ring).

To be sure, therefore, that the approximated solution f_1, f_2 is s.t. $f_i \in \mathbb{Z}^\wedge[X]$, further assumptions are required.

However it is easy to prove the following facts, which allow to recover the original result of Hensel Lemma (actually, the reader can easily verify that the classical constructive proof of Hensel Lemma, and the one which can be extracted by the theorem in section 4.7 and the remarks below, are essentially the same):

(1) $\deg_X(H_1) + \deg_X(H_2) = \deg_X(a)$
(2) given $h \in G_\gamma$, there is a s.t. $\operatorname{in}(a) = h$ and $\deg_X(a) = \deg_X(h)$
(3) if $h \in G_{-n}$ for some n, with $\deg_X(h) < \deg_X(a)$, there are $h_1, h_2 \in G_{-n}$, s.t. $H_2 h_1 + H_1 h_2 = h$ and $\deg_X(h_i) < \deg_X(H_i)$

(4) if $(a_n: n \in N)$ is a Cauchy sequence in $\underline{A}$ s.t. for some D, $\deg_X(a_n) < D$ for all n, and f is its limit in $\underline{A}^\wedge$, then $f \in \mathbb{Z}^\wedge[X]$.

Therefore, if one chooses an initial approximation a_{11}, a_{12} with $\deg_X(a_{1i}) = \deg_X(H_i)$, and, inductively, one refines an approximate solution $((a_{1n}: n \leqslant N), (a_{1n}: n \leqslant N))$ with $\deg_X(a_{ni}) = \deg_X(H_i)$ by computing a_{1N+1}, a_{2N+1} s.t. they also satisfy $\deg_X(a_{Ni} - a_{N+1i}) < \deg_X(H_i)$, the Cauchy sequence so obtained has its limit in $\mathbb{Z}^\wedge[X]$, as required.

5.2. OTHER HENSEL LIFTING CONSTRUCTIONS

The use of Hensel lifting techniques were proposed by Yun (1974; 1976) to solve a variety of polynomial equations in $\mathbb{Z}[X]$ (for instance $z_1^n = a$, $b_1 z_1 + b_2 z_2 = b$, etc.).

Yun's approach is essentially a specialization of our method to the case of $\mathbb{Z}[X]$ with the p-adic topology. However, the regularity conditions we require can help to understand the possible limits of this approach: for instance $(RC1)$ is not satisfied if $P(Z_1) = Z_1^n - a$, with n multiple of p, since then $D_1(P) = nZ_1^{n-1}$ and $v(n) \leqslant -1$, so $v(D_1(P)) + v(Z_1) \leqslant v(P) - 1$.

One can verify that such an equation cannot be straightforwardly solved by Yun's method, if p is given.

$(RC2)$ is not satisfied, for instance, if $P(Z_1) = Z_1^2 - 2aZ_1 + a^2$ with $v(a) = v(Z_1) = 0$, for then the initial approximation is got solving $z_1^2 - 2\,\mathrm{res}(a) z_1 + \mathrm{res}(a)^2 = 0$, whose solution $H_1 := \mathrm{res}(a)$ obviously satisfies $\mathrm{in}(D_1(P)) = 2Z_1 - 2\,\mathrm{res}(a) = 0$.

Also, the fact that solutions are looked for in $\mathbb{Z}[X]^\wedge$, which strictly contains $\mathbb{Z}^\wedge[X]$, means that additional requirements on the shape of a solution must be added if solutions are required to be in $\mathbb{Z}^\wedge[X]$ or even in $\mathbb{Z}[X]$.

For instance, if $a, b \in \mathbb{Z}[X]$, and $GCD(a, b) = 1 - pX$, the equation $aZ_1 + bZ_2 = 1$ has obviously no solution neither in $\mathbb{Z}^\wedge[X]$ nor in $\mathbb{Z}[X]$; however if $c_1, c_2 \in \mathbb{Z}[X]$ are s.t. $c_1 a + c_2 b = 1 - pX$, then

$$a_1 := c_1/(1 - pX) = c_1 \sum_{i=0,\infty} p^i X^i \quad \text{and} \quad a_2 := c_2/(1 - pX) = c_2 \sum_{i=0,\infty} p^i X^i$$

are s.t. $aa_1 + ba_2 = 1$ in $\mathbb{Z}[X]^\wedge$.

In this example, both our and Yun's construction will give a γ-approximate solution, which is the initial segment of a Cauchy sequence approximating such a solution, unless one imposes a degree bound on the solution, since for n sufficiently large, no $(-n)$-approximate solution exists satisfying such a degree bound.

5.3. GRÖBNER REDUCTION

Let $\underline{A}$ be the graded structure of example 6, i.e.

$$\underline{A} := (k[X_1, \ldots, X_n], \mathbb{Z}^n, \{V_m: m \in Z^n\}) = (k[X_1, \ldots, X_n], \mathbb{Z}^n, k[X_1, \ldots, X_n], v, \mathrm{in}),$$

where $\mathbb{Z}^n$ is endowed with a positive ordering, which implies $\underline{A}^\wedge = \underline{A}$.

Let I be an ideal in $k[X_1, \ldots, X_n]$, $(c_1, \ldots, c_r)$ a Gröbner basis of I, i.e. a basis s.t. $\{\mathrm{in}(c_1), \ldots, \mathrm{in}(c_r)\}$ generates the ideal $\mathbb{N}(I) := (\mathrm{in}(c): c \in I)$. It is known that the following conditions are equivalent:

(i) $(c_1, \ldots, c_r)$ is a Gröbner basis of I.
(ii) $c \in I$ iff there are $f_1, \ldots, f_r \in k[X_1, \ldots, X_n]$ s.t. $c = \Sigma\, c_i f_i$ and $v(c) \geqslant v(f_i) v(c_i)$ for all i.

To find such a representation for a given c means to find solutions of the equation $P(z_1, \ldots, z_r) = 0$ with $P := \Sigma c_i Z_i - c \in \underline{A}[Z_1, \ldots, Z_r]$, where $\underline{A}[Z_1, \ldots, Z_r]$ denotes the graded structure induced by $\underline{A}$, assigning $v(Z_i) := v(c_i)$.

We remark explicitly that, because of (ii) above, $c \in I$ if and only if a solution of $P(z_1, \ldots, z_r) = 0$ over $A^\wedge = A$ exists. We have $\text{in}(P) = \Sigma \, \text{in}(c_i)Z_i - \text{in}(c)$, $D_i(P) = c_i$, so $(RC1)$ and $(RC2)$ are obviously satisfied.

The initial approximation is got by solving the equation $\Sigma \, \text{in}(c_i)z_i - \text{in}(c) = 0$ in $G_{\text{in}(c)}$; since $(c_1, \ldots, c_r)$ is a Gröbner basis of I, if no solution $(H_1, \ldots, H_r)$ exists, then $c \notin I$. Remark that in this example one can always choose a solution s.t. just one of the H_i's is not zero; this means that each f_i solving the equation except one is such that $v(f_i) < v(Z_i)$. Once an approximation $(a_1, \ldots, a_r)$ is obtained, it is refined by solving the equation $\Sigma \, \text{in}(c_i)z_i - \text{in}(d) = 0$ in $G_{v(d)}$, where $d := c - \Sigma \, c_i a_i = -P(a_1, \ldots, a_r)$.

Remark again that if this equation has no non-trivial solution, since $(c_1, \ldots, c_r)$ is a Gröbner basis of I, then $d \notin I$, and so $c \notin I$.

In this example therefore, theoretical knowledge about Gröbner bases, allows to rule out the possibility that the lifting technique fails.

5.4. STANDARD SET REDUCTION

As it is well known, if we relax the condition that $\mathbb{Z}^n$ is endowed with a positive ordering, while (ii) implies (i), the converse is false, and in the case the ordering on $\mathbb{Z}^n$ is compatible with the valuation induced by the $(X_1, \ldots, X_n)$-adic topology (e.g., when it is the inverse of the usual total-degree ordering), (i) is equivalent with a weaker version of (ii) in which the f_i's are required to be in $K[[X_1, \ldots, X_n]]$, which is the completion of $\underline{A}$ w.r.t. the $(X_i, \ldots, X_n)$-adic topology and also w.r.t. the topology induced by the filtration (V_m); this is consistent with the approach of section 4.

This result holds in general (Mora, 1987): if $\underline{A}$ is a graded structure and I is an ideal of A, for a set $(c_1, \ldots, c_r)$ (which is then called a standard set of I), the following two conditions are equivalent:

(i) $\{\text{in}(c_1), \ldots, \text{in}(c_r)\}$ generates $\text{in}(I) := (\text{in}(c) \colon c \in I)$
(ii) $c \in I^\wedge$ iff there are $f_1, \ldots, f_r \in A^\wedge$ s.t. $c = \Sigma \, c_i f_i$ and $v(c) \geqslant v(f_i) + v(c_i)$ for all i,

where $A^\wedge$ is the completion of A w.r.t. the topology induced by $\{V_\gamma\}$ and $I^\wedge$ is the closure of I in $A^\wedge$.

The proof as given in Mora (1987) is essentially the generalization of the construction we have just given above. If, however we try to apply our algorithm to this case, some failure conditions could be met, as shown by the following discussion.

As above, to find a representation as in (ii) for a given c, means to find solutions of the equation $P(z_1, \ldots, z_r) = 0$ with $P := \Sigma \, c_i Z_i - c \in \underline{A}[Z_1, \ldots, Z_r]$, where $\underline{A}[Z_1, \ldots, Z_r]$ denotes the graded structure induced by $\underline{A}$, assigning $v(Z_i) := v(c_i)$. Again, $(RC1)$ and $(RC2)$ are obviously satisfied.

The initial approximation is got by solving the equation $\Sigma \, \text{in}(c_i)z_i - \text{in}(c) = 0$ in $G_{v(c)}$; and again, since $(c_1, \ldots, c_r)$ is a standard set of I, if no solution $(H_1, \ldots, H_r)$ exists, then $c \notin I^\wedge$.

Once an approximation $(a_1, \ldots, a_r)$ is obtained, it is refined by solving the equation $\Sigma \, \text{in}(c_i)z_i - \text{in}(d) = 0$ in $G_{v(d)}$, where $d := c - \Sigma \, c_i a_i = -P(a_1, \ldots, a_r)$.

Remark again that if this equation has no non-trivial solution, since $(c_1, \ldots, c_r)$ is a Gröbner basis of I, then $d \notin I^\wedge$, and so $c \notin I^\wedge$. This means that, if it is impossible to refine any given approximation, one can conclude that no solution of the problem exists over $A^\wedge$; however, from the existence of a γ-approximation, for any given γ, one cannot conclude that a solution of the problem exists, i.e. that $c \in I^\wedge$.

5.5. CONSTRUCTIVE LIFTING OF SYZYGIES AND BUCHBERGER'S ALGORITHM

Let us consider here another equivalent definition of the concept of Gröbner basis, which provides an algebraic interpretation of Buchberger's algorithm (Buchberger 1965; 1985), and also clarifies the strict relation with the usual unconstructive lifting techniques in local algebra (e.g. Hironaka, 1964).

The following holds for an ideal I in $k[X_1, \ldots, X_n]$ (Zacharias, 1978; Möller, 1985):

> $(c_1, \ldots, c_r)$ is a Gröbner basis of I iff for each element $(h_1, \ldots, h_r)$ in a homogeneous basis of the G-module of syzygies of $(\mathrm{in}(c_1), \ldots, \mathrm{in}(c_r))$, there is an element $(d_1, \ldots, d_r)$ in the A-module of syzygies of $(c_1, \ldots, c_r)$ s.t. if $h_i \neq 0$ then $\mathrm{in}(d_i) = h_i$.

Again the same result generalizes to graded structures (Robbiano, 1986; Mora, 1987):

> $(c_1, \ldots, c_r)$ is a standard set of an ideal I in $\underline{A}$ iff for each element $(h_1, \ldots, h_r)$ in a homogeneous basis of the G-module of syzygies of $(\mathrm{in}(c_1), \ldots, \mathrm{in}(c_r))$, there is an element $(d_1, \ldots, d_r)$ in $(A^\wedge)^r$ s.t. $\Sigma\, d_i h_i = 0$ and if $h_i \neq 0$ then $\mathrm{in}(d_i) = h_i$.

Buchberger's algorithm can be interpreted as follows:

- —given a basis $(c_1, \ldots, c_r)$ of I, try to lift each element $(h_1, \ldots, h_r)$ in a homogeneous basis of the module of syzygies of $(\mathrm{in}(c_1), \ldots, \mathrm{in}(c_r))$, to an element $(d_1, \ldots, d_r)$ in the module of syzygies of $(c_1, \ldots, c_r)$ s.t. if $h_i \neq 0$ then $\mathrm{in}(d_i) = h_i$;
- —if the lifting is successful, then the basis is Gröbner;
- —if for some $(h_1, \ldots, h_r)$ in a homogeneous basis of the module of syzygies of $(\mathrm{in}(c_1), \ldots, \mathrm{in}(c_r))$, no element $(d_1, \ldots, d_r)$ in the module of syzygies of $(c_1, \ldots, c_r)$ exists, s.t. if $h_i \neq 0$ then $\mathrm{in}(d_i) = h_i$, then the basis is not Gröbner; then update the original basis, adding $\Sigma\, \mathrm{in}^*(h_i)c_i$, and restart the procedure.

To lift an element $(h_1, \ldots, h_r)$ in a homogeneous basis of the module of syzygies of $(\mathrm{in}(c_1), \ldots, \mathrm{in}(c_r))$, to an element $(d_1, \ldots, d_r)$ in the module of syzygies of $(c_1, \ldots, c_r)$ s.t. if $h_i \neq 0$ then $\mathrm{in}(d_i) = h_i$, means to solve the polynomial equation $P(z_1, \ldots, z_r) = 0$ with $P := \Sigma\, c_i Z_i \in \underline{A}[Z_1, \ldots, Z_r]$, where $\underline{A}[Z_1, \ldots, Z_r]$ denotes the graded structure induced by $\underline{A}$, assigning $v(Z_i) := v(c_i)$, by refining an initial approximation $\mathrm{in}^*(h_1), \ldots, \mathrm{in}^*(h_r)$.

The same discussion as in section 5.3 implies that either a solution over $A^\wedge = A$ is found after a finite number of steps, or some approximation cannot be refined, implying that $(h_1, \ldots, h_r)$ cannot be lifted.

If Γ_0 is not well-ordered, ad hoc techniques to compute a standard set have been proposed, which use properties of graded structures over a polynomial ring, which cannot be applied in general (Lazard, 1983; Mora, 1987). In the general case of standard sets for graded structures, the only known computational result is that the computation of a "truncated" standard set is possible by the technique of this paper (Mora, 1987), where a γ-truncated standard set is set $(c_1, \ldots, c_r)$, $c_i \in I$, s.t.

$$c \in I^\wedge \text{ iff there are } g \in I^\wedge \cap V_\gamma, f_1, \ldots, f_r \in A^\wedge \text{ s.t. } c - g = \sum c_i f_i$$
$$\text{and } v(c) \geqslant v(f_i) + v(c_i) \text{ for all } i, v(c) \geqslant v(g).$$

Both the computation of a truncated standard set and, given $c \in I^\wedge$ and a truncated standard set $(c_1, \ldots, c_r)$, of a representation of c, are fairly trivial generalizations of the approach discussed above.

References

Buchberger, B. (1965). *Ein Algorithmmus zum Auffinden der Basis-elemente des Restklassenringes nach einem null-dimensionalen Polynomideal*, Ph.D. Thesis, Innsbruck University.

Buchberger, B. (1985). Gröbner bases: an algorithmic method in polynomial ideal theory. In (Bose, N. J., ed.) *Recent Trends in Multidimensional System Theory*. Reidel.

Hensel, K. (1913). *Zahlentheorie*. Goschen.

Hermann, G. (1926). Die Frage die endlichen vielen Schritte in der Theorie der Polynomideale. *Math. Ann.* **95,** 736–788.

Hironaka, H. 1964. Resolution of singularities of an algebraic variety over a field of characteristic zero, *Ann. Math.* **79,** 109–326.

Lazard, D. Gröbner bases, Gaussian elimination, and resolutions of systems of algebraic equations. *Proc. EUROCAL '83 L.N.C.S.* **162,** 142–156.

Miola, A. & Yun, D. Y. Y. (1974). The computational aspects of Hensel-type univariate polynomial greatest common divisor algorithms. *Proc. EUROSAM '74*, 46–54.

Möller, H. M. (1987). A reduction strategy for the Taylor resolution, *Proc. EUROCAL '85, L.N.C.S.* **204,** 526–534.

Mora, T. (1987). *Seven variations on standard bases.*

Renschuch, B. (1976). *Elementäre und praktische Idealtheorie*. VEB Deutscher Verlag D. Wiss.

Robbiano, L. (1986). On the theory of graded structures, *J. Symb. Comp.* **2**.

Yun, D. Y. Y. (1974). *The Hensel Lemma in Algebraic Manipulation*, Ph.D Thesis, M.I.T.

Yun, D. Y. Y. (1976). Algebraic algorithms using *p*-adic constructions, *Proc. ACM Symsac '76*, 248–259.

Zacharias, G. (1978). Generalized Gröbner bases in commutative polynomial rings, Bachelor Th., M.I.T.

Zassenhaus, H. (1969). On Hensel factorization, I, *J. Number Theory* **1,** 269–311.

On the Computation of Generalized Standard Bases

MICHELA BRUNDU AND FABIO ROSSI

Dip. di Scienze Matematiche, Piazzale Europa 1, 34100 Trieste, Italy

In the context of graded structures, we give algorithms to compute generalized standard bases of a given module and of the associated modules of syzygies, generalizing known results and applying them to some computational problems

Introduction

The purpose of this paper is two-fold: first, to give a set up for various notions already present in the literature concerning computational algebra and algebraic geometry; secondly, to give applications of general facts to specific problems in this area.

In particular, in Section 0 we set notations and recall known results (see Robbiano, 1986; Buchberger, 1965), focussing on the notion of graded structure, introduced by Robbiano (1986), which generalizes and unifies different objects, like the polynomial ring as graded ring w.r.t. the usual degree or w.r.t. the monomial decomposition, the tangent cone to an algebraic variety, etc.

In Section 1 we introduce the concept of reduction procedure in a graded structure, giving some properties, relevant from a computational point of view and generalizing result of Buchberger (1965; 1976; 1985), Möller (1985), Mora & Möller (1986), concerning the polynomial case.

In Section 2, we exhibit an algorithm to calculate generalized standard bases (i.e. particular system of generators of graded structures that in the 'classical' cases become either Gröbner bases or Macaulay bases or standard bases, etc.). The notion of Gröbner basis for submodules of free modules on the polynomial ring was introduced in Buchberger (1965) and in Mora & Möller (1986), but it had a poor description until now. We introduce a suitable graded structure on them and, hence, a notion of base; in addition our algorithm gives a way to compute the modules of syzygies and their generators, getting a free resolution of a chosen module.

There are still many problems related to the effectiveness of this algorithm, which will be examined by the authors in successive papers.

Finally, in Section 3, we study the quotients of graded structures and, in particular, the notion of Gröbner base for an ideal of a quotient of the polynomial ring, naturally arising in this context.

In the second part of Section 3 we show that the Lazard's idea to find $\mathfrak{M}$-standard bases (where $\mathfrak{M}$ is the maximal ideal generated by all the variables of a polynomial ring) by the computation of a Gröbner base of a suitable ideal, can be extended, under some assumption, to a graded structure.

Using this fact we describe a constructive way to find I-standard bases of an ideal, where I is any ideal, possibly different from $\mathfrak{M}$, i.e. a way to find the tangent cone to a scheme in a point, possibly not closed.

0. Preliminaries

All rings are assumed to be commutative with identity. Let us describe some examples we will recall throughout this paper.

EXAMPLE 1. Let A be a noetherian ring, I an ideal of A such that $\cap I^n = (0)$.

We denote by $G_I(A)$ the associated graded ring $\oplus I^n/I^{n+1}$ and, for $x \in A$, we denote by in(x) the class $[x] \in I^n/I^{n+1}$, where $n = v_I(x) = \max\{p \mid x \in I^p\}$.

So we get a map in: $A \to G_I(A)$. For an ideal J of A, let us set in$(J) = (\text{in}(x) \mid x \in J)$. If $\{x_1, \ldots, x_r\} \subset J$, we say that $\{x_1, \ldots, x_r\}$ is an *I-standard base* of J if in$(J) = (\text{in}(x_1), \ldots, \text{in}(x_r))$.

EXAMPLE 2. Let k be a field $P = k[X_1, \ldots, X_n]$ and h: $P \to P[X_0]$ be the homogenization map defined by ${}^hf = X_0^{\deg f} f(X_1/X_0, \ldots, X_n/X_0)$, where $\deg f$ is the usual total degree of the polynomial f.

If $J = (f_1, \ldots, f_r)$ is an ideal of P, denoting by ${}^hJ$ the ideal $({}^hf \mid f \in J)$, we say that $\{f_1, \ldots, f_r\}$ is a *Macaulay base* of J if ${}^hJ = ({}^hf_1, \ldots, {}^hf_r)$. i.e. if the intersection of the projective closures of the hypersurfaces $(f_i = 0)$ of $\mathbb{A}^n$ is the projective closure of their affine intersection.

EXAMPLE 3. Let $P = k[X_1, \ldots, X_n]$ and, for $f \in P$ let $M_H(f)$ be the homogeneous part of maximum degree of f. So we get a map M_H: $P \to P$ and we define, for every ideal J of P, $M_H(J) = (M_H(f) \mid f \in J)$. We say that $\{f_1, \ldots, f_r\} \subset J$ is an *H-base* of J if $M_H(J) = (M_H(f_1), \ldots, M_H(f_r))$, i.e. if the points at infinity of the affine algebraic variety $V(J)$ are points at infinity of each $\{f_i = 0\}$.

EXAMPLE 4 Let $P = k[X_1, \ldots, X_n]$ and $T = \{$*terms* or *monic monomials* of $P\}$. T is a semigroup and the map log: $T \to \mathbb{Z}^n$ defined by $\log(X_1^{a_1} \ldots X_n^{a_n}) = (a_1, \ldots, a_n)$ induces a semigroup isomorphism $T \cong \mathbb{N}^n$.

Every total ordering on $\mathbb{Z}^n$ compatible with its group structure, or, shortly, every *group ordering* (g.o.) on $\mathbb{Z}^n$ induces a semigroup ordering on $\mathbb{N}^n$ and hence on T.

We say that a g.o. $<$ on $\mathbb{Z}^n$ is a *term-ordering* (t.o.) if $\mathbb{N}^n \subseteq (\mathbb{Z}^n)_+$, where $(\mathbb{Z}^n)_+$ is the set of elements of $\mathbb{Z}^n$ which are non negative w.r.t. $<$.

So a g.o. on $\mathbb{Z}^n$ is a t.o. if and only if for every $(a_1, \ldots, a_n) \in \mathbb{Z}^n$: $a_i \geqslant 0$ for $i = 1, \ldots, n \Rightarrow (a_1, \ldots, a_n) \geqslant (0, \ldots, 0)$ in the given ordering, or if and only if for every $X^A = X_1^{a_1} \ldots X_n^{a_n} \in T$, $X^A \geqslant 1$.

We can think of $k[X_1, \ldots, X_n]$ as a graded ring

$$\tilde{P} = \bigoplus_{A \in \mathbb{Z}^n} \tilde{P}_A$$

where $\tilde{P}_A = \{cX^A \mid c \in k\}$ is a k-vector space of dimension 1, when $A \in \mathbb{N}^n$; $\{0\}$ otherwise.

Let us define a map M: $P \to \tilde{P}$ by $M(f) = c_A X^A$, where X^A is the maximum term of f, w.r.t. the fixed t.o. $<$.

If $J = (f_1, \ldots, f_r)$ is an ideal of P, denoting by $M(J) = (M(f) \mid f \in J)$, we say that $\{f_1, \ldots, f_r\}$ is a Gröbner base of J if $M(J) = (M(f_1), \ldots, M(f_r))$.

EXAMPLE 5. Let P and $\tilde{P}$ be as before and $<$ be a t.o. on $\mathbb{Z}^n$. We define a map L: $P \to \tilde{P}$ by $L(f) = c_A X^A$, where X^A is the least term of f, w.r.t. the fixed t.o. $<$. With obvious notations, we say that $\{f_1, \ldots, f_r\}$ is an *L-base* of J if $L(J) = (L(f_1), \ldots, L(f_r))$.

Let us recall known results linking together some of these notions.

PROPOSITION 1. *Let J be an ideal of P. Then* $\{f_1, \ldots, f_r\}$ *is a Macaulay base of J if and only if it is an H-base of J.*

DEFINITION. We say that a t.o. $<$ on $\mathbb{Z}^n$ is *compatible with the usual degree* or *homogeneous* if, for every $X^A, X^B \in T$, $\deg(X^A) > \deg(X^B) \Rightarrow X^A > X^B$, i.e. for $A = (a_1, \ldots, a_n)$, $B = (b_1, \ldots, b_n) \in \mathbb{N}^n$, $|A| = \Sigma\, a_i > |B| = \Sigma\, b_i \Rightarrow A > B$.

PROPOSITION 2. *Let* $J = (f_1, \ldots, f_r)$ *be an ideal of P and* $<$ *be a homogeneous t.o. on* $\mathbb{Z}^n$. *Then*:

(1) *If* $\{f_1, \ldots, f_r\}$ *is a Gröbner base of J, then it is an H-base of J*

(2) *If* $\{f_1, \ldots, f_r\}$ *is an L-base of J, then it is an* $\mathfrak{M}$*-standard base of J, where* $\mathfrak{M} = (X_1, \ldots, X_n) \subset P$.

DEFINITION. A *graded structure* $\mathfrak{A}$ is a quintuple (A, Γ, v, G, F), where A is a ring, $(\Gamma, <)$ is a totally ordered group, $v: A - (0) \to \Gamma$ is a function, G is a Γ-graded ring, i.e. $G = \bigoplus_{\gamma \in \Gamma} G_\gamma$, $F: A \to G$ is a function, all these data satisfying nine axioms $A_1, \ldots, A_9$ (see Robbiano, 1986; Section 1).

Without writing them, let us just remark that they describe v as a valuation and F as a "leading term" map, such that $F(a)$ lies in the part of degree $v(a)$ of G and F is a surjective on the homogeneous elements of G.

An $\mathfrak{A}$-*module* $\mathfrak{M}$ is a quintuple (M, Γ, w, T, Φ), where M is an A-module, $w: M - (0) \to \Gamma$ is a function, T is a Γ-graded G-module, i.e. $T = \bigoplus_{\gamma \in \Gamma} T_\gamma$, $\Phi: M \to T$ is a function, all these data satisfying nine axioms $M_1, \ldots, M_9$ (see Robbiano, 1986).

Let $\mathfrak{M} = (M, \Gamma, w, T, \Phi)$ and $\mathfrak{M}' = (M', \Gamma, w', T', \Phi')$ be two $\mathfrak{A}$-modules.

An $\mathfrak{A}$-*morphism* from $\mathfrak{M}$ to $\mathfrak{M}'$ is a couple (λ, Λ), where $\lambda: M \to M'$ is an A-module homomorphism and $\Lambda: T \to T'$ is a graded G-module homomorphism such that $w'(\lambda(m)) \leqslant w(m)$, for every $m \in M$ such that $\lambda(m) \neq 0$ and

$$\Lambda(\Phi(m)) = \begin{cases} \Phi'(\lambda(m)) & \text{if} \quad w'(\lambda(m)) = w(m) \\ 0 & \text{if} \quad w'(\lambda(m)) < w(m) \quad \text{or} \quad \lambda(m) = 0 \end{cases}$$

Then the $\mathfrak{A}$-modules and $\mathfrak{A}$-morphisms give rise to a category, denoted by $\mathfrak{G}_{\mathfrak{A}}$.

DEFINITION. A *filtered structure* $\mathfrak{A}^*$ is a triple $(A, \Gamma, \mathscr{F}_A)$, where A is a ring, $(\Gamma, <)$ is a totally ordered group, $\mathscr{F}_A = \{F_\gamma A\}_{\gamma \in \Gamma}$ is a filtration i.e. $F_\gamma A$ are additive subgroups of A such that $F_\gamma A \subseteqq F_{\gamma'} A$ if $\gamma < \gamma'$ and $F_\gamma A \,.\, F_{\gamma'} A \subseteqq F_{\gamma + \gamma'} A$; furthermore for every $a \in A - (0)$, there exists a minimum γ such that $a \in F_\gamma A$.

With obvious notations, $\mathfrak{M}^* = (M, \Gamma, \mathscr{F}_M)$ will denote an $\mathfrak{A}^*$ *module*.

A homomorphism of A-modules $\lambda: M \to M'$ is an $\mathfrak{A}$-morphism from $\mathfrak{M}$ to $\mathfrak{M}^*$ if $\lambda(F_\gamma M) \subseteqq F_\gamma M'$ for every $\gamma \in \Gamma$.

We denote the category of $\mathfrak{A}^*$-modules and $\mathfrak{A}^*$-morphisms by $\mathfrak{G}_{\mathfrak{A}^*}$.

It is possible to get a filtered structure $\mathfrak{A}^*$ over A, from a graded structure $\mathfrak{A}$, defining a filtration by $F_\gamma A = \{a \in A \mid v(a) \leqslant \gamma\}$.

Conversely, given $\mathfrak{A}^* = (A, \Gamma, \mathscr{F}_A)$ it is possible to define a graded structure over A setting $v(a) = \min\{\gamma \mid a \in F_\gamma A\}$ for every $a \in A - (0)$ and $G = \bigoplus_{\gamma \in \Gamma} G_\gamma$, where $G_\gamma = (F_\gamma A)/\bigcup_{\gamma' < \gamma} (F_{\gamma'} A)$.

Furthermore, there exists a category equivalence between $\mathfrak{G}_{\mathfrak{A}}$ and $\mathfrak{G}_{\mathfrak{A}^*}$ (see Robbiano, 1986 thm. 1.2).

DEFINITION. A graded structure $\mathfrak{A}$ is called *noetherian* if both A and G are noetherian rings. An $\mathfrak{A}$-module $\mathfrak{M}$ is called *finitely generated* (f.g.) if both M and T are f.g. modules over A and G, respectively.

DEFINITION. Let $\mathfrak{A}$ and $\mathfrak{M}$ be as usual. For every set $\{m_1, \ldots, m_r\} \subset M - (0)$, setting $w_i = w(m_i)$, we define $\mathfrak{L}(w_1, \ldots, w_r) = (A^r, \Gamma, w^+, \bigoplus_i G(-w_i), F^+)$, where A^r is the free module of rank r over A, $w^+: A^r - (0) \to \Gamma$ is defined by $w^+(a_1, \ldots, a_r) = \max\{v(a_i) + w_i \mid a_i \neq 0\}$, $F^+: A^r \to \bigoplus_i G(-w_i)$ is defined by $F^+(a_1, \ldots, a_r) = (\tilde{F}(a_1), \ldots, \tilde{F}(a_r))$,

$$\text{where } \tilde{F}(a_i) = \begin{cases} 0 & \text{if } a_i = 0 \text{ or } v(a_i) + w_i < w^+(a_1, \ldots, a_r) \\ F(a_i) & \text{if } \qquad v(a_i) + w_i = w^+(a_1, \ldots, a_r). \end{cases}$$

So there exists a canonical morphism (λ, Λ): $\mathfrak{L}(w_1, \ldots, w_r) \to \mathfrak{M}$, where $\lambda: A^r \to M$ sends $(a_1, \ldots, a_r)$ to $\sum_i a_i m_i$ and $\Lambda: \bigoplus_i G(-w_i) \to T$ sends $(\alpha_1, \ldots, \alpha_r)$ to $\sum_i \alpha_i \Phi(m_i)$.

Let us notice that, if M is generated by $\{m_1, \ldots, m_r\}$, then (λ, Λ) is the first step of a presentation of $\mathfrak{M}$.

We say that an $\mathfrak{A}$-module $\mathfrak{M}$ is *finite free* if it is isomorphic to an $\mathfrak{A}$-module $\mathfrak{L}(w_1, \ldots, w_r)$.

DEFINITION. Let $\mathfrak{A}$ be a graded noetherian structure, $\mathfrak{M}$ be a f.g. module and $m_1, \ldots, m_r \in M - (0)$. We say that $\{m_1, \ldots, m_r\}$ is a *generalized standard base* of $\mathfrak{M}$ (g.s.b.) if every element m of $M - (0)$ can be written by

$$m = \sum_i a_i m_i, \quad \text{where} \quad a_i \in A \text{ and } w(m) \geqslant v(a_i) + w(m_i) \text{ for every } i \text{ such that } a_i \neq 0.$$

We say that $\{m_1, \ldots, m_r\}$ is a *standard set* of $\mathfrak{M}$ if T is generated by $\Phi(m_1), \ldots, \Phi(m_r)$ as G-module. (see Mora, preprint).

THEOREM 3. *In the previous notation, consider the following conditions*:

(*i*) $\{m_1, \ldots, m_r\}$ *is a g.s.b. of* $\mathfrak{M}$
(*ii*) $\{m_1, \ldots, m_r\}$ *is a standard set of* $\mathfrak{M}$.

Then (i) $\Rightarrow$ (ii) and, if $\mathfrak{M}$ is a Krull module, they are equivalent (Robbiano, 1986, thm. 3.1). (We will not give the definition of Krull module, which is really technical. Anyway since it generalizes in some way the Krull intersection theorem, it is satisfied in many examples.)

REMARK 4. Almost all the above examples are graded structures.

EXAMPLE 1. $\mathfrak{A} = (A, \mathbb{Z}, -v_I, G, \text{in})$, where $\mathbb{Z}$ has the usual ordering and $G_n = (G_I(A))_{-n}$, is a graded structure. Here we need to use $-v_I$ instead of v_I to satisfy the axioms $A_1, \ldots, A_9$. In fact A_7 says: $v(a+b) \leqslant \text{Max}\{v(a), v(b)\}$, while the relation $v_I(a+b) \geqslant \text{Min}\{v_I(a), v_I(b)\}$ holds.

Now, if J is an ideal of A, the structure $\mathfrak{J}$ induced on J by $\mathfrak{A}$, putting $F_\gamma J = F_\gamma A \cap J$, is obviously $\mathfrak{J} = (J, \mathbb{Z}, -v_I, \text{in}(J), \text{in})$.

So and *I-standard base* of J is a standard set for the $\mathfrak{A}$-submodule $\mathfrak{J}$. Let us remark that both $\mathfrak{A}$ and $\mathfrak{J}$ are negatively graded structures.

EXAMPLE 2. $\mathfrak{A} = (P, \mathbb{Z}, \deg, P[X_0], h)$, where $\mathbb{Z}$ has the usual ordering, is not a graded structure since X_0 is not in $\mathrm{Im}(h)$, as it is required by the axiom A.3.

EXAMPLE 3. $\mathfrak{A} = (P, \mathbb{Z}, \deg, P, M_H)$ is a positively graded structure and an H-base for an ideal J is a g.s.b. for the associated structure $\mathfrak{J}$.

EXAMPLE 4. $\mathfrak{A} = (P, \mathbb{Z}^n, \log(M(-)), \tilde{P}, M)$ is a positively graded structure and Gröbner bases for ideals of P are g.s.b. for the induced substructure of $\mathfrak{A}$.

EXAMPLE 5. $\mathfrak{A} = (P, \mathbb{Z}^n, -\log(L(-)), \tilde{P}, L)$ is a negatively graded structure and the notions of L-base of an ideal of P and standard set for the associated structure coincide.

REMARK 5. Let us remark that all the structures above are noetherian; so every substructure is a f.g. $\mathfrak{A}$-module.

Furthermore the structures in Ex. 3 and 4 are Krull structures.

The structure $\mathfrak{A}$ in Ex. 1 is a Krull structure if A is local (see Robbiano, 1986); with a similar argument one can prove that it is still Krull if I is contained in the Jacobson radical of A.

DEFINITION. Let Γ and Δ be two ordered groups and $\mathfrak{A}_\Gamma = (A, \Gamma, v_\Gamma, G(\Gamma), F(\Gamma))$, $\mathfrak{A}_\Delta = (A, \Delta, v_\Delta, G(\Delta), F(\Delta))$ be two graded structures over the same ring A. If $\alpha: \Gamma \to \Delta$ is an ordered group homomorphism such that $\alpha(v_\Gamma(a)) = v_\Delta(a)$ for every a, then the triple $\mathscr{A}_\alpha = (\mathfrak{A}_\Gamma, \mathfrak{A}_\Delta, \alpha)$ is called a *double structure*. The definition of $\mathscr{A}_\alpha$-modules $\mathscr{M}_\alpha = (\mathfrak{M}_\Gamma, \mathfrak{M}_\Delta)$ is the natural one.

In this case, one can consider the ordering on Γ to be a refinement of Δ.

Let us recall some results about orderings.

PROPOSITION 6. *Let $\mathfrak{A} = (A, \Gamma, v, G, F)$ be a noetherian structure. Then*:

(*a*) $\Gamma \cong \mathbb{Z}^n$ *for a suitable* n;
(*b*) $\Gamma^0(\mathfrak{A}) = \mathrm{Im}\, v$ *generates a finitely generated semigroup of* Γ (*see* Robbiano, 1986, *proposition* 1.3).

PROPOSITION 7. *Let $<$ be a group ordering on $\mathbb{Z}^n$.*

(*i*) *If $\Gamma \subset (\mathbb{Z}^n)_+ \cup \{0\}$ is a finitely generated sub-semigroup of $\mathbb{Z}^n$, then Γ is well-ordered.*
(*ii*) *Let $u_1, \ldots, u_t$ be elements of $\mathbb{Z}^n$ and $\Gamma_1, \ldots, \Gamma_t$ be a finitely generated sub-semigroups of $\mathbb{Z}^n$. If each Γ_i is well-ordered, then $E = \bigcup_i (u_i + \Gamma_i)$ is well-ordered.*

PROOF. (i) By Robbiano (1986), corollary 2.6. (ii) Obvious.

DEFINITION. Let $\mathfrak{A}$ be as before. We say that $\mathfrak{A}$ is a *positively graded structure* if $\Gamma^0(\mathfrak{A}) \subseteq \Gamma^+ = \{\gamma \in \Gamma \mid \gamma \geqslant 0\}$ and we write $\Gamma^0(\mathfrak{A}) \geqslant 0$. We say that a graded $\mathfrak{A}$-module $\mathfrak{M}$ is *well-ordered* (resp. *term-ordered* or *positively graded*) if $\Gamma^0(\mathfrak{M})$ is well-ordered (resp. $\Gamma^0(\mathfrak{M}) \geqslant 0$).

COROLLARY 8. *Let $\mathfrak{A}$ be a noetherian structure.*

(*i*) *If $\Gamma^0(\mathfrak{A}) \geqslant 0$, then $\Gamma^0(\mathfrak{A})$ is well-ordered.*
(*ii*) *If $\mathfrak{M}$ is a finite $\mathfrak{A}$-module and $\Gamma^0(\mathfrak{A})$ is well-ordered, then $\Gamma^0(\mathfrak{M})$ is well-ordered.*

PROOF. (i) By proposition 7, (i). (ii) If $u_1, \ldots, u_t$ are the degrees of the generators of $\mathfrak{M}$, then $\mathfrak{M}$ is a quotient of the free $\mathfrak{A}$-module $\mathfrak{L}(u_1, \ldots, u_t)$. So $\Gamma^0(\mathfrak{M}) \subseteq \Gamma^0(\mathfrak{L}(u_1, \ldots, u_t)) = \bigcup_i (u_i + \Gamma^0(\mathfrak{A}))$ and we are done by prop. 7, (ii).

In the case $A = k[X_1, \ldots, X_n]$, $\Gamma = \mathbb{Z}^n$, and $v(f) = \log(\text{max term of } f)$ (see Example 4), $\mathfrak{A}$ well-ordered $\Rightarrow$ $\mathfrak{A}$ term-ordered. In a general structure, this may be false.

EXAMPLE 9. Let $A = k[X]/(X^2) \cong k \oplus B$, where B is the k-vector space generated by $\bar{X}$. Let $v\colon A \to \mathbb{Z}$ be such that $v(c) = 0$, $v(c\bar{X}) = -1$, $v(b + c\bar{X}) = 0$ for every $b, c \in k - (0)$. Let us consider the associated graded structure $\mathfrak{A}$, where $\mathbb{Z}$ has the usual ordering. Here $\Gamma^0 = \{0, -1\}$, which is well-ordered, but $\Gamma^0 \not\subset \Gamma_+$. Let us remark that A has nilpotent elements.

REMARK 10. Let $\mathfrak{A} = (A, \ldots)$ be a well-ordered structure such that A is reduced. Then $\mathfrak{A}$ is term-ordered. In fact, let us assume there exists $\gamma \in \Gamma^0$, $\gamma < 0$; then let us consider $a \in A$ such that $v(a) = \gamma$. Since A is reduced $a^n \neq 0$ for every n and since v is a valuation, $v(a^{n+1}) \leqslant v(a) + v(a^n)$ for every n. Therefore the sequence $\gamma_n = v(a^n)$ is strictly decreasing, and it is a contradiction, being Γ^0 well-ordered.

REMARK 11. $\Gamma^0(\mathfrak{A}) \geqslant 0$ does not imply that $\Gamma^0(\mathfrak{M}) \geqslant 0$, since the generators of M as A-module can have valuations negative "enough".

1. Reduction Procedure

1.1. FIRST CHARACTERIZATION OF GENERALIZED STANDARD BASES

Let $\mathfrak{A} = (A, \Gamma, v, G, F)$ be a graded structure and $\mathfrak{M} = (M, \Gamma, w, T, \Phi)$ be an $\mathfrak{A}$-module. Let us start with a technical result we will need in this section.

Let $\{b_1, \ldots, b_s\} \subset M - (0)$ and $m = \sum_1^s a_i b_i$, $a_i \in A$, $(a_1, \ldots, a_s) \neq (0, \ldots, 0)$.

Moreover, we set $I = \{1, \ldots, s\}$ and $I' = \{i \in I / v(a_i) + w_i = \mu\}$, where $\mu = \max\{v(a_i) + w_i / a_i \neq 0\}$ and $w_i = w(b_i)$.

PROPOSITION 1.1.1. *In the above notation* $\sum_{I'} F(a_i)\Phi(b_i) = 0$ *if and only if either* $m = 0$ *or* $w(m) < \mu$.

PROOF. Let $(\lambda, \Lambda)\colon \mathfrak{L}(w_1, \ldots, w_s) \to \mathfrak{M}$ be the morphism associated to $\{b_1, \ldots, b_s\}$; then, by definition $\Phi^+(a_1, \ldots, a_s) = \tilde{F}(a_1), \ldots, \tilde{F}(a_s))$ (see Section 0).

Furthermore

$$\Lambda(\Phi^+(a_1, \ldots, a_s)) = \sum_{I'} F(a_i)\Phi(b_i) \quad \text{and} \quad m = \lambda(a_1, \ldots, a_s).$$

If $\sum_{I'} F(a_i)\Phi(b_i) = 0$, assuming $m \neq 0$ and $w(m) = \mu$, we get $w(\lambda(a_1, \ldots, a_s)) = \mu = w^+(a_1, \ldots, a_s)$ and so $\Lambda(\Phi^+(a_1, \ldots, a_s)) = \Phi(\lambda(a_1, \ldots, a_s))$; hence $0 = \Phi(m)$: contradiction.

Conversely, if $m = 0$, the thesis easily holds. If $m \neq 0$ and $w(m) < \mu$, then $w(\lambda(a_1, \ldots, a_s)) < w^+(a_1, \ldots, a_s)$ and so $\Lambda(\Phi^+(a_1, \ldots, a_s)) = 0$.

DEFINITION. Let m and m' be in M, $m \neq 0$, and B be a finite subset of $M - \{0\}$ we say that *m reduces to m' modulo B* and we write $m \underset{B}{\rightarrow} m'$ if there exist $b_1, \ldots, b_s$ in B and $a_1, \ldots, a_s$ in A such that

(i) $m' = m - \sum_1^s a_i b_i$, $a_i b_i \neq 0$ for all i;

(ii) $\Phi(m) = \sum_1^s F(a_i)\Phi(b_i)$;

(iii) $w(m) = v(a_i) + w(b_i)$ for all i.

LEMMA 1.1.2. *If $m \neq 0$ and $m \underset{B}{\rightarrow} m'$, then either $m' = 0$, or $w(m') < w(m)$.*

PROOF. Let us remark that $\sum_I a_i b_i \neq 0$, by proposition 1.1.1. It is enough to show that $\Phi(m) = \Phi(\sum_i a_i b_i)$. Let $I'' = \{i \in I / w(a_i b_i) = v(a_i) + w(b_i) = w(m)\}$. We have: $0 \neq \Phi(m) = \sum_{I''} F(a_i)\Phi(b_i) = \sum_{I''} \Phi(a_i b_i)$, then, by proposition 1.1.1., $\sum_{I''} a_i b_i \neq 0$. So $\Phi(m) = \Phi(\sum_{I''} a_i b_i) = \Phi(\sum_I a_i b_i)$.

NOTATION. We denote by "$\underset{B}{\rightarrow}*$" the reflexive and transitive closure of the relation "$\underset{B}{\rightarrow}$".

THEOREM 1.1.3. *Let $m \in M - (0)$. If $m \underset{B}{\rightarrow}* 0$ then there exist $b_1, \ldots, b_s$ in B and $a_1, \ldots, a_s$ in A such that*

(i) $m = \sum_1^s a_i b_i$, $a_i b_i \neq 0$ for all i;

(ii) $w(m) \geqslant v(a_i) + w(b_i)$ for all i.

PROOF. By assumption, there exist $m_1, \ldots, m_n$ such that

$$m \underset{B}{\rightarrow} m_1 \underset{B}{\rightarrow} m_2 \underset{B}{\rightarrow} \cdots \underset{B}{\rightarrow} m_{n-1} \underset{B}{\rightarrow} m_n = 0.$$

Let us prove the theorem by induction on n. The case $n = 1$ is obvious. Now let $n > 1$. Then $m \underset{B}{\rightarrow} m_1 \neq 0$, so $m_1 = m - \sum_I c_i b_i$, where $c_i b_i \neq 0$ and $w(m) \geqslant v(c_i) + w(b_i)$ for all i and by lemma 1.1.2, $w(m) > w(m_1)$. Since $m_1 \underset{B}{\rightarrow}* 0$ by the inductive hypothesis, $m_1 = \sum_J d_j b_j$, where $d_j b_j \neq 0$ and $w(m_1) \geqslant v(d_j) + w(b_j)$ for all j. Then, with obvious notations:

$$m = \sum_I c_i b_i + \sum_J d_j b_j = \sum_{J \cup I} a_h b_h$$

and also (ii) is easily verified.

COROLLARY 1.1.4. *If for every $m \in M$, $m \underset{B}{\rightarrow}* 0$, then B is a g.s.b. for $\mathfrak{M}$.*

PROPOSITION 1.1.5. *Let $\Gamma^0(\mathfrak{M})$ be well-ordered. Then, for each $m \in M - (0)$, there exists (at least) an element $\mathrm{red}_B(m) \in M$ such that:*

(i) $m \underset{B}{\rightarrow} * \mathrm{red}_B(m)$;
(ii) either $\mathrm{red}_B(m) = 0$ or $\Phi(\mathrm{red}_B(m)) \notin (\Phi(b) / b \in B)$.

PROOF. If $\Phi(m) \notin (\Phi(b)/b \in B)$, let us set $\mathrm{red}_B(m) = m$. Otherwise $\Phi(m) = \sum_1^s \alpha_i \Phi(b_i)$, $\alpha_i \Phi(b_i) \neq 0$, for suitable b_i's in B and homogeneous α_i's in G. Without loss of generality, we may assume $w(m) = \deg(\alpha_i) + \deg \Phi(b_i)$, for all i.

Let $\{a_1, \ldots, a_s\}$ be a fixed lifting of $\{\alpha_1, \ldots, \alpha_s\}$. Setting $m_1 = m - \Sigma\, a_i b_i$, we have $m \xrightarrow[B]{} m_1$. Then, if $m_1 \neq 0$, $w(m_1) < w(m)$.

In this case, either $\Phi(m_1) \notin (\Phi(b)/b \in B)$ (and so $m_1 = \mathrm{red}_B(m)$), or we can repeat the procedure on m_1, getting some m_2. If $m_2 \neq 0$, $w(m_2) < w(m_1)$, and so on. Since $\Gamma^0(\mathfrak{M})$ is well-ordered, the sequence of $\{w(m_n)\}$ must be finite.

THEOREM 1.1.6. *Let us assume that $\Gamma^0(\mathfrak{M})$ is well-ordered. The following conditions are equivalent*:

(*i*) *B is a g.s.b. for $\mathfrak{M}$*;
(*ii*) *for every $m \in M$, $m \xrightarrow[B]{}{}^* 0$.*

PROOF. It is a consequence of corollary 1.1.4, theorem 3, and proposition 1.1.5.

1.2. SECOND CHARACTERIZATION OF GENERALIZED STANDARD BASES

Let $\mathfrak{A}$, $\mathfrak{M}$ and $B = \{b_1, \ldots, b_n\}$ be as before and let us assume $\mathfrak{A}$ is noetherian. Setting $w_i = w(b_i)$ for all i, we denote as usual by (λ, Λ): $\mathfrak{L}(B) = \mathfrak{L}(w_1, \ldots, w_n) \to \mathfrak{M}$ the morphism associated to B, where Λ: $\bigoplus_i G(-w_i) \to T$ is a homomorphism of Γ-graded G-modules.

So $\ker \Lambda$ is finitely generated; if it is non zero, let $\{L^{(1)}, \ldots, L^{(k)}\}$ be a system of homogeneous generators of $\ker \Lambda$ and δ_i = degree of $L^{(i)}$. Hence $L^{(i)} = (l_1^{(i)}, \ldots, l_n^{(i)})$, where $l_j^{(i)} \in G_{\delta_i - w_j}$.

There exist $e_j^{(i)} \in A$ such that $F(e_j^{(i)}) = l_j^{(i)}$, so $v(e_j^{(i)}) = \delta_i - w_j$. Setting $E^{(i)} = (e_1^{(i)}, \ldots, e_n^{(i)})$, we have $w^+(E^{(i)}) = \delta_i$ and $F^+(E^{(i)}) = L^{(i)}$.

We will always assume to fix, for each $B \subset M$, a basis $\{L^{(1)}, \ldots, L^{(k)}\}$ of $\ker \Lambda$ and a lifting $\{E^{(1)}, \ldots, E^{(k)}\}$ which satisfies the above conditions.

DEFINITION. We call *S-set of B* the k-uple $S(B) = (E^{(1)} . B, \ldots, E^{(k)} . B)$, where $E^{(i)} . B = e_1^{(i)} . b_1 + \cdots + e_n^{(i)} . b_n = \lambda(E^{(i)})$.

REMARK 1.2.1. The definition of S-set is the natural generalization of the notion of S-polynomial. In fact, when $\mathfrak{A} = \mathfrak{M}$, $A = k[X_1, \ldots, X_n]$ and the graduation is given by the monomial decomposition (see Example 4, Section 0), if $B = \{f, g\}$, then $\ker \Lambda$ is generated by the element $L = (l_1, l_2) = (\mathrm{l.c.m.}(M(f), M(g))/M(f), -\mathrm{l.c.m.}(M(f), M(g))/M(g))$, where $M(-)$ denotes the leading term. In this case, taking $e_i = l_i$ as lifting, $S(B) = \{l_1 f + l_2 g\}$ coincides with the S-polynomial of f and g.

The aim of this section is to characterize the g.s.b. in terms of S-sets.

More precisely, we will prove that it is sufficient that $x \xrightarrow[B]{}{}^* 0$ just for the x's which belong to $S(B)$ to verify that B is a g.s.b.

LEMMA 1.2.2. *Let $\mathfrak{A}$ be a noetherian structure and $\mathfrak{M}$ be any $\mathfrak{A}$-module. Let $B = \{b_1, \ldots, b_n\} \subset M - (0)$. Setting $I = \{1, \ldots, n\}$, let $m = \sum_I a_i b_i \neq 0$, $a_i \in A$.* Let us assume:

(*i*) $w(m) < \max\{v(a_i) + w_i / i \in I,\ a_i \neq 0\} = \gamma$,
(*ii*) *for every $x \in S(B)$, $x \xrightarrow[B]{}{}^* 0$.*

Then there exist $a'_1, \dots, a'_n \in A$ *such that* $m = \sum_I a'_i b'_i$ *and*

(*iii*) $\max\{v(a'_i) + w_i / i \in I, a'_i \neq 0\} < \gamma$.

PROOF. Let us consider (λ, Λ): $\mathfrak{L}(B) = \mathfrak{L}(w_1, \dots, w_n) \to \mathfrak{M}$. Let us fix a homogeneous system of generators $\{L^{(i)}, \dots, L^{(k)}\}$ of $\ker \Lambda$ and a lifting $\{E^{(1)}, \dots, E^{(k)}\}$, as before.

Since $w(\lambda(a_1, \dots, a_n)) = w(m) < \gamma = w^+(a_1, \dots, a_n)$, we have $0 = \Lambda(\Phi^+(a_1, \dots, a_n)) = \tilde{F}(a_1)\Phi(b_1) + \cdots + \tilde{F}(a_n)\Phi(b_n)$, where $\tilde{F}(a_i)$ is either 0 or $F(a_i)$ if, respectively, either $v(a_i) + w_i < \gamma$ or $v(a_i) + w_i = \gamma$.

Then $(\tilde{F}(a_1), \dots, \tilde{F}(a_n)) \in \ker \Lambda - (0)$ and is homogeneous of degree γ.

Without loss of generality, we may assume that $(\tilde{F}(a_1), \dots, \tilde{F}(a_n)) = \beta_1 L^{(1)} + \cdots + \beta_t L^{(t)}$ for some $t \leqslant k$, where $\beta_i L^{(i)} \neq 0$ and β_i are homogeneous in G, of degree $\gamma - \delta_i$.

$$\text{So } \tilde{F}(a_1) = \sum_1^t \beta_j l_1^{(j)}, \dots, \tilde{F}(a_n) = \sum_1^t \beta_j l_n^{(j)}.$$

Let $c_1, \dots, c_t \in A$ such that $F(c_i) = \beta_i$ for all i, and let us set

$$\tilde{a}_1 = \sum_1^t c_j e_1^{(j)}, \dots, \tilde{a}_n = \sum_1^t c_j e_n^{(j)}$$

and $a_i^* = a_i - \tilde{a}_i$.

So, denoting by $J = \{1, \dots, t\}$, we have

$$m = \sum_I a_i b_i = \sum_I a_i^* b_i + \sum_I \tilde{a}_i b_i = \sum_I a_i^* b_i + \sum_I \left(\sum_J c_j e_i^{(j)} \right) b_i = \sum_I a_i^* b_i + \sum_J c_j (E^{(j)} \,.\, B).$$

To prove the lemma it is enough to show that:

(a) either $a_i^* = 0$ or $v(a_i^*) < \gamma - w_i$;
(b) either $E^{(j)} \,.\, B = 0$ or $w(E^{(j)} \,.\, B) < \delta_j$.

In fact, assuming this, we get that m is splittable in two sums, the first one fulfilling (iii) by the claim (a).

Furthermore, by (ii), if $E^{(j)} \,.\, B \neq 0$, then $E^{(j)} \,.\, B \xrightarrow[B]{}^* 0$. Then, by theorem 1.1.3., there exist a subset H_j of I and suitable $\bar{a}$'s in A such that $E^{(j)} \,.\, B = \sum_{H_j} \bar{a}_h^{(j)} b_h$ and $w(E^{(j)} \,.\, B) \geqslant v(\bar{a}_h^{(j)}) + w_h$ for all $h \in H_j$.

By claim (b), $v(\bar{a}_h^{(j)}) < \delta_j - w_h$. So the second sum appearing in m is $\sum_J c_j (\sum_{H_j} \bar{a}^{(j)} b_h)$ and the coefficients of the b_i's are bounded by $\max\{v(c_j) + v(\bar{a}_h^{(j)})\} < \gamma - \delta_j + \delta_j - w_h = \gamma - w_h$. Hence also the second sum fulfils (iii) by (a). Now let us hint of the proof of (a) and (b):

(a) For each i, $v(a_i) + w_i \leqslant \gamma$. If $v(a_i) + w_i = \gamma$, then $F(a_i) = \tilde{F}(a_i) = \sum_1^t \beta_j l_i^{(j)}$. Furthermore $v(c_j e_i^{(j)}) \leqslant v(c_j) + v(e_i^{(j)}) = \gamma - w_i$. It is easy to verify that $\tilde{a}_i = \sum_J c_j e_i^{(j)}$ is such that $v(\tilde{a}_i) = \gamma - w_i = v(a_i)$ and $F(\tilde{a}_i) = F(a_i)$. Hence either $a_i^* = a_i - \tilde{a}_i = 0$ or $v(a_i^*) < \gamma - w_i$. On the other hand, if $a_i = 0$ or $v(a_i) + w_i < \gamma$, $0 = \tilde{F}(a_i) = \sum_J F(c_j) F(e_i^{(j)})$, so also $\Sigma\, F(c_j) F(e_i^{(j)}) = 0$, where the sum is on the j's such that $v(c_j) + v(e_i^{(j)}) = \gamma - w_i$. Then, by proposition 1.1.1, we get (a).
(b) Since $0 = \Lambda(L^{(j)}) = \sum_I F(e_i^{(j)}) \Phi(b_i)$ and $v(e_i^{(j)}) + w_i = \delta_j$, if $e_i^{(j)} \neq 0$, by proposition 1.1.1 either $\sum_I e_i^{(j)} \,.\, b_i = 0$ or $w(\sum_I e_i^{(j)} \,.\, b_i) < \delta_j$.

THEOREM 1.2.3. *Let $\mathfrak{A}$ be noetherian and $\Gamma^0(\mathfrak{U})$ be well-ordered; let $\mathfrak{M}$ be a f.g. $\mathfrak{A}$-module and $B = \{b_1, \ldots, b_n\}$ be a system of non zero generators of M. The following conditions are equivalent*:

(*i*) *B is a g.s.b. of $\mathfrak{M}$*;
(*ii*) *for every $x \in S(B)$, $x \xrightarrow[B]{} {}^* 0$.*

PROOF. (i) ⇒ (ii) by thm. 1.1.6. (ii) ⇒ (i) We want to show that, for every $m \in M - (0)$, we can write $m = \sum_I a_i b_i$ where $a_i \in A$ and $w(m) \geqslant v(a_i) + w(b_i)$, for all i such that $a_i \neq 0$. Since B generates M, we can write $m = \sum_I a_i b_i$.

If it happens that $w(m) < \max\{v(a_i) + w_i / i \in I, a_i \neq 0\} = \gamma_0$, then by lemma 1.2.2, there exist $a_1^{(1)}, \ldots, a_n^{(1)} \in A$ such that $m = \sum_I a_i^{(1)} b_i$ and $\max\{v(a_i^{(1)}) + w_i\} = \gamma_1 < \gamma_0$.

If $w(m) < \gamma_1$, we can apply again the lemma, getting a sequence $\gamma_0 > \gamma_1 > \ldots$ of elements of $\Gamma^0(\mathfrak{L}(B))$, which is well-ordered since $\Gamma^0(\mathfrak{A})$ is so (see cor. 8). Hence there exist $s \in \mathbb{N}$ such that $w(m) \geqslant \gamma_s$ and $m = \sum_I a_i^{(s)} b_i$.

2. Algorithms

2.1. EFFECTIVE STRUCTURES

T. Mora (1988) introduced the notion of effective structure.

Let $\mathfrak{A} = (A, \Gamma, v, G, F)$ be a graded noetherian effective structure and $\mathfrak{M} = (M, \Gamma, w, T, \Phi)$ be an $\mathfrak{A}$-module.

DEFINITION. We say that $\mathfrak{M}$ is *effective* if:

(E_1) M, Γ, T are effective;
(E_2) w and Φ are computable functions;
(E_3) for each $\alpha \in T$, α homogeneous, it is possible to compute $m \in M$ such that $\Phi(m) = \alpha$;
(E_4) if $\alpha, \alpha_1, \ldots, \alpha_n \in T - (0)$ are homogeneous, it is possible to decide whether α is in the submodule $G(\alpha_1, \ldots, \alpha_n)$ and, if so, to compute homogeneous $g_i \in G$ such that $\alpha = \Sigma\, g_i \alpha_i$, $g_i \alpha_i \neq 0$, $\deg(\alpha) = \deg(g_i) + \deg(\alpha_i)$ for all i;
(E_5) for each finite set $B \subset M$ it is possible to compute a finite homogeneous basis of $\ker \Lambda_B$, where (λ_B, Λ_B): $\mathfrak{L}(B) \to \mathfrak{M}$.

EXAMPLES. The graded structures in Ex. 1, 3, 4, 5 Section 0, are all effective. Other examples of effective structures (in the "Gröbner" situation) can be found for instance in Möller (1985).

Throughout this section, we will assume that $\mathfrak{A}$ and $\mathfrak{M}$ are effective.

ALGORITHM 2.1.1. Let $B = \{b_1, \ldots, b_n\} \subset M - (0)$, $n \geqslant 2$, and $B_t = \{b_1, \ldots, b_t\}$, $t \leqslant n$. We can canonically embed $\ker \Lambda_{B_t}$ in $\ker \Lambda_{B_{t+1}}$, adding zeros in the last coordinates. In particular the following algorithm produces a homogeneous basis L of ker Λ_B containing a base L_{B_t} of ker Λ_{B_t} for all $t \leqslant n$.

$t := 1$;

Compute L_{B_t} (By E_5);

$L := L_{B_t}$;

While $t < n$ do
  Compute $L_{B_{t+1}}$ (By E_5);
  $L' := \{(l_1, \ldots, l_{t+1}) \in L_{B_{t+1}} \mid l_{t+1} \neq 0\}$;
  $L := L \cup L'$;
  $t := t + 1$;

ALGORITHM 2.12. Let $\Gamma^0(\mathfrak{M})$ be w.o. and $B = \{b_1, \ldots, b_n\} \subset M - (0)$. The following procedure gives, for any $m \in M - (0)$, an element $\text{red}_B(m)$ in M, fulfilling the properties in prop. 1.1.5. Moreover, if $\text{red}_B(m) \neq m$, we get also suitable a_i's in A such that $m = \Sigma\, a_i b_i + \text{red}_B(m)$; if we have such representation, we will write $\text{rep}_B(m) = (a_1, \ldots, a_n)$.

$m' := m$;

For $i := 1, \ldots, n$ do

  $a_i := 0$;

While $m' \neq 0$ and $\Phi(m') \in (\Phi(b_1), \ldots, \Phi(b_n))$ do
  Compute $c_1, \ldots, c_n \in A$ such that $\Phi(m') = \Sigma\, F(c_i)\Phi(b_i)$
  $w(m') = v(c_i) + w(b_i)$ or $c_i = 0$
  $F(c_i)\Phi(b_i) = 0 \Rightarrow c_i = 0$;

  $m' := m' - \Sigma\, c_i b_i$;
  for $i := 1, \ldots, n$ do
    $a_i := a_i + c_i$;
$\text{red}_B(m) := m'$;
$\text{rep}_B(m) := (a_1, \ldots, a_n)$.

REMARK 2.1.3. If $m \neq \text{red}_B(m)$, then $w(m) \geqslant v(a_i) + w(b_i)$ for each i; if, in addition, $\text{red}_B(m) \neq 0$, then $w(m) > w(\text{red}_B(m))$.

2.2. COMPUTATION OF GENERALIZED STANDARD BASES

In this section both $\mathfrak{A}$ and $\mathfrak{M}$ are effective and well-ordered, and $\mathfrak{M}$ is f.g.; in this case if B is a finite subset of $M - (0)$ then $S(B) = (E^{(1)} . B, \ldots, E^{(k)} . B)$ is computable.

ALGORITHM 2.2.2. Let $B' = \{b_1, \ldots, b_n\}$ be a system of generators of M. The following algorithm produces elements $b_{n+1}, \ldots, b_r$ such that $B = \{b_1, \ldots, b_n, b_{n+1}, \ldots, b_r\}$ is a generalized standard base for $\mathfrak{M}$.

$t := n;$

$C_t := B';$

Compute L_{C_t} (By Algorithm 2.1.1);

$L := L_{C_t};$
$S := \phi;$

While $L \neq \phi$ do
  Choose $(l_1, \ldots, l_t) \in L$ and $E = (e_1, \ldots, e_t)$ where $F(e_i) = l_i$:

$L := L - \{(l_1, \ldots, l_t)\};$
$x := e_1 b_1 + \cdots + e_t b_t;$
$y := \text{red}_{C_t}(x);$
$(a_1, \ldots, a_t) := \text{rep}_{C_t}(X);$
$\tilde{a}_i := e_i - a_i;$
If $y = 0$ then $E(x) := (\tilde{a}_1, \ldots, \tilde{a}_t);$
$S := S \cup \{(\tilde{a}_1, \ldots, \tilde{a}_t)\};$
else $\tilde{a}_{t+1} := -1;$
$E(x) := (\tilde{a}_1, \ldots, \tilde{a}_{t+1});$
$S := S \cup \{(\tilde{a}_1, \ldots, \tilde{a}_{t+1})\};$
$b_{t+1} := y;$
$C_{t+1} := \{b_1, \ldots, b_{t+1}\};$
Compute $L_{C_{t+1}}$ (By E_5);
$L' := \{(l_1, \ldots, l_{t+1}) \in L_{C_{t+1}} \mid l_{t+1} \neq 0\};$
$L := L \cup L';$
$t := t + 1;$

$B := C_t$

Since T is f.g. over a noetherian ring, the algorithm stops. Furthermore $x \xrightarrow[B]{}{}^* 0$ for every x in $S(B)$, so B is a g.s.b. by theorem 1.2.3.

REMARK 2.2.3. For each $x \in S(C_t)$, the algorithm 2.2.2 produces also the element $\tilde{E}(x) = (\tilde{a}_1, \ldots, \tilde{a}_r)$ of $\ker \lambda_B$ which arises from $E(x) = (\tilde{a}_1^{(j)}, \ldots, \tilde{a}_t^{(j)}) \in S$ adding $r - t$ zeros at the end.

In the next section, we prove that these elements are a g.s.b. for the first syzygies module $\ker \lambda_B$ of B w.r.t. a suitable graded structure on it.

2.3. MODULES OF SYZYGIES

Let $P = k[X_1, \ldots, X_n]$, T be the set of all terms of P and $<$ be a t.o. on T (see Example 4, Section 0). If P^r denotes the free module of rank r over P, the set T_r of the "terms" of P^r is $\{(0, \ldots, 0, \varphi_j, 0, \ldots, 0)/\varphi_j \in T\}$.

So T_r can be identified either with $T \times \{1, \ldots, r\}$ or with $\mathbb{N}^n \times \{1, \ldots, r\}$.

If $w = (\varphi_1, \ldots, \varphi_r)$, $\varphi_i \in T$, it is possible to define a t.o. $<_w$ on T_r, called the t.o. *induced by* $<$ *and* w by setting $(\alpha, i) <_w (\beta, j) \Leftrightarrow \alpha\varphi_i < \beta\varphi_j$ or $\alpha\varphi_i = \beta\varphi_j$ and $i < j$. (see, for instance, Mora & Möller, 1986; Bayer, 1982).

Let $J = (g_1, \ldots, g_n)$ be an ideal of P; a generalization of Buchberger's algorithm gives further elements $g_{n+1}, \ldots, g_r$ such that $\{g_1, \ldots, g_n, g_{n+1}, \ldots, g_r\}$ is a Gröbner basis of

J and also produces a Gröbner basis $\{\Phi_{ij}/1 \leq i < j \leq r\}$ for $K = Syz^1(g_1, \ldots, g_r) \subset P^r$ w.r.t. the t.o. on T_r induced by $<$ and $w = (M(g_1), \ldots, M(g_r))$ (to find the definitions and properties see Mora & Möller, 1986; Spear, 1977; Trinks, 1978; Zacharias, 1978).

We will prove that, in the context of graded structures, the G-base $\{\Phi_{ij}\}$ is a g.s.b. for a suitable graded structure on K (see also Remark 2.2.3). The aim of this section is to describe this structure and to show that the alg. 2.2.2 is a good generalization of the polynomial situation.

REMARK 2.3.1. Let $\mathfrak{A}$ and $\mathfrak{M}$ be as usual, $B = \{b_1, \ldots, b_n\} \subset M - (0)$, $w_i = w(b_i)$. The morphism associated to B is $(\lambda, \Lambda): \mathfrak{L}(B) \to \mathfrak{M}$, where $\mathfrak{L}(B) = (A^n, \Gamma, w^+, \bigoplus_i G(-w_i), \Phi^+)$ (see Section 0). Let us consider the diagram

$$\begin{array}{ccccccc} 0 \to K & \to & A^r & \xrightarrow{\lambda} & M \\ \downarrow & & \downarrow{\scriptstyle \Phi+} & & \downarrow{\scriptstyle \Phi} \\ 0 \to H & \to & \bigoplus_i G(-w_i) & \xrightarrow{\Lambda} & T \end{array}$$

where $K = \ker \lambda = Syz^1(b_1, \ldots, b_n)$ and $H = \ker \Lambda = Syz^1(\Phi(b_1), \ldots, \Phi(b_n))$.

Of course the graded structure on K induced by $\mathfrak{L}(B)$ is $(K, \Gamma, w_K^+, H, \Phi_K^+) := \mathfrak{Ker}(B)$; unfortunately, a g.s.b. for this structure becomes, in the polynomial case described at the beginning of this section, a T-base (see Mora & Möller, 1986). The suitable structure is slightly more complicated and will be described later (see theorem 2.3.4.c), in which it is denoted by $\mathfrak{K}_{\Gamma_1}$).

DEFINITION—PROPOSITION 2.3.2. *Let $\mathfrak{A}$ be as before. Let $\Gamma_1 = \Gamma \oplus \mathbb{Z}$ with the following group ordering: $(\gamma, i) <_{\Gamma_1} (\delta, j) \Leftrightarrow \gamma < \delta$ or $\gamma = \delta$ and $i < j$, where $\mathbb{Z}$ has the usual ordering. Furthermore, defining $v_{\Gamma_1}: A \to \Gamma_1$ by $v_{\Gamma_1}(a) = (v(a), 0)$; $G(\Gamma_1)_{(\gamma, i)} = G_\gamma$ if $i = 0$ and $\{0\}$ otherwise; $G(\Gamma_1) = \bigoplus_{\Gamma_1} G(\Gamma_1)_{(\gamma, i)}$ and $F_{\Gamma_1}: A \to G(\Gamma_1)$ by $F_{\Gamma_1}(a) = F(a)$, we get a graded structure $\mathfrak{A}_{\Gamma_1} = (A, \Gamma_1, v_{\Gamma_1}, G(\Gamma_1), F_{\Gamma_1})$.*

Moreover, if $p: \Gamma_1 \to \Gamma$ is the first projection, then $\mathscr{A}_p = (\mathfrak{A}_{\Gamma_1}, \mathfrak{A}, p)$ is a double structure on A. (see Section 0)

DEFINITION—PROPOSITION 2.3.3. *Let $\mathfrak{M}$ be an $\mathfrak{A}$-module, as usual, and $B = \{b_1, \ldots, b_n\}$ be a basis of M, $w_i = w(b_i)$. We denote by $\mathfrak{N}(B)_{\Gamma_1} = (A^n, \Gamma_1, w_{\Gamma_1}, T(\Gamma_1), \Phi_{\Gamma_1})$ the following $\mathfrak{A}_{\Gamma_1}$-structure on A^n: $w_{\Gamma_1}(a_1, \ldots, a_n) = (w^+(a_1, \ldots, a_n), \bar{i})$, where $\bar{i} = \max\{i = 1, \ldots, n / v(a_i) + w_i = w^+(a_i, \ldots, a_n)\}$, $T(\Gamma_1)_{(\gamma, i)} = G_{\gamma - w_i}$ if $1 \leq i \leq n$ and $\{0\}$ otherwise; $\Phi_{\Gamma_1}(a_1, \ldots, a_n) = 0$ if $(a_1, \ldots, a_n) = (0, \ldots, 0)$ and $F(a_{\bar{i}})$ otherwise. The couple $(\mathfrak{N}(B)_{\Gamma_1}, \mathfrak{L}(B))$ is an $\mathscr{A}_p$-module on A^n.*

THEOREM 2.3.4. *Let $\mathfrak{A}$ and $\mathfrak{M}$ be as above and $B' = \{b_1, \ldots, b_n\}$ be a basis for M. Let us extend B' to a g.s.b. $B = \{b_1, \ldots, b_n, \ldots, b_r\}$ by the algor. 2.2.2, which produces also the elements $\{\tilde{E}(x) \mid x \in S(B)\}$. Let $\{L^{(1)}, \ldots, L^{(k)}\}$ be the homogeneous basis of $H = \ker \Lambda_B$ as in Rmk. 2.1.1. Now let us fix an element $x^{(j)} = e_1^{(j)} b_1 + \cdots + e_r^{(j)} b_r$ in $S(B)$, denote $\tilde{E}^{(j)} := \tilde{E}(x^{(j)})$ and set $t = \max\{i = 1, \ldots, r / e_i^{(j)} \neq 0\}$. Then we have the following facts:*

(*a*) *$w^+(\tilde{E}^{(j)}) = \deg L^{(j)} = v(e_i^{(j)}) + w_i$ for all i such that $e_i^{(j)} \neq 0$;*
(*b*) *$F^+(\tilde{E}^{(j)}) = L^{(j)}$ for all $j = 1, \ldots, k$;*
(*c*) *denoting by $\mathfrak{K}_{\Gamma_1}$ the $\mathfrak{A}_{\Gamma_1}$-structure on $K = \ker \lambda_B$, induced by $\mathfrak{N}(B)_{\Gamma_1}$, we have $w_{\Gamma_1}(\tilde{E}^{(j)}) = (\deg L^{(j)}, t)$ and $\Phi_{\Gamma_1}(\tilde{E}^{(j)}) = F(e_t^{(j)})$;*
(*d*) *$\{\tilde{E}^{(1)}, \ldots, \tilde{E}^{(k)}\}$ is a g.s.b. for $\mathfrak{K}_{\Gamma_1}$ and also for $\mathfrak{Ker}(B)$.*

PROOF. The parts (a), (b), (c) easily follow from the previous algorithms. (d) Keeping the usual notation $\mathfrak{K}_{\Gamma_1} = (K, \Gamma_1, w_{\Gamma_1}, \Phi_{\Gamma_1}(K), \Phi_{\Gamma_1})$.

Since the "Krull-type property" holds for $\mathfrak{K}_{\Gamma_1}$, it is enough to show that $\Phi_{\Gamma_1}(K)$ is generated by $\{\Phi_{\Gamma_1}(\tilde{E}^{(1)}), \ldots, \Phi_{\Gamma_1}(\tilde{E}^{(k)})\}$.

Writing the homogeneous components of $\Phi_{\Gamma_1}(K)$, one can check that these are generated by elements like $F(e_i^{(j)})$, then we are done by (c). The last part of the statement follows from proposition 4.6 (Robbiano, 1986) (it will be recalled in proposition 3.2.1) and from the fact that $(\mathfrak{K}_{\Gamma_1}, \mathfrak{Ker}(B))$ is a double structure.

Let us remark that we can define, by recursion, a structure $\mathfrak{A}_{\Gamma_n}$ on A, for every $n \in \mathbb{N}$, starting from $\mathfrak{A}_{\Gamma_0} = \mathfrak{A}_\Gamma$ and the $\mathfrak{A}_{\Gamma_n}$-module $\mathfrak{N}(B)_{\Gamma_n}$, starting from a module $\mathfrak{M}$ and a set B. It is easy to see that, if we start from an effective structure, we produce, in this way, only effective structures. Applying the previous algorithms, we get the following result:

PROPOSITION 2.3.5. *Let* $\Gamma = \Gamma_0, \Gamma_1, \ldots, \Gamma_n$, $\mathfrak{A} = \mathfrak{A}_\Gamma$, $\mathfrak{M} = \mathfrak{M}_\Gamma$ *be as usual.*

(*i*) *If* B' *is a basis of* M, *the algorithm* 2.2.2 *produces a g.s.b.* $B = B_{r_1}$ *of* $\mathfrak{M}_\Gamma$ *(where* $|B_{r_1}| = r_1$*) and a g.s.b.* $\{\tilde{E}_1^{(j)}\}_{1 \leq j \leq r_2}$ *of* $\mathfrak{K}_{\Gamma_1} = (\ker \lambda^{(1)}, \Gamma_1, \ldots)$, *where* $\lambda^{(1)}: A^{r_1} \to M$.

(*ii*) *If* $n > 1$, *starting from a g.s.b.* $\{\tilde{E}_{(n-1)}^{(j)}\}_{1 \leq j \leq r_n}$ *of* $\mathfrak{K}_{\Gamma_{n-1}}$, *the algor.* 2.2.2 *produces a g.s.b.* $\{\tilde{E}_n^{(j)}\}_{1 \leq j \leq r_{n+1}}$ *of* $\mathfrak{K}_{\Gamma_n} = (\ker \lambda^{(n)}, \Gamma_n, \ldots)$, *where* $\lambda^{(n)}: A^{r_n} \to A^{r_{n-1}}$ *is the nth boundary of this free resolution of M.*

3. Applications

3.1. QUOTIENTS OF GRADED STRUCTURES

The aim of this section is to define a Gröbner base for an ideal of a quotient of a polynomial ring and to compute such bases.

Let $P = k[X_1, \ldots, X_n]$, T be the set of all terms of P and log: $T \to \mathbb{Z}^n$ be as usual. We will identify T and $\mathrm{Im}(\log) = \mathbb{N}^n$.

Given an ideal I of P, the problem to determine a "canonical" base of the k-vector space P/I was solved by Buchberger:

THEOREM 3.1.1. (Buchberger, 1965) *Let us fix a t.o.* $<$ *in* $\mathbb{Z}^n$; *denoting by* $N(I) = T - (T \cap M(I))$, *the set* $\{[x] \bmod I / x \in N(I)\}$ *is a base of* P/I *as* k*-vector space.*

Furthermore Buchberger's algorithm to compute Gröbner bases allows one to find, given $[f]$ in P/I, a canonical representative of $[f]$.

In fact there exists $R \subset P$ and a computable map $\sigma: P \to R$ such that $\sigma(f)$ is the unique element of R such that $f \equiv \sigma(f) \bmod I$ and $0, 1 \in R$.

Since Buchberger's reduction "$\xrightarrow[B]{}*$" is confluent (this means: if $f \xrightarrow[B]{}* f'$ and $f \xrightarrow[B]{}* f''$, then there exists f° such that $f' \xrightarrow[B]{}* f^\circ$ and $f'' \xrightarrow[B]{}* f^\circ$), if B is a Gröbner base for I, then for every $f \in P$, there exists a unique irreducible element g such that $f \xrightarrow[B]{}* g = \mathrm{red}_B(f)$.

So, the set of all canonical representatives is $R = \{g \in P / g \text{ irreducible}\}$.

Now let $\bar{P} = P/I$; setting $\tau = \{M(\sigma(f))/f \in P\}$, τ can be regarded as a subset of T; let $\overline{\log}$: $\tau \to \mathbb{N}^n$ be the restriction of the map "log".

REMARK 3.1.2.

(1) The maps $\bar{v}$: $\bar{P} \to \mathbb{N}^n$, where $\bar{v}([f]) = \overline{\log}(M(\sigma(f)))$ *and* g: $\bar{P} \to \tau$, where $g([f]) = M(\sigma(f))$ are well-defined;

(2) $\tau = T-(T\cap M(I))$;
(3) the map $g\colon \bar{P}\to P$ induces a map $\bar{M}\colon P/I\to P/M(I)$ defined by $\bar{M}([f])=[M(\sigma(f))]$; so, by 2), $\bar{M}([f])=0 \Leftrightarrow [f]=0$.

It is easily checked that $\bar{\mathfrak{A}}=(\bar{P},\Gamma,\bar{v},P/M(I),\bar{M})$ is a graded structure and an $\mathfrak{A}$-module, where $\mathfrak{A}=(P,\Gamma,v,P,M)$ (as in Section 0, Ex. 4). $P/M(I)$ is canonically graded over $\mathbb{Z}^n=\Gamma$, setting for $A\in\mathbb{N}^n$, $(P/M(I))_A=$ the k-vector space generated by X^A, if $X^A\notin M(I)$, or (0), if $X^A\in M(I)$.

Proposition 3.1.3. *Let* $\mathfrak{A}$ *be as above,* $\mathfrak{M}=(M,\Gamma,w,T,\Phi)$ *be a Krull module and* $\mathfrak{N}=(N,\Gamma,w_N,S,\phi_N)$ *be a submodule.*

Then the quotient $\mathfrak{M}/\mathfrak{N}$ *exists and has the following form:* $(\bar{M},\Gamma,\bar{w},\bar{T},\bar{\Phi})$, *where:* $\bar{M}=M/N$, $\bar{w}\colon \bar{M}-(0)\to\Gamma$ *is defined by* $\bar{w}([x])=\min\{w(y)/[y]=[x]\}$, $\bar{T}=T/S$ *and* $\bar{\Phi}\colon \bar{M}\to\bar{T}$ *is defined by* $\bar{\Phi}([x])=[\Phi(y)]$, *where* y *is such that* $[x]=[y]$ *and* $w(y)=\bar{w}([x])$.

Proof. The existence of $\mathfrak{M}/\mathfrak{N}$ follows from Prop. 1.4 (Robbiano, 1986). The last part of the statement easily follows from the construction of the equivalence between the categories $\mathfrak{G}_{\mathfrak{A}}$ and $\mathfrak{G}_{\mathfrak{A}^*}$ (see Section 0), recalling that $\mathfrak{N}$ is a submodule of $\mathfrak{M}$ if N is a submodule of M and $F_\gamma N=(F_\gamma M)\cap N$, for all $\gamma\in\Gamma$.

Theorem 3.1.4. *Let* $\mathfrak{A}$, $\mathfrak{M}$ *and* $\mathfrak{N}$ *be as above and assume* $\mathfrak{M}$ *is a Krull module. Let* $B=\{m_1,\ldots,m_r\}$ *be a g.s.b. of* $\mathfrak{M}$. *Then* $\bar{B}=\{[m_1],\ldots,[m_r]\}$, *where* $[m_i]$ *is the class of* m_i *in* M/N, *is a g.s.b. for* $\mathfrak{M}/\mathfrak{N}$.

Proof. Let $[x]\in M/N-(0)$ and $y\in M$ such that $[y]=[x]$ and $w(y)=\bar{w}([x])$. Since B is a g.s.b. of M, there exist $a_1,\ldots,a_r$ in A such that $y=\sum_i a_i m_i$ and $w(y)\geqslant v(a_i)+w(m_i)$, for every i such that $a_i\neq 0$. Hence

$$[x]=[y]=\sum_i a_i[m_i] \quad\text{and}\quad \bar{w}([x])\geqslant v(a_i)+w(m_i)\geqslant v(a_i)+\bar{w}([m_i]).$$

Remark 3.1.5. Let $\mathfrak{A}$ and $\bar{\mathfrak{A}}$ as in Rmk. 3.1.2; if $\mathfrak{J}$ is the graded structure induced by $\mathfrak{A}$ on the ideal I, then $\bar{\mathfrak{A}}=\mathfrak{A}/\mathfrak{J}$.

Since the description of terms and valuations in P/I can be set up in the category $\mathfrak{G}_{\mathfrak{A}}$, the following definition arises in a natural way:

Definition. Let $J\supset I$ be two ideals of P. We say that $\{[f_1],\ldots,[f_p]\}$ is a *Gröbner base* for J/I as ideal of P/I if it is a g.s.b. for the graded structure $\mathfrak{J}/\mathfrak{J}=(J/I,\ldots)$.

Remark 3.1.6. By thm. 3.1.4, we can find a Gröbner base for J/I just computing a Gröbner base $\{f_1,\ldots,f_p\}$ for J and taking the elements $\mathrm{red}_B(f_1),\ldots,\mathrm{red}_B(f_p)$, where B is a Gröbner base of I; these are canonical representatives for $[f_1],\ldots,[f_p]$, which are a Gröbner base for J/I.

3.2. Generalized Lazard's Algorithm

Proposition 3.2.1. (Robbiano, 1986, proposition 4.6) *Let* $\mathscr{A}_\alpha=(\mathfrak{A}_\Gamma,\mathfrak{A}_\Delta,\alpha)$ *be a double structure over a ring* A *and* $\mathscr{M}_\alpha=(\mathfrak{M}_\Gamma,\mathfrak{M}_\Delta)$ *be a finitely generated* $\mathscr{A}_\alpha$*-module. If* $\{m_1,\ldots,m_r\}$ *is a g.s.b. for* $\mathfrak{M}_\Gamma$, *then it is a g.s.b. for* $\mathfrak{M}_\Delta$.

REMARK 3.2.2. If $A = k[X_1, \ldots, X_n]$, $\Gamma = \mathbb{Z}^n$, $\Delta = \mathbb{Z}$, $<_\Gamma$ is a homogeneous term-ordering on Γ and $\alpha: \Gamma \to \Delta$ is defined by $\alpha(a_1, \ldots, a_n) = \sum_i a_i$, then:

(1) denoting by $\mathfrak{A}_\Gamma$ and $\mathfrak{A}_\Delta$ the two structures described in Section 0, Remark 4 (3 and 4 respectively), it turns out that $(\mathfrak{A}_\Gamma, \mathfrak{A}_\Delta, \alpha)$ is a double structure.
So the fact that every Gröbner base is an H-base, with a suitable term-ordering (Section 0, prop. 2) is an obvious corollary of prop. 3.2.1.

(2) Denoting by $\mathfrak{A}_\Gamma$ and $\mathfrak{A}_\Delta$ the two structures defined in (Section 0, Remark 4 and 1 respectively), where the ideal I is $\mathfrak{M} = (X_1, \ldots, X_n)$, it is easy to verify that $(\mathfrak{A}_\Gamma, \mathfrak{A}_\Delta, \alpha)$ is a double structure. It is well-known that an L-base is also an $\mathfrak{M}$-standard base; to prove this one cannot use Proposition 3.2.1 since an L-base is not a g.s.b. for $\mathfrak{A}_\Delta$. We will give a proof in a slightly more general setting in 3.3.

It is well-known that, in the case of L-bases, the Buchberger algorithm does not necessarily converge.

Lazard's algorithm (Lazard, 1983) uses a homogenization map from $k[X_1, \ldots X_n]$ to $k[X_0, \ldots X_n]$ to do computations in this new ring, in which the order of the terms is really a suitable term-ordering; of course, since $k[X_1, \ldots X_n]$ is already a graded ring itself, the homogenization makes sense.

Another approach to do computations in $G_{\mathfrak{M}}(P)$ is to start from a local ring A, e.g. $A = P_{\mathfrak{M}}$. In this case Mora (1982) was able to find an L-base of an ideal I, following the hint of the Buchberger algorithm but avoiding the possibility of infinitely many steps.

The two above algorithms are not tricks: they use some intrinsic properties of the rings they deal with, which guarantee the finiteness of the computation, not fulfilled in an arbitrary negatively graded structure.

We will show that it is quite natural to set up the Lazard construction inside the theory of graded structures, assuming some "good properties".

DEFINITION. Let $A = \bigoplus_{\gamma\in\Gamma} A_\gamma$ be a Γ-graded ring (Γ an ordered group). We can consider the quintuple $\mathbf{A}_\Gamma = (A, \Gamma, v_\Gamma, A, F_\Gamma)$, where for every $a \in A - (0)$, if $a = a_{\gamma_1} + \cdots + a_{\gamma_p}$ is the Γ-homogeneous decomposition and $\gamma_1 <_\Gamma \gamma_2 <_\Gamma \cdots <_\Gamma \gamma_p$, then $v_\Gamma: A - (0) \to \Gamma$ is defined by $v_\Gamma(a) := \gamma_p$, and $F_\Gamma: A \to A$ is defined by $F_\Gamma(a) := a_{\gamma_p}$.

Then $\mathbf{A}_\Gamma$ is a graded structure and we will say that it is the *special graded structure* (s.g.s.) associated to A.

DEFINITION. Let $\mathscr{A}_\alpha$ be a double structure over a ring A, where $\alpha: \Gamma \to \mathbb{Z}$ is an ordered group homomorphism. We say that $\mathscr{A}_\alpha$ is a *special double structure* (s.d.s.) if $\mathscr{A}_\alpha = (\mathbf{A}_\Gamma, \mathfrak{A}_\mathbb{Z}, \alpha)$, where $\mathbf{A}_\Gamma$ is the s.g.s. over A. We say that $\mathscr{A}_\alpha$ is a *very special double structure* (v.s.d.s.) if $\mathscr{A}_\alpha = (\mathbf{A}_\Gamma, \mathbf{A}_\mathbb{Z}, \alpha)$, where $\mathbf{A}_\Gamma$ and $\mathbf{A}_\mathbb{Z}$ are both s.g.s. over A.

EXAMPLES. Let $A = k[X_1, \ldots, X_n]$ be a polynomial ring. The double structures in Remark 3.2.2, (1) and (2) are both v.s.d.s.

REMARK 3.2.3. Let $A = \bigoplus_{\gamma\in\Gamma} A_\gamma$ be a Γ-graded ring and $\alpha: \Gamma \to \mathbb{Z}$ be any ordered group homomorphism. Then A is also a $\mathbb{Z}$-graded ring, writing $A = \bigoplus_{\delta\in\mathbb{Z}} A_\delta$ and $A_\delta = \bigoplus_{\alpha(\gamma)=\delta} A_\gamma$ and $(\mathbf{A}_\Gamma, \mathbf{A}_\mathbb{Z}, \alpha)$ is a v.s.d.s.

Now we introduce a polynomial structure associated to a special double structure and the homogenization procedure.

DEFINITION. Let $\mathscr{A}_\alpha = (\mathbf{A}_\Gamma, \mathfrak{A}_\mathbb{Z}, \alpha)$ be a s.d.s. over A and $\Theta = \Gamma \oplus \mathbb{Z}$ with the total ordering $<$ defined by $(\gamma, n) < (\gamma', n') \Leftrightarrow \alpha(\gamma) + n < \alpha(\gamma') + n'$ or $\alpha(\gamma) + n = \alpha(\gamma') + n'$ and $\gamma >_\Gamma \gamma'$. Then Θ is an ordered group and the polynomial ring $A[t]$ is naturally Θ-graded via $A[t] = \oplus_{\theta \in \Theta} A[t]_\theta$, where $A[t]_\theta = \{at^i \mid a \in A$ is Γ-homogeneous and $(-v_\Gamma(a), i) = \theta\}$. We will denote by $\mathbf{A[t]}_\Theta$ the s.g.s. $(A[t], \Theta, v^*_\Theta, A[t], F^*_\Theta)$ associated to $A[t]$ w.r.t. the Θ-degree.

DEFINITION. Let $\mathscr{A}_\alpha = (\mathbf{A}_\Gamma, \mathfrak{A}_\mathbb{Z}, \alpha)$ be a s.d.s. over A and β: $\Theta = \Gamma \oplus \mathbb{Z} \to \mathbb{Z}$ be defined by $\beta(\gamma, n) = \alpha(\gamma) + n$. Since β is an ordered group homomorphism, by Remark 3.2.3, $A[t]$ is a Θ-graded ring and also a $\mathbb{Z}$-graded ring. We will denote by $\mathscr{A}[\ell]_\beta$ the v.s.g.s. $(\mathbf{A[t]}_\Theta, \mathbf{A[t]}_\mathbb{Z}, \beta)$, where $\mathbf{A[t]}_\mathbb{Z} = (A[t], \mathbb{Z}, v^*_\mathbb{Z}, A[t], F^*_\mathbb{Z})$.

REMARK 3.2.4. If $y \in A$ is a $\mathbb{Z}$-homogeneous then $v^*_\Theta(yt^s) = (-v_\Gamma(t), s)$ and $v^*_\mathbb{Z}(yt^s) = -v_\mathbb{Z}(y) + s$.

DEFINITION. Let $\mathscr{A}_\alpha = (\mathbf{A}_\Gamma, \mathfrak{A}_\mathbb{Z}, \alpha)$ be a s.d.s., $\mathscr{A}[\ell]_\beta = (\mathbf{A[t]}_\Theta, \mathbf{A[t]}_\mathbb{Z}, \beta)$ the associated v.s.d.s.. We define the *dehomogenization morphism* a: $A[t] \to A$ via $a(p(t)) = p(1_A)$ and the *homogenization map* h: $A \to A[t]$ in the following way: for every $x \in A$, let $x = x_0 + \cdots + x_r$ by the $\mathbb{Z}$-homogeneous decomposition, i.e. $x_i \in A_{d_i}$, $d_i \in \mathbb{Z}$, $d_0 < \cdots < d_r$. Let us define $h(x) = x_0 + x_1 t^{i_1} + \cdots + x_r t^{i_r}$, where $i_1 = d_1 - d_0, \ldots, i_r = d_r - d_0$.

In this way $h(x)$ is $\mathbb{Z}$-homogeneous in $A[t]$ of degree $-d_0$. It is easy to verify that the following properties hold:

PROPOSITION 3.2.5.

(1) *For every x in A, $a(h(x)) = x$.*

(2) *For every $g(t)$ $\mathbb{Z}$-homogeneous in $A[t]$: $a(F^*_\Theta(g(t))) = F_\Gamma(a(g(t)))$.*

(3) *If $x = \sum_i x_i f_i$ in A, then there exist $n, n_1, \ldots, n_r \in \mathbb{N}$ such that $t^n h(x) = \sum_i t^{n_i} h(x_i) h(f_i)$.*

Let us consider the following situation: let $\mathscr{A}_\alpha = (\mathbf{A}_\Gamma, \mathfrak{A}_\mathbb{Z}, \alpha)$ be a s.d.s. and $\mathscr{A}[\ell]_\beta = (\mathbf{A[t]}_\Theta, \mathbf{A[t]}_\mathbb{Z}, \beta)$ be as above.

Let $J = (f_1, \ldots, f_p)$ be an ideal of A and $\tilde{J} = (h(f_1), \ldots, h(f_p)) \subset A[t]$.

With obvious notations $\mathscr{J}_\alpha = (\mathfrak{J}_\Gamma, \mathfrak{J}_\mathbb{Z}, \alpha)$ denotes the corresponding substructure of $\mathscr{A}_\alpha$ induced on J and $\tilde{\mathscr{J}}_\beta = (\tilde{\mathfrak{J}}_\Theta, \tilde{\mathfrak{J}}_\mathbb{Z}, \beta)$ is the substructure of $\mathscr{A}[\ell]_\beta$ induced on $\tilde{J}$.

THEOREM 3.2.6. *If $\{z_1, \ldots, z_s\}$ is a standard set of $\tilde{\mathfrak{J}}_\Theta$ and the z_i's are $\mathbb{Z}$-homogeneous in $A[t]$, then $\{a(z_1), \ldots, a(z_s)\}$ is a standard set for $\mathfrak{J}_\Gamma$.*

PROOF. Let us recall that $\tilde{\mathfrak{J}}_\Theta = (\tilde{J}, \Theta, v^*_\Theta, F^*_\Theta(\tilde{J}), F^*_\Theta)$. By assumption, $F^*_\Theta(\tilde{J}) = (F^*_\Theta(z_1), \ldots, F^*_\Theta(z_s))$. We have to show that $F_\Gamma(J) = (F_\Gamma(a(z_1)), \ldots, F_\Gamma(a(z_s)))$. Let $x \in J$, so $x = \sum_i x_i f_i$. By prop. 3.2.5, 3), there exist $n, n_1, \ldots, n_p \in \mathbb{N}$ such that $t^n h(x) = \sum_i t^{n_i} h(x_i) h(f_i)$. So $F^*_\Theta(t^n h(x)) \in F^*_\Theta(\tilde{J})$ i.e. $t^n F^*_\Theta(h(x)) = \sum_j y_j F^*_\Theta(z_j)$, for suitable $y_j \in A[t]$, so $a(t^n F^*_\Theta(h(x))) = a(\sum_j y_j F^*_\Theta(z_j)) = \sum_j a(y_j) a(F^*_\Theta(z_j))$. But $h(x), z_1, \ldots, z_s$ are $\mathbb{Z}$-homogeneous, so by prop. 3.2.5, 1) and 2) we have $F_\Gamma(x) = F_\Gamma(a(h(x))) = \sum_j a(y_j) F_\Gamma(a(z_j))$.

REMARK 3.2.7. Now the problem is: how to construct a st. set for $\tilde{\mathfrak{J}}_\Theta$ whose elements are $\mathbb{Z}$-homogeneous? Since $\mathscr{A}[\ell]_\beta$ is a v.s.d.s. $\tilde{\mathscr{J}}_\beta = (\tilde{\mathfrak{J}}_\Theta, \tilde{\mathfrak{J}}_\mathbb{Z}, \beta)$ is a f.g. module on it and $\tilde{\mathfrak{J}}_\mathbb{Z} = \mathbf{J}_\mathbb{Z}$, i.e. it is a s.g.s., our problem can be stated in the following situation:

$\mathscr{A}_\alpha = (\mathbf{A}_\Gamma, \mathbf{A}_\mathbb{Z}, \alpha)$ is a noetherian v.s.d.s. and $\mathscr{M}_\alpha = (\mathfrak{M}_\Gamma, \boldsymbol{M}_\mathbb{Z})$ is a f.g. $\mathscr{A}_\alpha$-module, $\mathbf{M}_\mathbb{Z}$ special module on the A-module M, which is $\mathbb{Z}$-graded.

THEOREM 3.2.8. *With the above notations, let* $\{m_1, \ldots, m_r\}$ *be a* $\mathbb{Z}$*-homogeneous system of generators of* M. *Assume that* $\Gamma^0(\mathfrak{M}_\Gamma)$ *is well-ordered. Then, applying the algorithm* 2.2.2 *to* $\{m_1, \ldots, m_r\}$, *we get a g.s.b. of* $\mathfrak{M}_\Gamma$ *whose elements are* $\mathbb{Z}$*-homogeneous in* M.

PROOF. It is easy to prove this fact, reading carefully the steps of the algorithm. Let us remark just the key points.

(i) In the following commutative diagram the second row is a sequence of Γ-graded A-modules:

$$\begin{array}{ccccc} 0 \leftarrow K & \leftarrow & A^r & \overset{\lambda}{\leftarrow} & M \\ \downarrow & & \downarrow \Phi^+ & & \downarrow \Phi \\ 0 \leftarrow H \leftarrow & & \bigoplus_i A(-w_i) & \overset{\Lambda}{\leftarrow} & T_\Gamma \end{array}$$

where $w_i = w_\Gamma(m_i)$ and λ, Λ are as in Section 0. Then it is possible to choose a lifting in A^r, via Φ^+, of any Γ homogeneous element of H, such that its components are $\mathbb{Z}$-homogeneous in A.

(ii) In this way, if the elements of B are $\mathbb{Z}$-homogeneous in M, then the elements of $S(B)$ are $\mathbb{Z}$-homogeneous in M.

(iii) Let $x \in M$, $B = \{b_1, \ldots, b_n\}$, $b_i \in M$, and $x, b_1, \ldots, b_n$ be $\mathbb{Z}$-homogeneous. If $x \xrightarrow[B]{}{}^* x'$, then x' is $\mathbb{Z}$-homogeneous in M.

THEOREM 3.2.9. *Let* $\mathscr{A}_\alpha = (\mathbf{A}_\Gamma, \mathfrak{A}_\mathbb{Z}, \alpha)$ *be a noetherian effective s.d.s. and let us assume that* $(\operatorname{Ker} \alpha) \cap \Gamma^0(\mathbf{A}_\Gamma) = (0)$ *and* $\Gamma^0(\mathbf{A}_\Gamma) \leqslant 0$.

Let $J = (f_1, \ldots, f_p)$ *be an ideal of* A *and* $\mathscr{J}_\alpha = (\mathfrak{J}_\Gamma, \mathfrak{J}_\mathbb{Z}, \alpha)$ *be the induced structure of* J. *The following procedure give rise a standard set for* $\mathfrak{J}_\Gamma$:

(1) Homogenize *the elements* $\{f_1, \ldots, f_p\}$; *let* $(h(f_1), \ldots, h(f_p)) = \tilde{J}$ *ideal of* $A[t]$.

(2) Compute *a* $\mathbb{Z}$*-homogeneous g.s.b., say* $\{z_1, \ldots, z_s\}$, *of* $\tilde{\mathfrak{J}}_\Theta$, *the substructure of* $\mathbf{A}[\mathbf{t}]_\Theta$ *induced on* $\tilde{J}$, starting from *the generators* $\{h(f_1), \ldots, h(f_p)\}$.

(3) Dehomogenize $z_1, \ldots, z_s$, *getting* $\{a(z_1), \ldots, a(z_s)\}$, *which is a standard set for* $\mathfrak{J}_\Gamma$.

PROOF. Step 2. The computation can be done with the algorithm 2.2.2.: since $\Gamma^0(\mathbf{A}_\Gamma) \leqslant 0$ and $(\operatorname{Ker} \alpha) \cap \Gamma^0(\mathbf{A}_\Gamma) = (0)$ then $\mathbf{A}[\mathbf{t}]_\Theta$ is an effective positively graded structure, therefore it is well-ordered. Furthermore, if we start from $\mathbb{Z}$-homogeneous generators of $\tilde{J}$, by thm. 3.2.8 we get a $\mathbb{Z}$-homogeneous g.s.b. of $\tilde{\mathfrak{J}}_\Theta$. Step 3 follows from theorem 3.2.6.

3.3. *I*-STANDARD BASES

The aim of this section is to give a constructive procedure furnishing an I-standard base for an ideal K of a polynomial ring with coefficients in a field. This is well-known when I is the ideal generated by the variables.

The problem is then the following: let $P = k[X_1, \ldots, X_n]$, $I = (g_1, \ldots, g_s)$ and K be any ideal of P. Let $\mathfrak{P} = (P, \mathbb{Z}, -v_I, G_I(P), \mathrm{in}_I)$ be the associated graded structure (Ex. 1, Section 0) and $\mathfrak{K} = (K, \mathbb{Z}, -v_I, in_I(K), in_I)$ its substructure.

Let us recall that the notion of I-standard base of K and standard set of $\mathfrak{K}$ are equivalent. Let us consider the following commutative diagram

$$\begin{array}{ccccccccc} 0 & \to & L & \to & P[T_1, \ldots, T_s] & \xrightarrow{\pi} & P & \to & 0 \\ & & \| \uparrow & & \cup\uparrow & & \cup\uparrow & & \\ 0 & \to & L & \to & J & \to & K & \to & 0 \end{array}$$

where $\pi(T_i) = g_i$ and $\pi(a) = a$, for each a in P; $L = \ker \pi = (g_1 - T_1, \ldots, g_s - T_s)$; $J = \pi^{-1}(K)$.

Setting $A = P[T_1, \ldots, T_s]$ and $\mathfrak{M} = (T_1, \ldots, T_s) \subset A$, we can define a valuation on A $v_{\mathbb{Z}} = -v_{\mathfrak{M}}$: $A \to \mathbb{Z}$ and an "initial form" $F_{\mathbb{Z}} = \text{in}_{\mathfrak{M}}$: $A \to G_{\mathfrak{M}}(A)$. So $\mathfrak{A}_{\mathbb{Z}} = (A, \mathbb{Z}, v_{\mathbb{Z}}, G_{\mathfrak{M}}(A), F_{\mathbb{Z}})$ is a graded structure. Let $\mathfrak{J}_{\mathbb{Z}} = (J, \mathbb{Z}, v_{\mathbb{Z}}, F_{\mathbb{Z}}(J), F_{\mathbb{Z}})$ and $\mathfrak{L}_{\mathbb{Z}} = (L, \mathbb{Z}, v_{\mathbb{Z}}, F_{\mathbb{Z}}(L), F_{\mathbb{Z}})$ be induced substructures on J and L, respectively.

PROPOSITION 3.3.1. *With the previous notations*:

(*i*) $\mathfrak{A}_{\mathbb{Z}}/\mathfrak{L}_{\mathbb{Z}}$ *exists and is isomorphic to* $\mathfrak{P}$;

(*ii*) $\mathfrak{J}_{\mathbb{Z}}/\mathfrak{L}_{\mathbb{Z}}$ *exists and is isomorphic to* $\mathfrak{K}$;

(*iii*) *if* $\{m_1, \ldots, m_p\}$, *is a standard set for* $\mathfrak{J}_{\mathbb{Z}}$, *then* $\{\bar{m}_1, \ldots, \bar{m}_p\}$, *with* $\bar{m}_i \in J/L \cong K$, *is an I-standard base of K.*

PROOF.

(i) The valued filtration on A induced by $v_{\mathbb{Z}}$ is $F_{-n}A = \mathfrak{M}^n$; so the induced filtered structure on L is $F_{-n}L = \mathfrak{M}^n \cap L$. Therefore $\{F_{-n}(A/L)\}$ is the I-filtration on $P \cong A/L$.

(ii) Analogous.

(iii) Let $\hat{A}$ be the $\mathfrak{M}$-adic completion of A. Mora in Mora (1988, Part I) defines, starting from $\mathfrak{A}_{\mathbb{Z}}$, a graded structure $\hat{\mathfrak{A}}$ on $\hat{A}$. It is easy to see that $\hat{\mathfrak{A}}$ is just the structure $(\hat{A}, \mathbb{Z}, -v_{\mathfrak{M}^\wedge}, G_{\mathfrak{M}^\wedge}(\hat{A}), \text{in}_{\mathfrak{M}^\wedge})$ (as in example 1, Section 0). He shows also that (under some assumptions on Γ^0, which are here fulfilled) B is a standard set of $\mathfrak{A}_{\mathbb{Z}}$ if and only if B is a standard set of $\mathfrak{A}$. Since $\mathfrak{M}^\wedge \subseteq \text{rad}(\hat{A})$ (Atiyah & Macdonald, 1969; prop. 10.15), $\hat{\mathfrak{A}}$ is Krull (Remark 5, Section 0). Hence each substructure of $\hat{\mathfrak{A}}$ is Krull and the quotients of substructures of $\hat{\mathfrak{A}}$ are defined. In particular a g.s.b. of $\hat{\mathfrak{J}}$ projects on a g.s.b. of $\hat{\mathfrak{J}}/\hat{\mathfrak{L}}$, by thm. 3.1.4. It is easy to prove, using ii), that $\hat{\mathfrak{J}}/\hat{\mathfrak{L}} \cong \hat{\mathfrak{K}}$, where $\hat{\mathfrak{K}}$ is the structure on $\hat{K} = I$-adic completion on K. From being $\{m_1, \ldots, m_p\}$ a st. set for $\hat{\mathfrak{J}}_{\mathbb{Z}}$, we get it is a st. set also for $\hat{\mathfrak{J}}$. But $\hat{\mathfrak{J}}$ is Krull, then $\{m_1, \ldots, m_p\}$ is a g.s.b. for $\hat{\mathfrak{J}}$; so $\{\bar{m}_1, \ldots, \bar{m}_p\}$ is a g.s.b. for $\hat{\mathfrak{K}} \cong \hat{\mathfrak{J}}/\hat{\mathfrak{L}}$. Therefore $\{\bar{m}_1, \ldots, \bar{m}_p\}$ is a standard set for $\mathfrak{K}$, i.e., it is an I-standard base of K.

In this way we have reduced ourselves to find a st. set for $\mathfrak{J}_{\mathbb{Z}}$, i.e. an $\mathfrak{M} = (T_1, \ldots, T_s)$ – st. base for $J = (K, g_1 - T_1, \ldots, g_s - T_s)$ in $A = P[T_1, \ldots, T_s]$. This is not the known setting, since P is not a field, but this can be solved using the "generalized Lazard algorithm" in 3.2.

Let $\Gamma = \mathbb{Z}^s$ with any homogeneous ordering $<_\Gamma$. Let F_Γ: $A \to A$ be defined by $F_\Gamma(f(X_i, T_i)) =$ [minimum (w.r.t. $<_\Gamma$) monomial in the T_i's occurring in f] and v_Γ: $A \to \Gamma$ be defined by $v_\Gamma(f(X_i, T_i)) = -\log F_\Gamma(f)$.

In this way A is a Γ-graded ring and $\mathbf{A}_\Gamma = (A, \Gamma, v_\Gamma, A, F_\Gamma)$ is a s.g.s. Furthermore

$\mathscr{A}_\alpha = (\mathbf{A}_\Gamma, \mathfrak{A}_\mathbb{Z}, \alpha)$ is a s.d.s., where $\alpha\colon \mathbb{Z}^s \to \mathbb{Z}$ sends $(n_1, \ldots, n_s)$ to $n_1 + \cdots + n_s$. With obvious notations $\mathscr{J}_\alpha = (\mathfrak{J}_\Gamma, \mathfrak{J}_\mathbb{Z}, \alpha) \subset \mathscr{J}_\alpha$.

REMARK 3.3.2. The procedure of theorem 3.2.9, giving a standard set of $\mathfrak{J}_\Gamma$, starting from any system of generators of J, can be here applied. In fact both the requirements $(\ker \alpha) \cap \Gamma^0(\mathbf{A}_\Gamma) = (0)$ and $\Gamma^0(\mathbf{A}_\Gamma) \leqslant 0$ hold.

At this point, we have just to show that a standard set of $\mathfrak{J}_\Gamma$ is a standard set of $\mathfrak{J}_\mathbb{Z}$. After this all the procedure is completed and constructive. In fact we get a standard set $\{m_1, \ldots, m_p\}$ of $\mathfrak{J}_\mathbb{Z}$ and we can compute $\bar{m}_1, \ldots, \bar{m}_p$ in P, by substituting the T_i's with the g_i's, getting an I-standard base of K. Let us finally remark that, in a general double structure $(\mathfrak{J}_\Gamma, \mathfrak{J}_\mathbb{Z}, \alpha)$, it is not true that a standard set of $\mathfrak{J}_\Gamma$ is a standard set of $\mathfrak{J}_\mathbb{Z}$, unless $\mathfrak{J}_\Gamma$ is a Krull structure. This is not our case; however the particularity of our situation allows us to prove the result.

THEOREM 3.3.3. *Let P be a commutative ring, $A = P[T_1, \ldots, T_s]$, J be an ideal of A, $\mathfrak{M} = (T_1, \ldots, T_s)$, $\Gamma = \mathbb{Z}^s$ with a homogeneous ordering and $\mathfrak{J}_\Gamma = (J, \Gamma, v_\Gamma, F_\Gamma(J), F_\Gamma)$, $\mathfrak{J}_\mathbb{Z} = (J, \mathbb{Z}, v_\mathbb{Z} = -v_\mathfrak{M}, F_\mathbb{Z}(J), F_\mathbb{Z} = \mathrm{in}_\mathfrak{M})$ be as above. If $\{y_1, \ldots, y_r\}$ is a standard set of $\mathfrak{J}_\Gamma$, then it is a standard set of $\mathfrak{J}_\mathbb{Z}$.*

PROOF. Let $y \in J$. By assumption $F_\Gamma(y) = \sum_i a_i F_\Gamma(y_i)$, where $a_i = F_\Gamma(a_i) \neq 0$ are Γ-homogeneous and $v_\Gamma(y) = v_\Gamma(a_i) + v_\Gamma(y_i)$. Therefore $v_\Gamma(y) = v_\Gamma(\sum_i a_i y_i)$. Let $y' = y - \sum_i a_i y_i$; obviously $v_\mathbb{Z}(y) = v_\mathbb{Z}(a_i) + v_\mathbb{Z}(y_i) = v_\mathbb{Z}(\sum_i a_i y_i)$. If $y' = 0$, then $F_\mathbb{Z}(y) = \sum_i F_\mathbb{Z}(a_i) F_\mathbb{Z}(y_i)$ and we are done. If $y' \neq 0$, then $v_\Gamma(y') < v_\Gamma(y)$, so $v_\mathbb{Z}(y') \leqslant v_\mathbb{Z}(y)$, Then either $v_\mathbb{Z}(y') < v_\mathbb{Z}(y)$ and we are done, or $v_\mathbb{Z}(y') = v_\mathbb{Z}(y)$. In this case $F_\mathbb{Z}(y) = \sum_i F_\mathbb{Z}(a_i) F_\mathbb{Z}(y_i) + F_\mathbb{Z}(y')$. Since $y' \in J$, we can repeat the procedure of y' and write $y'' = y' - \sum_i b_i y_i$. If $y'' = 0$ or $v_\mathbb{Z}(y'') < v_\mathbb{Z}(y')$, then $F_\mathbb{Z}(y') = \sum_i F_\mathbb{Z}(b_i) F_\mathbb{Z}(y_i)$, so we can stop. Otherwise we get an y''' and so on. Therefore we construct a sequence $y', y'', y'''\ldots$ of elements of J such that $v_\Gamma(y') > v_\Gamma(y'') > \cdots$ and $v_\mathbb{Z}(y') = v_\mathbb{Z}(y'') = \cdots$. So we have a set $V = \{\gamma, \gamma', \gamma'', \ldots\} \subset \Gamma^0(\mathbf{A}_\Gamma)$ with $\alpha(\gamma) = \alpha(\gamma') = \alpha(\gamma'') = \cdots$. For every $n \in \mathbb{Z}$, $\alpha^{-1}(n) \cap \Gamma^0(\mathbf{A}_\Gamma) = \{$solutions in $-\mathbb{N}^s$ of the equation $X_1 + \cdots + X_s = n\}$ is a finite set, so V must be finite and the procedure stops.

Acknowledgement

We are indebted to Corrado De Concini and Michael Stillman for their warm support, insightful ideas and critical help with this paper. We also thank Teo Mora and Lorenzo Robbiano for useful discussions.

References

Atiyah, M. F., Macdonald, I. G. (1969). *Introduction to Commutative Algebra*. Addison Wesley Publ. Comp.

Buchberger, B. (1965). *Ein Algorithmus zum Auffinden der Basiselemente*. Ph.D. Thesis, Innsbruck.

Buchberger, B. (1976). A theoretical basis for the reduction of polynomials to canonical form. *ACM SIGSAM Bull.* **10**(3), 19–29.

Buchberger, B. (1985). Gröbner bases: an algorithmic method in polynomial. In (Bose, N. K., ed.) *Recent Trends in Multidimensional Systems Theory*. Reidel.

Bayer, D. A. (1982). *The Division Algorithm and the Hilbert Scheme*. Ph.D. Thesis, Harvard.

Lazard, D. (1983). Gröbner bases, Gaussian elimination and resolution of systems of algebraic equations. *Proc. EUROCAL 83 Springer, L.N.C.S.* 162, 146–156.

Mora, T. (1988). *Seven variations on standard bases*. Preprint.

Mora, T. (1982). An algorithm to compute the equations of tangent cones. *Proc. EUROCAM 82 Springer, L.N.C.S.* 144, 24–31.

Mora, T., Möller, H. M. (1986). New constructive methods in classical ideal theory. *J. Algebra* **100**, 138–178.

Möller, H. M. (1985). *On the computation of Gröbner bases in commutative rings*. Preprint.
Robbiano, L. (1986). On the theory of graded structures. *J. Symb. Comp.* **2**, 139–170.
Schreyer, F. O. (1980). Die Berechnung von Syzygien mit dem verallgemeinerten Weierstrasschen Divisionsatz. Dipolmarbeit Hamburg.
Spear, D. (1977). A constructive approach to commutative ring theory. *Proc. MACSYMA Users Conf.* 369–376.
Trinks, W. (1978). Über B. Buchbergers Verfahern, Systeme algebraischer Gleichungen zu lösen. *J. Number Th.* **10**, 475–488.
Zacharias, G. (1978). *Generalized Gröbner Bases on Commutative Polynomial Rings*. Bachelor Th. M.I.T.

On the Construction of Gröbner Bases Using Syzygies

H. MICHAEL MÖLLER

FB Mathematik und Informatik, Fern Universität, Lützowstr. 125, D-5800 Hagen, Fed. Rep. of Germany

(*Received* 21 *October* 1986)

Gröbner bases are useful for analysing multivariate polynomial ideals. For different coefficient domains R, it is shown how to construct (weak) Gröbner bases using bases of modules of syzygies, and under constructibility conditions on R an algorithm for finding the required module bases is given. These methods are described in detail for principal ideal rings R. This leads to strong Gröbner bases and in case of fields R the construction is the known Buchberger algorithm.

1. Introduction

For multivariate polynomials over fields, the concept of Gröbner bases was introduced by Buchberger (1965) and up to now solutions using Gröbner bases have been developed for many problems connected with polynomial ideals. For a survey, see Buchberger (1985). In Möller and Mora (1986) some of the properties characterising Gröbner bases are summarised and some connections with syzygies are pointed out.

An ordering in the set of polynomials and a simplification procedure (reduction) w.r.t. this ordering by a finite set F of polynomials are two essential notions in this context. The algorithm of Buchberger for constructing a Gröbner basis uses the fact that it is sufficient to test only so-called S-polynomials for being reducible to zero. Here, each S-polynomial depends on exactly two polynomials of F.

Gröbner bases are generalised by some authors to polynomials over rings R. Buchberger (1983) introduced reduction rings R, such that in $R[X_1, \ldots, X_n]$ the Buchberger algorithm can be easily adopted. These rings and the Buchberger algorithm have also been studied by Winkler (1984) and Stifter (1985). If R is a Euclidean ring, Kandri-Rody & Kapur (1984) extended Gröbner bases to $R[X_1, \ldots, X_n]$ using the natural ordering in the Euclidean ring. Considering principal ideal rings R with a special ordering, Pan (1985) generalised Gröbner bases to polynomial rings over such R.

For commutative Noetherian rings R (with additional conditions concerning constructibility and uniqueness of some elements), Gröbner bases are introduced in $R[X_1, \ldots, X_n]$ by Trinks (1978), Zacharias (1978) and Schaller (1979), partially based on papers of Spear (1977) and Lauer (1976). By means of an explicit or implicit use of syzygies, they interpreted S-polynomials anew. Here, S-polynomials depend in general on more than two polynomials. Also, the reduction had to be generalised to what we will call weak reduction (and consequently we will call the reduction of Buchberger *et al.* strong reduction and the respective Gröbner bases weak and strong Gröbner bases). Weak Gröbner bases and weak reductions are studied in a broader context by Robbiano (1986). For non-commutative rings R, see Mora (1986).

The intention of this paper is to show the systematic use of syzygies in order to construct Gröbner bases starting with arbitrary commutative rings and ending with a presentation of Buchberger's algorithm for polynomials over Euclidean rings. If R is a commutative Noetherian ring, we give in terms of syzygies some equivalent conditions for weak Gröbner bases, show the connection with the different concepts of Trinks (1978), Zacharias (1978) and Schaller (1979) and give a new recursive procedure for constructing bases of modules of syzygies. This procedure allows a very efficient algorithm for constructing weak Gröbner bases.

Considering only unique factorisation rings, we show that Buchberger's algorithm works with S-polynomials depending on only two polynomials for arbitrary finite sets of input polynomials exactly if R is a principal ideal ring (with suitable constructibility conditions). This is an easy consequence of proposition 1 and the following example and shows that the concept of reduction rings is naturally restricted to principal ideal rings. Also strong reductions exist in general only if R is a principal ideal ring, as elementary arguments show. Therefore there is only sense to define strong Gröbner bases in rings $R[X_1, \ldots, X_n]$ with a principal ideal ring R.

In section 4 we specify the results on weak Gröbner bases and weak reductions to polynomials over PIR's and introduce the strong Gröbner bases and reductions. Strategies are presented for keeping the bases of the modules of syzygies restricted. This leads to a new Buchberger algorithm for polynomials over principal ideal rings, where, as with the criteria of Buchberger (1979) for detecting unnecessary S-polynomials, many S-polynomials have not to be considered. An installation of this algorithm already exists in SCRATCHPAD II for polynomials over Euclidean rings, see Gebauer & Möller (1987). An example concludes the paper.

In contrast with many of the articles quoted above, we did not deal with questions of uniqueness of special Gröbner bases in order to concentrate on the role of syzygies.

2. Characterisation of Weak Gröbner Bases

Let R be a commutative ring with identity, $\mathscr{P} := R[X_1, \ldots, X_n]$, and T the set of terms (power products) $\varphi = X_1^{i_1} \ldots X_n^{i_n}$ with $i_1, \ldots, i_n \in \mathbb{N}_0$. As usual, T is ordered by $<_T$ such that

$$\begin{aligned} &1 = X_1^0 \ldots X_n^0 \leq_T \varphi \\ &\varphi_i <_T \varphi_j \Rightarrow \varphi\, \varphi_i <_T \varphi\, \varphi_j \end{aligned} \qquad \text{for all } \varphi, \varphi_i, \varphi_j \in T.$$

For

$$f = \sum_{i=1}^{r} c_i \varphi_i$$

with $c_i \in R \backslash \{0\}$, $\varphi_1 <_T \ldots <_T \varphi_r$, let

$$lc(f) := c_r,\ lt(f) := \varphi_r,\ M_T(f) := c_r \varphi_r$$

be the *leading coefficient*, the *leading term* and the *maximal part* of f respectively. For avoiding a separate treatment of the zero polynomial, we define

$$lc(0) := lt(0) := M_T(0) := 0$$

and extend $<_T$ to $T \cup \{0\}$ by $0 <_T \varphi$ for all $\varphi \in T$.

DEFINITION. Let I be an ideal in $\mathscr{P}$. Then $F=\{f_1,\ldots,f_r\}\subset I$ is called a *weak Gröbner basis* of I if the ideal

$$M_T(I):=\left\{\sum_{k=1}^{m} M_T(g_k)\mid m\in\mathbb{N},\ g_k\in I\right\}$$

is generated by $\{M_T(f_1),\ldots,M_T(f_r)\}$.

In theorem 1 we will show that weak Gröbner bases are, in fact, ideal bases. Hence, if every ideal in $\mathscr{P}$ has a weak Gröbner basis, then $\mathscr{P}$ and its subring R are Noetherian. Conversely, if R is Noetherian, every ideal in $\mathscr{P}$ has a weak Gröbner basis, because $\mathscr{P}$ is in this case Noetherian by the Hilbert Basissatz and then any ideal $M_T(I)$ has a finite basis $\{M_T(f_1),\ldots,M_T(f_r)\}$. Thus, in the following, R will always denote a (commutative) Noetherian ring with identity.

DEFINITION. Let $I=(f_1,\ldots,f_r)$, i.e. $f_1,\ldots,f_r\in I$ and for all $f\in I$ polynomials $g_1,\ldots,g_r\in\mathscr{P}$ exist, such that f has the representation

$$f=\sum_{i=1}^{r} g_i f_i.$$

We call this representation a *weak Gröbner representation* of f (in terms of $f_1,\ldots,f_r$) if

$$lt(f)=\max_{i=1}^{r} lt(g_i)lt(f_i).$$

DEFINITION. For a preassigned tuple $M:=(M_T(f_1),\ldots,M_T(f_r))$, we call $G=(g_1,\ldots,g_r)\in\mathscr{P}^r$ a *syzygy* w.r.t. M, if

$$\sum_{i=1}^{r} g_i M_T(f_i)=0,$$

and call G is *homogeneous of degree* ϕ, if $\phi\in T$ and $g_i=0$ or

$$g_i=M_T(g_i),\quad lt(g_i)lt(f_i)=\phi,\quad i=1,\ldots,r.$$

The set $S=S(M)$ of all syzygies with respect to the same tuple $M=(M_T(f_1),\ldots,M_T(f_r))$ is called the module of syzygies (w.r.t. M). Clearly, S is homogeneous, i.e. any $G\in S$ splits into a sum of homogeneous r-tuples, all of them being in S, too.

DEFINITION. Let $F=\{f_1,\ldots,f_r\}\subset\mathscr{P}$ and $f,g\in\mathscr{P}$. We say *f reduces weakly to g modulo F*, for short $f\overset{w}{\underset{F}{\to}}g$, if $lt(g)<_T lt(f)$ and $f-g$ has a weak Gröbner representation in terms of F. $\overset{w}{\underset{F}{\to}}{}^{+}$ denotes the reflexive transitive closure of $\overset{w}{\underset{F}{\to}}$. The reduction $\overset{w}{\underset{F}{\to}}$ is Noetherian, because in

$$g_1\overset{w}{\underset{F}{\to}}g_2\overset{w}{\underset{F}{\to}}\cdots$$

we have $lt(g_1)>_T lt(g_2)>_T\ldots$ and hence no $lt(g_i)$ is a multiple of a preceding $lt(g_j)$, $j<i$. This gives an ascending chain of ideals

$$(lt(g_1))\subset(lt(g_1),lt(g_2))\subset(lt(g_1),lt(g_2),lt(g_3))\subset\ldots$$

which is finite because $\mathscr{P}$ is Noetherian.

THEOREM 1. *Let*

$$F=\{f_1,\ldots,f_r\}\subset\mathscr{P},\quad I=(f_1,\ldots,f_r)\quad\textit{and}\quad M=(M_T(f_1),\ldots,M_T(f_r)).$$

Then the following conditions are equivalent:

(*C*1) *F is a weak Gröbner basis of I.*

(*C*2) *Every $f \in I$ has a weak Gröbner representation in terms of F.*

(*C*3) *Let $G_1, \ldots, G_m$ be a basis of $S(M)$, $G_i = (g_{i1}, \ldots, g_{ir})$ homogeneous of degree ϕ_i, $i = 1, \ldots, m$. Then any so-called S-polynomial $\sum_{j=1}^{r} g_{ij} f_j$ has a weak Gröbner representation in terms of F.*

(*C*4) $f \xrightarrow[F]{w} {}^{+} 0$ *for every $f \in I$.*

(*C*5) *With $G_1, \ldots, G_m$ as in C3*

$$\sum_{j=1}^{r} g_{ij} f_j \xrightarrow[F]{w} {}^{+} 0, \quad i = 1, \ldots, m.$$

PROOF. C1$\Rightarrow$C2: Let $f \in I$ and let a weak Gröbner representation already exist for all $g \in I$, $lt(g) <_T lt(f)$. Then, by C1

$$M_T(f) = \sum_{i=1}^{r} h_i M_T(f_i), \quad h_i \text{ homogeneous},$$

and w.l.o.g. $lt(h_i) lt(f_i) = lt(f)$ or $h_i = 0$, $i = 1, \ldots, r$. Using a weak Gröbner representation for

$$f' := f - \sum_{i=1}^{r} h_i f_i$$

(by construction $lt(f') <_T lt(f)$!),

$$f' = \sum_{i=1}^{r} g'_i f_i, \quad lt(g'_i) lt(f_i) \leq_T lt(f') <_T lt(f),$$

we get for f the weak Gröbner representation

$$f = \sum_{i=1}^{r} (h_i + g'_i) f_i.$$

C2$\Rightarrow$C3: Each S-polynomial is in I.

C3$\Rightarrow$C1: Let $f \in I$. We will show $M_T(f) \in (M_T(f_1), \ldots, M_T(f_r))$. Consider an arbitrary representation

$$f = \sum_{i=1}^{r} g_i f_i, \quad \phi := \max lt(g_i) lt(f_i).$$

If $lt(f) = \phi$, we have

$$M_T(f) = \sum_{j \in J} M_T(g_j) M_T(f_j) \quad \text{with} \quad J := \{ j \mid lt(g_j) lt(f_j) = \phi \},$$

i.e. $M_T(f) \in (M_T(f_1), \ldots, M_T(f_r))$. Else $lt(f) <_T \phi$. But then

$$0 = \sum_{j \in J} M_T(g_j) M_T(f_j).$$

Hence

$$G := \sum_{j \in J} M_T(g_j) e_j \in S(M),$$

where e_j denotes the jth unit vector. Using the basis $\{G_1, \ldots, G_m\}$ of $S(M)$ and the weak Gröbner representations

$$\sum_{j=1}^{r} g_{ij} f_j = \sum_{j=1}^{r} h_{ij} f_j, \quad lt(h_{ij}) lt(f_j) <_T \phi_i, \quad i = 1, \ldots, m,$$

we get

$$G = \sum_{i=1}^{m} u_i G_i, \quad u_i = 0 \quad \text{or} \quad lt(u_i)\phi_i = \phi$$

and

$$\sum_{j \in J} M_T(g_j) f_j = \sum_{i=1}^{m} u_i \sum_{j=1}^{r} g_{ij} f_j = \sum_{j=1}^{r} \left\{ \sum_{i=1}^{m} u_i h_{ij} \right\} f_j.$$

The maximal term on the left-hand side is ϕ, on the right-hand side $<_T \phi$. Therefore, replacing in $f = \sum g_i f_i$ the summand

$$\sum_{j \in J} M_T(g_j) f_j \quad \text{by} \quad \sum_j \left\{ \sum_i u_i h_{ij} \right\} f_j,$$

we obtain a representation for f with a maximal term $<_T \phi$. An iterative application of this procedure gives finally a weak Gröbner representation of f and, as shown above, the maximal terms in this representation imply

$$M_T(f) \in (M_T(f_1), \ldots, M_T(f_r)).$$

C1 ⇒ C4 is proved as C1 ⇒ C2. C4 ⇒ C5 is obvious. For C5 ⇒ C3 take in

$$h_0 := \sum_{j=1}^{r} g_{ij} f_j \underset{F}{\overset{w}{\rightarrow}} h_1 \underset{F}{\overset{w}{\rightarrow}} \ldots \underset{F}{\overset{w}{\rightarrow}} h_S = 0$$

the weak Gröbner representations of $h_0 - h_1, h_1 - h_2, \ldots, h_{S-1} - h_S$ and sum it up obtaining a weak Gröbner representation for h_0. □

In theorem 1 we expressed the conditions characterising weak Gröbner bases in terms of $\mathscr{P}$, the ring of polynomials over R. This parallels the result known for polynomials over fields as in Möller & Mora (1986). The definitions and results for polynomials over commutative rings by Trinks (1978), Zacharias (1978) and Schaller (1979) are mostly given in terms of R. The relationship with these results is easily established. Consider, for instance, the weak reduction. A given $f \in \mathscr{P}$ is weakly reducible modulo $F = \{f_1, \ldots, f_r\}$ to a polynomial g, if $f - g$ has a weak Gröbner representation in terms of F. Because of $lt(g) <_T lt(f)$, the maximal part of f and $f - g$ coincide. This means that $lt(f) = lt(f - g)$ and that

$$lc(f) = \sum_{j \in J} d_j lc(f_j) \quad \text{with} \quad J = \{j \mid lt(f_j)/lt(f)\} \tag{1}$$

holds for suitable $d_j \in R$. But (1) is already sufficient for f to be weakly reducible modulo F,

$$f \underset{F}{\overset{w}{\rightarrow}} f - \sum_{j \in J} d_j \frac{lt(f)}{lt(f_j)} f_j.$$

Admitting only these reductions, Zacharias used C4 for defining (weak) Gröbner bases, and Trinks called a sequence of such reductions "D-Folge".

The connection with syzygies is mentioned by all three authors, but only Zacharias used the notion syzygy. We observe that in a homogeneous syzygy w.r.t. $M = (M_T(f_1), \ldots, M_T(f_r))$, all components can be divided by a term, such that $(c_1 \varphi_1, \ldots, c_r \varphi_r)$, a syzygy w.r.t. M of degree ϕ, is obtained with

$$\phi = lcm\{lt(f_i) \mid c_i \neq 0\}, \tag{2}$$

and necessary and sufficient for being a syzygy of degree ϕ is

$$0 = \sum_{i=1}^{r} c_i lc(f_i).$$

Obviously at most 2^r different ϕ's satisfying (2) exist. Considering only homogeneous syzygies with ϕ as in (2), and using that all such syzygies of the same degree ϕ constitute a linear space $V(\phi)$. Trinks remarked that instead of requiring $f \xrightarrow[F]{w} {}^{+}0$ for all $f \in I$ it suffices to consider only S-polynomials depending on syzygies which belong to a basis of one $V(\phi)$, i.e. he used C5 implicitly for defining (weak) Gröbner bases. Schaller used the notion "complete basis" for weak Gröbner bases and "simple representation" for weak Gröbner representation and C2 for defining weak Gröbner bases. Whereas Zacharias proposed a construction of a basis of $S(M)$ similar to Trinks, Schaller showed that a basis of $S(M)$ is obtained by constructing the module of syzygies $S \subset R^r$,

$$\left\{(c_1, \ldots, c_r) \in R^r \,\middle|\, \sum_{i=1}^{r} c_i lc(f_i) = 0\right\}$$

say $C_1, \ldots, C_m$ with $C_i = (c_{i1}, \ldots, c_{ir})$, $i = 1, \ldots, m$, and then a basis of $S(M)$ is given by $G_1, \ldots, G_m$ with $G_i = (g_{i1}, \ldots, g_{ir})$,

$$g_{ij} = c_{ij} \frac{lcm\{lt(f_k) \mid c_{ik} \neq 0\}}{lt(f_j)}, \quad j = 1, \ldots, r, \quad i = 1, \ldots, m.$$

3. An Iterative Construction of Weak Gröbner Bases

To construct weak Gröbner bases using C3 or C5 of theorem 1, some bases of modules of syzygies have to be considered until a weak Gröbner basis is found. Therefore it is more practical not to calculate each single module basis separately but to construct the basis of the next module by using the previous one.

Denoting

$$M_k := (M_T(f_1, \ldots, M_T(f_k)), \quad k = 1, 2, \ldots,$$

we will construct a basis for $S(M_r)$ using a basis of $S(M_{r-1})$. The initialisation of this procedure is easy because $S(M_1)$ is generated by $\{c \in R \mid c\, lc(f_1) = 0\}$, which is $\{0\}$ in case R is an integral domain and $f_1 \neq 0$. For convenience of notation we write

$$c_i := lc(f_i), \quad \varphi_i := lt(f_i), \quad i = 1, \ldots, r.$$

Our main tool in this paragraph will be the mapping

$$\pi_r : S(M_r) \to R, \quad \pi_r : (g_1, \ldots, g_r) \to lc(g_r).$$

We have

$$im\pi_r = (c_1, \ldots, c_{r-1}) : (c_r)$$

and

$$\ker \pi_r = (S(M_{r-1}), 0) = \{(g_1, \ldots, g_{r-1}, 0) \mid (g_1, \ldots, g_{r-1}) \in S(M_{r-1})\}.$$

(The representation for the kernel is obvious, the identity for $im\pi_r$ will be clear from the following theorem.)

Definition. For $J \subseteq \{1, \ldots, r\}$ let $\phi(J) := lcm\{\phi_j \mid j \in J\}$ and J is *maximal* for $\phi \in T$ if $\phi_i / \phi \Leftrightarrow i \in J$. We call a set $J \subseteq \{1, \ldots, r\}$ *basic*, if $r \in J$ and J is maximal for $\phi(J)$. If J is basic, $J^\times := J \backslash \{r\}$ and $b_r \in (c_j \mid j \in J^\times) : (c_r)$, i.e.

$$\sum_{j \in J^\times} b_j c_j + b_r c_r = 0 \quad \text{for suitable } b_j \in R,$$

then

$$\sum_{j \in J^\times} b_j \frac{\phi(J)}{\phi_j} e_j + b_r \frac{\phi(J)}{\phi_r} e_r,$$

where e_i denotes the ith unit vector, is a homogeneous syzygy of $S(M_r)$ of degree $\phi(J)$, which we call *associated to* b_r *and* J.

THEOREM 2. *For any basic* J *let* $\{b(1, J), \ldots, b(s_J, J)\}$ *with* $b(i, J) \neq 0$ *for* $i = 1, \ldots, s_J$, *be a basis of* $(c_j | j \in J \setminus \{r\}) : (c_r)$. *Let* $B(i, J)$ *be associated to* $b(i, J)$ *and* $J, i = 1, \ldots, s_J$. *If* $\{A_1, \ldots, A_m\}$ *is a homogeneous basis of* $S(M_{r-1})$, *then*

$$\{(A_1, 0), \ldots, (A_m, 0)\} \cup \bigcup_{J \text{ basic}} \{B(i, J) | 1 \leq i \leq s_J\}$$

is a homogeneous basis of $S(M_r)$.

PROOF. Let $G = (g_1, \ldots, g_r) \in S(M_r)$, $g_r \neq 0$ and w.l.o.g. G homogeneous of degree ϕ. Let $J_1 := \{j | g_j \neq 0\}$. By homogeneity $j \in J_1 \Rightarrow \phi_j / \phi$, i.e. $\phi(J_1)/\phi$. Let $J = \{j | \phi_j / \phi(J_1)\}$. Then by construction $\phi(J)/\phi(J_1)$ but $J_1 \subseteq J$, i.e. $\phi(J_1) = \phi(J)$. Since $r \in J_1 \subseteq J$, J is basic.

Since G is homogeneous of degree ϕ,

$$g_j = lc(g_j) \cdot \frac{\phi}{\phi_j} \quad \text{for} \quad j \in J_1 \quad \text{and} \quad g_j = 0 \quad \text{for} \quad j \notin J_1.$$

$G \in S(M_r)$ means

$$\sum_{j \in J} lc(g_j) \frac{\phi}{\phi_j} \cdot c_j \phi_j = 0.$$

This gives $c_r lc(g_r) \in (c_j | j \in J \setminus \{r\})$ or, equivalently,

$$lc(g_r) \in (c_j | j \in J \setminus \{r\}) : (c_r).$$

Hence, a representation

$$lc(g_r) = \sum_{i=1}^{s_J} u_i b(i, J) \quad \text{with} \quad u_i \in R$$

exists. The syzygies $B(i, J)$ are homogeneous of degree $\phi(J)$. Therefore

$$G - \sum_{i=1}^{s_J} u_i \frac{\phi}{\phi(J)} B(i, J) \in S(M_r)$$

is homogeneous of degree ϕ and belongs to $\ker \pi_r$. Hence,

$$G = \sum_{i=1}^{s_J} u_i \frac{\phi}{\phi(J)} B(i, J) + \sum_{i=1}^{m} h_i(A_i, 0)$$

with appropriate $h_i \in \mathscr{P}$. Since G was arbitrary, the assertion is proven. □

In the basis of $S(M_r)$ as given in theorem 2, only syzygies occur which are homogeneous of a degree $\phi(J)$, $J \subseteq \{1, \ldots, r\}$. But usually some of them are redundant. This is caused by two reasons.

First, there may be some $(A_i, 0)$ of degree $\phi(J)$, $J \subseteq \{1, \ldots, r-1\}$, and ϕ_r divides $\phi(J)$, such that in the basis of $S(M_r)$ also some $B(j, J_1)$ of degree $\phi(J)$ exist ($J \subset J_1$, $r \in J_1$, $\phi(J_1) = \phi(J)$). These $(A_i, 0)$ and $B(j, J_1)$ may be linearly dependent.

Second, if J_1 and J_2 are maximal but $J_1 \subset J_2$, then $\phi(J_1)/\phi(J_2)$ and

$$(c_j | j \in J_1) : (c_r) \subseteq (c_j | j \in J_2) : (c_r).$$

Therefore, $B(i, J_1)\phi(J_2)/\phi(J_1)$ is associated to b_r and J_2 if $B(i, J_1)$ is associated to b_r and J_1. Hence, enlarging a basis of $(c_j | j \in J_1) : (c_r)$, say $b(1, J_1), \ldots, b(s_1, J_1)$, to a basis of $(c_j | j \in J_2) : (c_r)$ and taking the corresponding $B(i, J_1) \cdot \phi(J_2)/\phi(J_1)$ associated to $b(i, J_2)$ and J_2 for $B(i, J_2)$, then $B(i, J_2)$ being a multiple of $B(i, J_1)$ is redundant in the basis of $S(M_r)$.

If in the basis of $S(M_r)$ one element is an r-tuple with exactly one non-zero component, say the jth one, then $f_j = 0$ or c_j is a zero divisor. For excluding these trivial cases, we assume $f_j \neq 0, j = 1, \ldots, r$, and R is an integral domain. Then the sparsest elements are those with two non-zero components.

DEFINITION. We call a homogeneous basis of $S(M_r)$ a *principal basis* if every basis element has exactly two non-zero components.

PROPOSITION 1. *If R is a principal ideal ring and if*

$$M_r := (M_T(f_1), \ldots, M_T(f_r)) \in (\mathscr{P}\backslash\{0\})^r,$$

then $S(M_r)$ has a principal basis.

PROOF. With respect to theorem 2 it is sufficient to show that for any basic J a basis $\{b(1, J), \ldots, b(s_J, J)\}$ of $(c_j \mid j \in J, j \neq r) : (c_r)$ exists, such that there are associated $B(i, J), i = 1, \ldots, s_J$, having exactly two non-vanishing components. Let $J^\times := J\backslash\{r\}$.

Since R is a principal ideal ring,

$$(c_j \mid j \in J^\times) = (gcd\{c_j \mid j \in J^\times\}),$$
$$(c_j) \cap (c_r) = (lcm\{c_j, c_r\}) \quad \text{for } j \in J^\times$$

holds. In particular, R is a unique factorisation ring. Therefore,

$$(lcm\{gcd\{c_j \mid j \in J^\times\}, c_r\}) = (gcd\{lcm\{c_j, c_r\} \mid j \in J\})$$

holds. This gives

$$(c_j \mid j \in J^\times) \cap (c_r) = \sum_{j \in J^\times} (c_j) \cap (c_r),$$

or equivalently

$$(c_j \mid j \in J^\times) : (c_r) = \sum_{j \in J^\times} (c_j) : (c_r).$$

Hence, if u_j generates $(c_j) : (c_r)$, then $\{u_j \mid j \in J^\times\}$ is a basis of $(c_j \mid j \in J^\times) : (c_r)$. Associated to u_j and $\{j\}$ and hence to u_j and J is

$$B_j := -\frac{u_j c_r}{c_j} \frac{\phi(j, r)}{\phi_j} e_j + u_j \frac{\phi(j, r)}{\phi_r} e_r. \tag{3}$$

□

If R is a unique factorisation ring (and if no f_i is zero), we can show that each module $S(M_r)$ has a principal basis only if R is a principal ideal ring. For this converse of proposition 1, it is sufficient to give an example of an R-module $S(M_3)$ which has no principal basis, and R is a unique factorisation ring but not a principal ideal ring.

In that case, $c_1, c_2 \in R$ exist, such that (c_1, c_2) is not the principal ideal $(gcd\{c_1, c_2\})$. Dividing out common factors and denoting the remainders again c_1, c_2, we have $c_1, c_2 \in R$ without common divisor and $1 \notin (c_1, c_2)$. Let $c_3 := c_1 + c_2$. Then c_3 has no divisor in common with c_1 and no one with c_2. Let $\varphi \in T$ be arbitrary and

$$M_3 := (c_1 \varphi, c_2 \varphi, c_3 \varphi).$$

If $(g_1, 0, g_3)$ and $(0, h_2, h_3)$ are homogeneous basis elements of $S(M_3)$, then

$$\pi_3(g_1, 0, g_3) = g_3 \in (c_1) : (c_3) = (c_1),$$
$$\pi_3(0, h_2, h_3) = h_3 \in (c_2) : (c_3) = (c_2).$$

Hence, for any $G \in S(M_3)$ with

$$G = \sum u_i(g_{1i}, 0, g_{2i}) + \sum v_i(0, h_{2i}, h_{3i}) + \sum w_i(k_{1i}, k_{2i}, 0),$$

we have $\pi_3(G) \in (c_1, c_2)$. But $(-1, -1, 1) \in S(M_3)$ and

$$\pi_3(-1, -1, 1) = 1 \notin (c_1, c_2).$$

Hence, this $S(M_3)$ has no principal basis.

For the computation of a basis of $S(M_r)$ by theorem 2 it is sufficient to be able to compute syzygies associated to a $b \in R$ and a $J \subseteq \{1, \ldots, r\}$. This is guaranteed by the following two computability requirements:

(1) Ideals in R are detachable. That is there exists an algorithm, which, given $c \in R$ and a finite set $\{c_1, \ldots, c_r\} \subset R$, decides whether $c \in (c_1, \ldots, c_r)$ and if so, produces $u_1, \ldots, u_r \in R$ such that $c = u_1 c_1 + \ldots + u_r c_r$.
(2) Ideal quotients in R are computable. That is, there exists an algorithm, which, given $c \in R$ and a finite set $\{c_1, \ldots, c_r\} \subset R$, produces a (finite) basis of the ideal

$$(c_1, \ldots, c_r):(c) = \{d \in R \mid dc \in (c_1, \ldots, c_r)\}.$$

These two conditions are essentially due to Zacharias (1978) who required in place of our second condition, that syzygies in R are solvable. Our condition, together with theorem 2, gives her second condition, but conversely using the projection π_r we see whenever syzygies in R are solvable, then also ideal quotients are computable, i.e. both second conditions are equivalent by theorem 2.

For Noetherian rings R with the two computability conditions the weak reduction can be performed too, because by the first condition, $h_1, \ldots, h_r$ can be found, such that $M_T(f) = \Sigma h_i M_T(f_i)$ and hence $f \overset{w}{\rightarrow} f - \Sigma h_i f_i$. Then condition C5 of theorem 1 gives an algorithm for computing Gröbner bases:

If an S-polynomial $\sum g_i f_j$ is not weakly reducible to 0 modulo $F = \{f_1, \ldots, f_r\}$, its 0-reducibility can be forced by enlarging F. Let $\sum g_{ij} f_j \overset{w}{\underset{F}{\rightarrow}}{}^{+} h_i \neq 0$. Then $\sum g_{ij} f_j$ is weakly reducible to 0 modulo $F' := F \cup \{h_i\}$,

$$\sum_{j=1}^{r} g_{ij} f_j \overset{w}{\underset{F'}{\rightarrow}}{}^{+} h_i \overset{w}{\underset{F'}{\rightarrow}} 0.$$

Then the S-polynomials corresponding to a basis of $S(M_T(f_1), \ldots, M_T(f_r), M_T(f_{r+1}))$ with $f_{r+1} := h_i$ have to be weakly reduced. Some of the new S-polynomials may weakly reduce to 0 mod. F', but when a reduction to an $h \neq 0$ occurs, F' is enlarged again. The procedure indicated here is the translation of Buchberger's algorithm, which was given by Buchberger (1965) for polynomials over fields. Trinks (1978), Zacharias (1978) and Schaller (1979) elaborated this procedure to algorithms for obtaining weak Gröbner bases in polynomial rings over computable commutative Noetherian rings. However, none of these authors mentioned a recursive computation of the module bases as in theorem 2. Such recursion increases the efficiency of the computation: Since a basis element A_i in $S(M_{r-1})$ and the basis element $(A_i, 0)$ in $S(M_r)$ have the same S-polynomial, it is sufficient to test the S-polynomials once and not for each basis separately. Hence, if the polynomial set F is enlarged, the only new S-polynomials are those corresponding to the basis elements denoted by $B(i, J)$ in theorem 2. In addition, as indicated in the remark following theorem 2, not all S-polynomials corresponding to the $B(i, J)$ and $(A_i, 0)$ have to be used. We will study this in more detail in the next paragraph for polynomials over principal ideal rings.

4. Gröbner Bases in Principal Ideal Rings

In the following, R is always a principal ideal ring and F a finite set of polynomials $f_i \neq 0$ in $\mathscr{P} = R[X_1, \ldots, X_n]$ with $c_i = lc(f_i)$, $\phi_i = lt(f_i)$, $i = 1, 2, \ldots$. For any subset $\{f_j \mid j \in J\} \subseteq F$, let

$$c_J = lcm\{c_j \mid J\}, \quad \phi(J) = lcm\{\phi_j \mid j \in J\}, \quad T(J) = c_J \phi(J).$$

Using

$$(c_j):(c_r) = (c_{jr}/c_r),$$

proposition 1 shows that $S(M_r)$ with

$$M_r = (M_T(f_1), \ldots, M_T(f_r))$$

has the principal basis $\{B_{jk} \mid 1 \le j < k \le r\}$,

$$B_{jk} = \frac{T(j,k)}{T(j)} e_j - \frac{T(j,k)}{T(k)} e_k.$$

If R is a field, $T(j, k) = \phi(j, k)$ because of $c_{jk} = 1$, and then the S-polynomial corresponding to B_{jk} is

$$\frac{1}{c_j} \frac{\phi(j,k)}{\phi_j} f_j - \frac{1}{c_k} \frac{\phi(j,k)}{\phi_k} f_k.$$

This is the original S-polynomial $S(f_j, f_k)$ of Buchberger (1965).

As we have already remarked, the basis for $S(M_r)$ as constructed in theorem 2 (possibly) contains redundant elements. In the special instance considered here, we can detect such elements using the identity

$$\frac{T(i,j,k)}{T(i,j)} B_{ij} - \frac{T(i,j,k)}{T(i,k)} B_{ik} + \frac{T(i,j,k)}{T(j,k)} B_{jk} = 0, \tag{4}$$

which holds for arbitrary $i, j, k \in \{1, \ldots, r\}$. In case one coefficient, say $T(i, j, k)/T(i, j)$, is 1 we see using (4), that B_{ij} is a redundant basis element and $T(i, k)$ and $T(j, k)$ both divide $T(i, j)$.

DEFINITION. We say *criterion* B_r holds for (i, j) if $i < j < r$, $T(r)/T(i, j)$ and

$$T(i, r) \neq T(i, j) \neq T(j, r).$$

We say *criterion M* holds for (i, r) if $i < r$ and

$$\exists j < r: T(j, r)/T(i, r) \neq T(j, r).$$

We say *criterion F* holds for (i, r) if $i < r$ and

$$\exists j < i: T(j, r) = T(i, r).$$

We also write briefly in these cases $B_r(i, j)$, $M(i, r)$, or $F(i, r)$ respectively.

These criteria were formulated by Gebauer & Möller (1988) for polynomials over fields and used for removing redundant elements in the principal basis $\{B_{ij} \mid 1 \le i < j \le r\}$. We state explicitly that these criteria require only *lcm* computations of ring elements and terms.

THEOREM 3. *A basis of* $S(M_r)$ *is given by*

$$\{B_{ij} \mid 1 \le i < j \le r, \neg M(i, j), \neg F(i, j), \forall k > j: \neg B_k(i, j)\}.$$

PROOF. We order the basis $\{B_{ij} \mid 1 \le i < j \le r\}$ partially by $B_{ij} < B_{kl}$ if

$$T(i, j)/T(k, l) \neq T(i, j)$$

or if $T(i, j) = T(k, l)$ and $j < l$ or if $T(i, j) = T(k, l)$, $j = l$ and $i < k$.

If $B_k(i, j)$ holds, then in any basis representation

$$G = \sum g_{\mu\nu} B_{\mu\nu}, \quad g_{\mu\nu} \in \mathscr{P},$$

B_{ij} can be replaced using (4) by B_{ik} and B_{jk}, and $B_{ik} < B_{ij}$, $B_{jk} < B_{ij}$ holds. This can be done for arbitrary B_{ij} satisfying a criterion B_k, i.e. for all $G \in S(M_r)$ basis representations exist where no such B_{ij} occurs. The same arguments hold for criteria M and F. □

DEFINITION. Let $f, g \in \mathscr{P}$. We say f *reduces strongly to g modulo F*, for short $f \overset{S}{\underset{F}{\to}} g$, if $lt(g) <_T lt(f)$ and $\exists f_i \in F$, $\exists h \in \mathscr{P} : g = f - hf_i$. The reflexive transitive closure of $\overset{S}{\underset{F}{\to}}$ is denoted by $\overset{S}{\underset{F}{\to}}^+$.

The same arguments as for $\overset{w}{\underset{F}{\to}}$ give that $\overset{S}{\underset{F}{\to}}$ is also Noetherian. Since hf_i is already a weak Gröbner representation in terms of F, a polynomial f reduces weakly (to another polynomial) modulo F, if it reduces strongly.

A polynomial f reduces strongly if and only if $M_T(f)$ is a multiple of an $M_T(f_i)$, $f_i \in F$. In that case

$$f \overset{S}{\underset{F}{\to}} f - \frac{M_T(f)}{M_T(f_i)} f_i.$$

This criterion is considerably simpler than the corresponding one for the weak reduction as the derivation of equation (1) shows. But there are more polynomials weakly reducible than strongly. For instance, let $\mathscr{P} = \mathbb{Z}[x, y]$, $F = \{2x, 3y\}$, $f = xy$. Then $f \overset{w}{\underset{F}{\to}} 0$, because $f - 0$ has the weak Gröbner representation $2y \cdot 2x - x \cdot 3y$, and f does not reduce strongly modulo F.

In order to be able to reduce strongly all polynomials which are weakly reducible mod. F, we enlarge F to a set $C(F)$ and use $\overset{S}{\underset{C(F)}{\longrightarrow}}$. A similar procedure was used by Schaller (1979) in simplification rings by introducing the set of kernel polynomials.

DEFINITION. Let $F = \{f_1, \ldots, f_r\}$ and I the ideal generated by F. We call a set M a *completion* of F, briefly $C(F)$, if it contains for every J, maximal for $\phi(J)$, a polynomial $f_J \in I$ with $lt(f_J)/\phi(J)$ and $lc(f_J)/gcd\{c_j \mid j \in J\}$ and f_J has a weak Gröbner representation in terms of F.

If no $lt(f_i)$ is a multiple of another $lt(f_j)$, then we see by considering the singletons J, that we may assume $F \subset C(F)$. So we have $C(F) = \{2x, 3y, xy\}$ in the above-mentioned instance.

A completion of $F = \{f_1, \ldots, f_r\}$ can be computed easily, similar to the basis construction in theorem 2. If for $F^\times := F \setminus \{f_r\}$ a completion $C(F^\times)$ is known, we may take for each maximal J not containing r the corresponding $f \in C(F^\times)$ for element f_J in $C(F)$. Hence, let J be maximal and contain r, i.e. J a basic set. Then $J^\times = J \setminus \{r\}$ is maximal w.r.t. $F^\times$ and $\phi(J^\times)/\phi(J)$. Let $f^\times \in C(F^\times)$ correspond to $J^\times$, i.e.

$$lc(f^\times)/gcd\{c_j \mid j \in J^\times\}, \ lt(f^\times)/\phi(J^\times).$$

Let

$$u_1 \, lc(f^\times) + u_2 c_r = gcd\{lc(f^\times), c_r\}, \quad u_1, u_2 \in R.$$

Then the so-called *T-polynomial*

$$T(f^{\times}, f_r) := u_1 \frac{\phi(J)}{lt(f^{\times})} f^{\times} + u_2 \frac{\phi(J)}{\phi_r} f_r$$

satisfies

$$lc(T(f^{\times}, f_r)) = gcd\{lc(f^{\times}), c_r\}/gcd\{c_j \mid j \in J\}, \quad lt(T(f^{\times}, f_r)) = \phi(J)$$

and inserting the weak Gröbner representation of $f^{\times}$ in terms of $\{f_1, \ldots, f_{r-1}\}$ into the definition $T(f^{\times}, f_r)$ we read off the weak Gröbner representation in terms of F. Hence, we may take it for f_J.

Therefore, a completion $C(F)$ is obtained by taking $C(F^{\times})$ and for every basic set the corresponding T-polynomial. An element f of $C(F)$ can be cancelled if a $g \in C(F)$ exists, s.t. $lt(g)/lt(f)$ and $lc(g)/lt(f)$. This allows a reduction of the just constructed completion $C(F)$. An instance of this construction is contained in the final example.

PROPOSITION 2. *Let $C(F)$ be a completion of $F = \{f_1, \ldots, f_r\}$ and $f, g \in \mathscr{P}$. If $f \underset{F}{\overset{w}{\rightarrow}} g$, then $f \underset{C(F)}{\overset{S}{\rightarrow}} g'$ for a $g' \in \mathscr{P}$. If $f \underset{C(F)}{\overset{S}{\rightarrow}} g$, then $f \underset{F}{\overset{w}{\rightarrow}} g'$ for a $g' \in \mathscr{P}$.*

PROOF. Equation (1) shows that $f \underset{F}{\overset{w}{\rightarrow}} g$ implies

$$lc(f) = \sum_{j \in J} d_j lc(f_j)$$

with $d_j \in R$ and J maximal for $lt(f)$. Then $lc(f) \in (c_j \mid j \in J)$. This ideal in R is generated by

$$c \cdot lc(f_J) = gcd\{c_j \mid j \in J\}.$$

By construction $\phi(J)/lt(f)$. Hence,

$$f \underset{C(F)}{\overset{S}{\rightarrow}} f - \frac{lc(f)}{lc(f_J)} \frac{lt(f)}{lt(f_J)} f_J.$$

Conversely, $f \underset{C(F)}{\overset{S}{\rightarrow}} g$ implies $M_T(f_J)/M_T(f)$ for a maximal J. Using the weak Gröbner representation $f_J = \Sigma h_i f_i$, we get

$$f \underset{F}{\overset{w}{\rightarrow}} f - \sum_{i=1}^{r} \frac{M_T(f)}{M_T(f_J)} h_i f_i. \quad \square$$

DEFINITION. Let I be an ideal in $\mathscr{P}$. Then $F = \{f_1, \ldots, f_r\} \subset I$ is called a *strong Gröbner basis* of I, if for each $f \in I$ an $f_i \in F$ exists, such that $M_T(f_i)/M_T(f)$.

THEOREM 4. *Let I be generated by $F = \{f_1, \ldots, f_r\}$. Then the following conditions are equivalent.*

(C1)′ *F is a strong Gröbner basis of I.*
(C4)′ *$f \underset{F}{\overset{S}{\rightarrow}}{}^{+} 0$ for every $f \in I$.*
(C5)′ *Let $\{G_1, \ldots, G_m\}$ be a basis of $S(M_T(f_1), \ldots, M_T(f_r))$ and each $G_i = (g_{i1}, \ldots, g_{ir})$ be homogeneous. Then*

$$\sum_{j=1}^{r} g_{ij} f_j \underset{F}{\overset{S}{\rightarrow}}{}^{+} 0, \quad i = 1, \ldots, m.$$

PROOF. C1′ ⇒ C4′: Let $f \in I$ and $M_T(f_k)$ divide $M_T(f)$. Then

$$f \underset{F}{\overset{S}{\rightarrow}} f - \frac{M_T(f)}{M_T(f_k)} f_k =: f'.$$

Because of $f' \in I$ and $lt(f') <_T lt(f)$ we may use an inductive argument.

C4′⇒C5′: Clear because of $\sum g_{ij} f_j \in I$.
C5′⇒C1′: Essentially like C5⇒C3⇒C1 in theorem 1.

We have only to replace weak Gröbner representations by representations

$$\sum_{i=1}^{r} g_i f_i$$

with $k \in \{1, \ldots, r\}$, such that

$$lt(f) = lt(g_k f_k) >_T lt(g_i f_i) \square$$

for all $i \neq k$.

Corollary. *F is a weak Gröbner basis of I if and only if a completion C(F) is a strong Gröbner basis of I.*

Proof. Proposition 2 shows that f is weakly irreducible mod. F if and only if f is strongly irreducible mod $C(F)$. Condition C4 and C4′ mean that 0 is the only irreducible element of I in the respective setting. □

The algorithm sketched at the end of section 3 for computing weak Gröbner bases can be modified to obtain also strong Gröbner bases.

Algorithm. Input: $\{f_1, \ldots, f_r\} \subset R[X_1, \ldots, X_n]$, R a principal ideal ring.

Step 1: Calculate

$$D := \{(i, j) \mid 1 \leq i < j \leq r, \neg M(i, j), \neg F(i, j), \forall k > j : \neg B_k(i, j)\},$$

calculate a $C(F)$ for $F := \{f_1, \ldots, f_r\}$ and define $R := r$.

Step 2: Let $(i, j) \in D$. Calculate h, such that

$$\frac{lcm\{M_T(f_i), M_T(f_j)\}}{M_T(f_i)} f_i - \frac{lcm\{M_T(f_i), M_T(f_j)\}}{M_T(f_j)} f_j \xrightarrow[C(F)]{s} h$$

and h is no longer strongly reducible modulo $C(F)$. Remove (i, j) from D.

Step 3: If $h \neq 0$ enlarge F by $f_{R+1} := h$, D by

$$\{(k, R+1) \mid 1 \leq k \leq R, \neg M(k, R+1), \neg F(k, R+1)\},$$

remove all (k, l) with $B_{R+1}(k, l)$ from D, calculate a $C(F)$ for this new F and enlarge finally R by 1.

Step 4: If $D \neq \emptyset$ go to step 2.

Output: F, a weak Gröbner basis of $(f_1, \ldots, f_r)$,

$C(F)$, a strong Gröbner basis of $(f_1, \ldots, f_r)$.

Since by construction of h, each $R > r$, is no multiple of an $M_T(f)$, $f \in C(f_1, \ldots, f_{R-1})$, or equivalently as shown in proposition 2,

$$M_T(f_R) \notin (M_T(f_1), \ldots, M_T(f_{R-1})),$$

the algorithm produces a strictly increasing chain of ideals

$$(M_T(f_1), \ldots, M_T(f_r)) \subset (M_T(f_1), \ldots, M_T(f_{r+1})) \subset \ldots.$$

This chain is finite because $\mathscr{P}$ is Noetherian. Therefore only finitely many $h \neq 0$ are produced in the algorithm. This shows termination.

For the correctness, we remark that when step 2 is performed, $\{B_{kl} \mid (k, l) \in D\}$ is a basis of $S(M_T(f_1), \ldots, M_T(f_R))$ by theorem 3. At termination, condition C5′ holds for $C(F)$, i.e. $C(F)$ is a strong Gröbner basis and by the last corollary F is a weak one.

For the computation of weak and strong Gröbner bases by means of this algorithm, we need that in the principal ideal ring R all ring operations can be performed and that an algorithm exists which, given $r_1, r_2 \in R \setminus \{0\}$, produces an *lcm* of r_1, r_2 and elements $u_1, u_2 \in R$ with

$$u_1 r_1 + u_2 r_2 = gcd\{r_1, r_2\} \left(= \frac{r_1 r_2}{lcm\{r_1, r_2\}} \right).$$

The most time and space consuming parts of the algorithm are the reduction in step 2 and the calculation of the next $C(F)$, even if the latter is computed as proposed by enlarging the previous $C(F)$, by T-polynomials. Superfluous $C(F)$ computations are made, when the new $M_T(f_{R+1})$ is a multiple of an $M_T(f_i)$, $i > R+1$, appearing later in the algorithm. This happens, for instance, when, as proposed by some authors, T-polynomials are also inserted into F. (They are needed only for the strong reduction!) To avoid unnecessary reductions in step 2, we also do not enlarge F by T-polynomials and keep the module basis restricted by the criteria M, F and B. The decision not to take a pair (i, j) for reduction, when $M(i, j)$ or $F(i, j)$ or $B_k(i, j)$ for a $k > j$ holds, parallels criteria of Buchberger (1979) for the case of polynomials over fields.

The algorithm is with minor modifications installed in the computer algebra system SCRATCHPAD II for polynomials in $X_1, \ldots, X_n$ over Euclidean rings R and tested by various examples, see Gebauer & Möller (1987).

EXAMPLE. Let R be the ring of integers, $\mathscr{P} = R[x, y]$, $<_T$ the graduated lexicographical ordering

$$1 <_T x <_T y <_T x^2 <_T xy <_T y^2 <_T x^3 <_T \ldots$$

and $F = \{f_1, f_2\}$,

$$f_1 = 2x^2 y - 17y, \quad f_2 = 5xy^2 - 3x.$$

The algorithm gives first $D = \{(1, 2)\}$ and $C(F) = \{f_1, f_2, f_{12}\}$,

$$f_{12} = 3yf_1 - xf_2 = x^2 y^2 - 51 y^2 + 3x^2.$$

Taking the pair (1, 2) for reduction in step 2,

$$f_3 = 5yf_1 - 2xf_2 = -85y^2 + 6x^2.$$

Then, because of $M(1, 3)$ we have $D = \{(2, 3)\}$ and $C(F) = \{f_1, f_2, f_3, f_{12}\}$, because we may take f_{12} for f_{123} and f_2 for f_{23}. The pair (2, 3) gives

$$f_4 = -17f_2 - xf_3 = -6x^3 + 51x.$$

Because of $M(2, 4)$ and $M(3, 4)$, we have $D = \{(1, 4)\}$, and $C(F) = \{f_1, f_2, f_3, f_4, f_{12}\}$, because we may take f_1 for f_{14} and f_{12} for f_{1234}. The pair (1, 4) gives

$$-3xf_1 - yf_4 = 0.$$

Thus, the algorithm terminates giving the weak Gröbner basis $\{f_1, f_2, f_3, f_4\}$ and the strong Gröbner basis $\{f_1, f_2, f_3, f_4, f_{12}\}$.

References

Buchberger, B. (1965). Ein Algorithmus zum Auffinden der Basiselemente des Restklassenrings nach einem nulldimensionalen Polynomideal. PhD thesis, Universität Innsbruck.

Buchberger, B. (1979). A criterion for detecting unnecessary reductions in the construction of Gröbner bases. *EUROSAM 79. Springer Lec. Notes Comp. Sci.* **72,** 3–21.

Buchberger, B. (1983). A critical-pair/completion algorithm for finitely generated ideals, *Symp. Rekursive Kombinatorik. Springer Lec. Notes Comp. Sci.* **171,** 137–155.

Buchberger, B. (1985). Gröbner bases: An algorithmic method in polynomial ideal theory. In: (Bose, N. K., ed.) *Progress, Directions and Open Problems in Multidimensional Systems Theory*, pp. 184–232. Dordrecht: Reidel.

Gebauer, R., Möller, H. M. (1988). On an installation of Buchberger's algorithm. *J. Symbol. Comput.* **6,** 275–286.

Gebauer, R., Möller, H. M. (1987). Algebra snapshot: Gröbner bases. In: (Sutor, R. S., ed.) *The Scratchpad II Newsletter*. Yorktown Heights, NY: IBM Thomas J. Watson Research Center.

Kandri-Rody, A., Kapur, D. (1984). Algorithms for computing Gröbner bases of polynomial ideals over various Euclidean rings. *EUROSAM 84. Springer Lec. Notes Comp. Sci.* **174,** 193–206.

Lauer, M. (1976). Canonical representatives for residue classes of a polynomial ideal. *SYMSAC* **76,** 339–345.

Möller, H. M. (1985). A reduction strategy for the Taylor resolution. *EUROCAL 85. Springer Lec. Notes Comp. Sci.* **204,** 526–534.

Möller, H. M., Mora, F. (1986). New constructive methods in classical ideal theory. *J. Algebra* **100,** 138–178.

Mora, F. (1986). Gröbner and standard bases for noncommutative polynomial rings. *Proc. AAECC 3, Springer Lec. Notes Comp. Sci.* **223,** 353–362.

Pan, L. (1985). Applications of rewriting techniques. Dissertation. University of California at Santa Barbara.

Robbiano, L. (1986). On the theory of graded structure. *J. Symb. Computat.* **2,** 139–170.

Schaller, S. C. (1975). Algorithmic aspects of polynomial residue class rings. Thesis. University of Wisconsin at Madison.

Spear, D. A. (1977). A constructive approach to commutative ring theory. *Proc. MACSYMA User's Conference*, pp. 369–376.

Stifter, S. (1985). Computation of Gröbner bases over the integers and in general reduction rings. Dipl. thesis. Univ. Linz.

Trinks, W. (1978). Über B. Buchbergers Verfahren, Systeme algebraischer Gleichungen zu lösen. *J. Number Theory* **10,** 475–488.

Winkler, F. (1984). The Church–Rosser property in computer algebra and special theorem proving/An investigation of critical-pair/completion algorithms. Thesis Univ. Linz.

Zacharias, G. (1978). Generalized Gröbner bases in commutative polynomial rings. Thesis at M.I.T., Dept. Comp. Sci.

An Extension of Buchberger's Algorithm and Calculations in Enveloping Fields of Lie Algebras†

J. APEL AND W. LASSNER

Naturwiss.-Theoretisches Zentrum und Sektion Mathematik, Karl-Marx-Universität, DDR-7010 Leipzig

(*Received* 10 *April* 1986; *revised* 5 *December* 1986)

The powerful concept of Gröbner bases and an extension of the Buchberger algorithm for their computation have been generalised to enveloping algebras of Lie algebras. Algorithms for the computation of syzygies by use of Gröbner bases are given. This is the first method which allows the transformation of right fractions into left fractions in the Lie field of any finite dimensional Lie algebra. That enables CAS calculations in Lie fields. An AMP program LIEFIELD has been written for this purpose. Another AMP program SYZYGY produces a generating set of the syzygy module of any finite subset of an enveloping algebra. Examples for both programs are presented for the Weyl algebra and the so(3).

1. Introduction

Extending the Buchberger (1965) algorithm to certain types of non-commutative algebras we present an algorithm for CAS calculations in enveloping fields of Lie algebras, i.e. in "rational functions" of generators of Lie algebras.

The main difficulty of calculations in these non-commutative fields of fractions consists in transforming right fractions into left fractions. The algorithm presented solves this non-trivial task together with a mathematically more fundamental problem in abstract algebra—the calculation of the module of syzygies (Gröbner, 1949).

This result is not only of principal interest for algebraists. It also supports constructive ways in representation theory of Lie algebras (Kirillov, 1972; Božek, Havliček & Navrátil, 1985). The algorithm is of interest for physicists in connection with spectrum generating algebra methods (Cordero & Chirardi, 1972).

We present two algorithms and corresponding AMP (Drouffe, 1981) programs LIEFIELD and SYZYGY. The first one is thought for computer calculations in enveloping fields. SYZYGY calculates not only one syzygy but a generating set of the module of syzygies. We underline that the algorithm presented allows calculations in the enveloping field of any finite dimensional Lie algebra.

So the algorithm covers more than only the Weyl algebra for which also other algorithms are known (Lassner, 1980; Galigo, 1985) and more than the calculation of syzygies of only linear elements from enveloping algebras (Beckmann, 1984).

† This is a revised version of a contribution to the International Conference on Computer Algebra and its Applications in Theoretical Physics, Dubna, 17–20 September 1985.

2. Basic Notions and Problems

Let L be a finite dimensional Lie algebra over a field K. We recall that the tensor algebra T of L is $T^0 \oplus T^1 \oplus \ldots \oplus T^n \oplus \ldots$, where $T^n = L \otimes L \otimes \ldots \otimes L$ (n times): in particular, $T^0 = K \cdot 1$ and $T^1 = L$. Let I be the two-sided ideal of T generated by the tensors $x \otimes y - y \otimes x - [x, y]$, where $x, y \in L$. The associative algebra T/I is called enveloping algebra of L and denoted by $U(L)$ (Dixmier, 1974). The composition of the canonical mappings $L \to T \to U(L)$ is called canonical mapping δ of L into $U(L)$. We have

$$\delta(x)\delta(y) - \delta(y)\delta(x) = \delta([x, y])$$

for all $x, y \in L$. We denote the canonical image of $T^0 \oplus T^1 \oplus \ldots \oplus T^q$ in $U(L)$ by $U_q(L)$, i.e. $\deg(f) \leqslant q$ for $f \in U_q(L)$.

Lemma 1 (*Dixmier*, 1974). *Let $a_1, \ldots, a_p \in L$, σ the canonical mapping of L into $U(L)$, and π be a permutation of $\{1, \ldots, p\}$. Then*

$$\delta(a_1) \ldots \delta(a_p) - \delta(a_{\pi(1)}) \ldots \delta(a_{\pi(p)}) \in U_{p-1}(L).$$

From now we denote the canonical image of $x \in L$ in $U(L)$ by the same letter x. The Poincaré–Birkhoff–Witt theorem ensures the existence of the following basis in $U(L)$. Let $(x_1, \ldots, x_n)$ be a basis of the vector space L. Then the $x_1^{i_1} x_2^{i_2} \ldots x_n^{i_n}$, where $i_1, \ldots, i_n \in N u\{0\}$, form a basis of $U(L)$. We denote this basis by B and we assume the elements of $U(L)$ always to be represented with respect to this basis. We denote the coefficient of $t \in B$ in $f \in U(L)$ by $c(f, t)$. Now we sketch the definition of the enveloping field (or Lie field) $D(L)$ (Kirillov, 1972; Dixmier, 1974). Consider the set of formal expressions of the forms xy^{-1} and $y^{-1}x$, where $x, y \in U(L)$ and $y \neq 0$. A left fraction $y_1^{-1}x_1$ is defined to be equivalent to a right fraction $x_2 y_2^{-1}$ if $x_1 y_2 = y_1 x_2$. Two left (right) fractions are defined to be equivalent if they are equivalent to one and the same right (left) fraction. The enveloping field is defined to be the set of classes of equivalent fractions equipped with the following operations. Consider two fractions $y_1^{-1}x_1$ and $y_2^{-1}x_2$ which represent the classes a_1 and a_2. In $U(L)$ there always exists a common left multiple, i.e. for given y_1 and y_2 there exist two elements z_1 and z_2 such that $z_1 y_1 = z_2 y_2$ (Ore condition).

Clearly, $y_1^{-1}x_1$ is equivalent to $(z_1 y_1)^{-1}(z_1 x_1)$. The sum $a_1 + a_2$ is the equivalence class of $(z_1 y_1)^{-1}(z_1 x_1 + z_2 x_2)$. The class of the quotient $a_1^{-1}a_2$ is the class of $(z_1 x_1)^{-1}(z_2 x_2)$. The remaining arithmetic operations are defined through $a_1 - a_2 = a_1 + (-1)a_2$ and $a_1 a_2 = (a_1^{-1})^{-1} a_2$. In order to represent a product of two left fractions $a^{-1}b$ and $c^{-1}d$ again by a left fraction $f^{-1}g = a^{-1}bc^{-1}d$ one has to find a left fraction $c'^{-1}b'$ equivalent to the occurring right fraction bc^{-1} such that $bc^{-1} = c'^{-1}b'$, i.e. $f^{-1}g = (b'a)^{-1}(c'd)$.

Problem 1a. Given a right fraction ab^{-1}. Find a left fraction $c^{-1}d$ such that $c^{-1}d = ab^{-1}$.

Note that in general a solution of problem 1a needs more than some commutations according to rules of the type $[u, v^{-1}] = v^{-1}[v, u]v^{-1}$ etc. One has to solve the following task.

Problem 1b. Given $a, b \in U(L)$. Find a (non-zero) solution (c, d) such that $c, d \in U(L)$ and $ca - db = 0$.

This is a special case ($m = 2$) of the following problem.

PROBLEM 1. Given m elements $g_1, \ldots, g_m$ of $U(L)$. Find a (non-zero) syzygy $(h_1, \ldots, h_m)$ of $g_1, \ldots, g_m$, i.e. a solution of the equation

$$h_1 g_1 + \ldots + h_m g_m = 0, \quad h_1, \ldots, h_m \in U(L). \tag{1}$$

While finding only one syzygy of a finite set is trivial in the commutative case, this is in general a hard problem in the non-commutative case. We attack problem 1 inside a larger problem.

PROBLEM 2. Given a finite basis $(g_1, \ldots, g_m)$ of a left ideal of $U(L)$. Find a finite set

$$\{(h_{11}, \ldots, h_{1m}), \ldots, (h_{s1}, \ldots, h_{sm})\}, \; h_{ij} \in U(L), \; 1 \leqslant i \leqslant s, \; 1 \leqslant j \leqslant m,$$

which generates the $U(L)$-module of the syzygies of $g_1, \ldots, g_m$, i.e.

$$\sum_{j=1}^{m} h_{ij} g_j = 0, \quad 1 \leqslant i \leqslant s,$$

and any syzygy $(h_1, \ldots, h_m)$ is an $U(L)$-combination

$$(h_1, \ldots, h_m) = \sum_{i=1}^{s} k_i (h_{i1}, \ldots, h_{im}), k_1, \ldots, k_s \in U(L). \tag{2}$$

3. Extended Buchberger Algorithm

Starting from the notions and results for the commutative case as first introduced in Buchberger (1965) and tutorially described in Buchberger's summarising paper (1985) we present a generalisation of Gröbner bases and the Buchberger algorithm to enveloping algebras of Lie algebras. The class considered contains the commutative ring $K[x_1, \ldots, x_n]$ as special case, where the given algorithm specialises to the Buchberger algorithm, i.e. in its present version to that for non-reduced Gröbner bases.

The extension to enveloping algebras includes some new notions. The theoretical foundation is given in a series of lemmata. Since the complete proofs are rather long they are given in details in a forthcoming paper. Though our lemmata are proven for a rich class of non-commutative algebras the proofs benefit from the basic ideas in the commutative case.

Due to the non-commutativity various notions require the distinction between left-, right-, or two-sided ones. Since we consider only left ideals the attribute "left-" occurring in the definitions will be dropped if no confusion is possible.

For a special class of rings A the following lemma 2 originates from Zacharias (1978).

LEMMA 2. *Let A a ring with unit element, $F = (f_1, \ldots, f_m)$ and $G = (g_1, \ldots, g_2)$ two bases of the left ideal $I \subset A$, X and Y transformation matrices such that $G^T = XF^T$ and $F^T = YG^T$, and R be a matrix whose rows generate the syzygy module of G. Then the rows of*

$$Q = \begin{pmatrix} I_m - YX \\ RX \end{pmatrix} \tag{3}$$

generate the module of syzygies of F; I_m is the (m, m)-unit matrix.

PROOF. (1) (The Q-rows are syzygies)

$$QF^T = \begin{pmatrix} I_m - YX \\ RX \end{pmatrix} F^T = \begin{pmatrix} F^T - YXF^T \\ RXF^T \end{pmatrix} = \begin{pmatrix} F^T - YG^T \\ RG^T \end{pmatrix} = \begin{pmatrix} F^T - F^T \\ 0 \end{pmatrix} = 0.$$

(2) (Any syzygy is an A-combination of Q-rows). Assume h is a syzygy of F, i.e. $hF^T = 0$. Because of $F^T = YG^T$ this means hY is syzygy of G, i.e. hY is A-combination of R-rows $hY = kR$ with coefficients $k_i \in A$. So it follows $hYX = kRX$, $h(I_m - I_m + YX) = kRX$, and finally $h = kRX + h(I_m - YX)$, i.e. h is an A-combination of Q-rows.

As in commutative polynomial rings our extended Buchberger algorithm is a suitable tool for computing a basis G, matrices X, Y, R, and the syzygies of the given $(f_1, \ldots, f_m) = F$.

We define an ordering in the basis B of $U(L)$ by

$$x_1^{i_1} x_2^{i_2} \ldots x_n^{i_n} \prec x_1^{j_1} x_2^{j_2} \ldots x_n^{j_n}$$

iff the first non-zero component of

$$\left(\sum_{k=1}^{n} (i_k - j_k), i_1 - j_1, \ldots, i_n - j_n \right) \text{ is negative.}$$

We introduce the following notations with respect to $\prec$ for elements f of $U(L)$.

$LBE(f)$: (leading basis element of f) the maximal (*w.r.t.* $\prec$) basis element t such that $c(f, t) \neq 0$.

$LC(f)$: (leading coefficient of f) $LC(f) = c(f, LBE(f))$.

Note some properties of $\prec$ which are fundamental for the proofs:

(i) $1 = x_1^0 x_2^0 \ldots x_n^0 \prec u$ for all $u \in B \setminus \{1\}$.

(ii) Any sequence $u_1, u_2, \ldots, u_i, \ldots$ such that $u_{i+1} \prec u_i$ for all $i \geq 1$ is finite. ($\prec$ is a Noetherian ordering.)

(iii) For all $t \in B: u \prec v$ iff $LBE(tu) \prec LBE(tv)$.

Furthermore, we define a special "multiple property" for basis elements of $U(L)$.

DEFINITION 1. $u \in B$ is called *left top multiple* (LTM) of $v \in B$ if there exists a $t \in B$ such that $LBE(tv) = u$. $t \in B$ is called *common left top multiple* ($CLTM$) of $u \in B$ and $v \in B$ if there exist $t_1, t_2 \in B$ such that $LBE(t_1 u) = LBE(t_2 v) = t$. The minimal (*w.r.t.* $\prec$) $CLTM$ of $u \in B$ and $v \in B$ is called *least common left top multiple* and denoted by $LCLTM(u, v)$.

Note the analogy of the property

$$LCLTM(x_1^{i_1} \ldots x_n^{i_n}, x_1^{j_1} \ldots x_n^{j_n}) = x_1^{\max(i_1, j_1)} \ldots x_n^{\max(i_n, j_n)}$$

to that of LCM in $K[x_1, \ldots, x_n]$.

A first concept of Gröbner bases for a rich class of non-commutative algebras was developed by Mora (1985). This class is different from our one. While Mora saved the property that the basis elements form a free semigroup, we save Dickson's lemma. The existence of Gröbner bases for any finite generated left ideal is ensured and a constructive way for finding it is given for our class.

LEMMA 3. (*Analogue to Dickson's lemma.*) *Let $t_1, t_2, \ldots, t_i, \ldots$ be a sequence of elements of B such that t_i is not a LTM of t_j for $i > 1$ and $j < i$. Then this sequence is finite.*

The proof of Dickson (1913) can be used nearly without changes.

Let us deal now with the *reduction* of elements of $U(L)$ with respect to a finite subset of $U(L) \setminus \{0\}$.

DEFINITION 2. Let $F=\{f_1,\ldots,f_m\}$ be a finite subset of $U(L)\backslash\{0\}$. Let g and h be elements of $U(L)$. g is called *left reducing* to h modulo F (write $g \xrightarrow[F]{l} h$) if there exist $f_i \in F$, $u \in B$, and $a \in K$ such that $c(g, LBE(uf_i)) \neq 0$, $a = c(g, LBE(uf_i))/LC(uf_i)$ and $h = g - auf_i$.

$\sum_{i=1}^{m} h_i f_i + h$ is called a *left normal representation* of g modulo F if $g = \sum_{i=1}^{m} h_i f_i + h$, there is no h' such that $h \xrightarrow[F]{l} h'$, and $LBE(h_i f_i) \prec LBE(g)$, h is called a *left irreducible part* of g modulo F.

LEMMA 4. *Let $F=\{f_1,\ldots,f_m\}$ be a finite subset of $U(L)\backslash\{0\}$. Then there does not exist any infinite sequence $g_1, g_2, \ldots, g_i, \ldots$ such that $g_i \xrightarrow[F]{l} g_{i+1}$ for all $i \geqslant 1$.*

The proof of the lemma goes by induction according to $LBE(g_1)$.

PROBLEM 3. Given a set $F=\{f_1,\ldots,f_m\} \subset U(L)\backslash\{0\}$ and an element g of $U(L)$.
Compute a left normal representation of g modulo F.

ALGORITHM 1. $LNR(F, g; h, h_1, h_2, \ldots, h_m)$

1. $h := g$, $h_i := 0$ $(1 \leqslant i \leqslant m)$
2. *while* exist $f_j \in F$ and $t \in B$ such that $c(h, t) \neq 0$ and t is LTM of $LBE(f_j)$ *do*

2.1. choose a pair (f_j, t) where t is maximal (*w.r.t.* $\prec$) among all these pairs (f_j, t)
2.2. compute u and a such that $LBE(uf_j) = t$ and $a = c(h, t)/LC(f_j)$
2.3. $h := h - auf_j$
2.5. $h_j := h_j + au$.

PROOF OF CORRECTNESS. Obviously, all steps are computable. After each execution of loop 2 the equation

$$\sum_{i=1}^{m} h_i f_i + h = g$$

is satisfied. $LBE(h_i f_i) \prec LBE(g)$ holds in any step. After termination h is a left irreducible part of g modulo F.

PROOF OF TERMINATION. Let g_k be the value of h before the kth execution of loop 2. The sequence $g_1, g_2, \ldots, g_i, \ldots$ is finite according to lemma 4.

Now we transfer the ideas of "S-polynomials" (Buchberger, 1965) and Gröbner bases to enveloping algebras of Lie algebras.

DEFINITION 3. The element $SE(f_1, f_2) := c_2 u_1 f_1 - c_1 u_2 f_2$ of $U(L)$, where $u_i \in B$, $c_i = LC(f_i)$, and $LBE(u_i f_i) = LCLTM(LBE(f_1), LBE(f_2))$, $i = 1, 2$ is called the *left S-element* corresponding to the elements f_1 and f_2 of $U(L)\backslash\{0\}$.

DEFINITION 4. A finite subset $F=\{f_1,\ldots,f_m\}$ of $U(L)\backslash\{0\}$ is called *left Gröbner basis* of the left ideal $J \subset U(L)$ if F generates J and zero is a left irreducible part modulo F of any left S-element $SE(f_i, f_j)$ corresponding to two elements f_i and f_j of F.

Note that the so-defined Gröbner bases are just those finite subsets G of $U(L)\backslash\{0\}$ for which the left irreducible part of any $g \in U(L)$ modulo G is uniquely determined.

Remember that the syzygies of an arbitrary finite set F could be found according to lemma 2 if the module of syzygies of G would be given (by R). For a Gröbner basis G this gap will be closed by lemma 5.

LEMMA 5. *Let $G=(g_1, \ldots, g_m)$ be a Gröbner basis and $LC(g_k)=1$, $i \leqslant k \leqslant m$. For any pair (g_i, g_j) where $1 \leqslant i < j \leqslant m$ let u_1^{ij} and u_2^{ij} be basis elements of $U(L)$ such that*

$$LCLTM(LBE(g_i), LBE(g_j)) = LBE(u_1^{ij} g_i) = LBE(u_2^{ij} g_j),$$

and let h_k^{ij} be elements of $U(L)$ such that $LBE(h_k^{ij} g_k) \preccurlyeq LBE(SE(g_i, g_j))$ and

$$\sum_{k=1}^{m} h_k^{ij} g_k - (u_1^{ij} g_i - u_2^{ij} g_j) = 0. \tag{4}$$

Then the m-tuples

$$r^{ij} = (h_1^{ij}, \ldots, h_{i-1}^{ij}, h_i^{ij} - u_1^{ij}, h_{i+1}^{ij}, \ldots, h_{j-1}^{ij}, h_j^{ij} + u_2^{ij}, h_{j+1}^{ij}, \ldots, h_m^{ij}), \quad 1 \leqslant i < j \leqslant m \tag{5}$$

generate the module of syzygies of $g_1, \ldots, g_m$.

PROOF IDEA. That the r^{ij} are syzygies of $g_1, \ldots, g_m$ follows by (4). The proof that any syzygy $s=(s_1, \ldots, s_m)$ of $g_1, \ldots, g_m$ is $U(L)$-combination of the r^{ij} can be performed by induction according to the maximal (*w.r.t.* $\prec$) element among all $LBE(s_k g_k)$ $(1 \leqslant k \leqslant m)$.

This connection between Gröbner bases and syzygies in the commutative case can be found already in the thesis of Zacharias (1978). Later it was described and generalised by certain authors, for example by Moeller & Mora (1986).

Now we are ready to present the main algorithm which could be characterised as an extension of the Buchberger algorithm to enveloping algebras. In the present version the extension corresponds to non-reduced Gröbner bases. Since the elements of the finite set F appear as the first elements (up to a factor from K) in G, the first rows $I_m - YX$ of Q described in lemma 2 are zero. Hence, they can be dropped, i.e. $Q = RX$ may be used. Due to the lemmata 2 and 5 the central problem 2 goes over in the following one.

PROBLEM 4. Given a finite set $F=(f_1, \ldots, f_m)$, where $f_i \neq 0$. Compute a left Gröbner basis $G=(g_1, \ldots, g_l)$ such that F and G generate the same left ideal in $U(L)$, $LC(g_i)=1$ $(1 \leqslant i \leqslant l)$, and $f_i = LC(f_i) g_i$ $(1 \leqslant i \leqslant m)$. Furthermore, compute matrices X, R, and Q such that $G^T = XF^T$ and the rows of R and Q generate the modules of the syzygies of G and F, respectively.

Problem 4 will be solved by the following algorithm.

ALGORITHM 2. SYZYGY(F; G, X, R, Q)

1. $l := m$
2. $R := (0, 0, \ldots, 0)$, $Q := (0, 0, \ldots, 0)$ (rows of length m)
3. $c_k := LC(f_k)$ $(1 \leqslant k \leqslant m)$
4. $g_k := f_k / c_k$ $(1 \leqslant k \leqslant m)$
5. $X := \begin{pmatrix} 1/c_1 & & 0 \\ & \ddots & \\ 0 & & 1/c_m \end{pmatrix}$

6. $G := (g_1, \ldots, g_m)$
7. $P := \{(g_i, g_j) \mid 1 \leqslant i < j \leqslant m\}$
8. *while* $P \neq \emptyset$ *do*
8.1. take $(g_i, g_j) \in P$
8.2. $P := P \backslash \{(g_i, g_j)\}$
8.3. $h' := SE(g_i, g_j)$
8.4. $LNR(G, h'; h, h_1, \ldots, h_l)$
8.5. $h_i := h_i - u_1^{ij}$, $h_j := h_j + u_2^{ij}$ (u_1^{ij} and u_2^{ij} such that $SE(g_i, g_j) = u_1^{ij} g_i - u_2^{ij} g_j$)
8.6. $H := (h_1, \ldots, h_l)$
8.7. *if* $h = 0$
8.7.1. *then* $R := \begin{pmatrix} R \\ H \end{pmatrix}$
8.7.2. *else*
8.7.2.1. $R := \begin{pmatrix} R & 0 \\ H & LC(h) \end{pmatrix}$
8.7.2.2. $H := -1/LC(h) * H$
8.7.2.3. $X := \begin{pmatrix} X \\ HX \end{pmatrix}$
8.7.2.4. $g_{l+1} := h/LC(h)$
8.7.2.5. $G := (g_1, \ldots, g_l, g_{l+1})$
8.7.2.6 $P := P \cup \{(g_k, g_{l+1}) \mid 1 \leqslant k \leqslant l\}$
8.7.2.7. $l := l + 1$
8.8. $Q := \begin{pmatrix} Q \\ R_L X \end{pmatrix}$ (R_L is the last row of R).

Proof of correctness. Obviously all steps are computable. If $P = \emptyset$ then any S-element corresponding to two elements of G has 0 as a left irreducible part modulo G, hence G is a left Gröbner basis by definition 4. After each execution of loop 8 $G^T = XF^T$ and $RG^T = 0$ are satisfied. After termination the rows of R are tuples (5) of lemma 5. Therefore, they generate the syzygy module of G. That the Q-rows generate the module of syzygies of F follows from lemma 2.

Proof of termination. The only step which is of interest for the termination is loop 8. Let us consider the sequence $LBE(g_1)$, $LBE(g_2), \ldots, LBE(g_k), \ldots$. Since the g_k $(k > m)$ are left irreducible parts modulo $(g_1, \ldots, g_{k-1})$, there is no $1 \leqslant k' < k$ such that $LBE(g_k)$ is LTM of $LBE(g_{k'})$. According to lemma 3 such a sequence must be finite. Therefore, only a finite number of pairs (g_i, g_j) must be treated.

Remark. It is possible to avoid the initialisation of R and Q to abundant trivial syzygies $(0, 0, \ldots, 0)$ in step 2. It could be used as initialisation according to loop 8.

Of course, algorithm 2 solves not only problem 2 but also the subproblem 1. But a slight change makes the algorithm more efficient for this purpose. Since $G^T = XF^T$ and $RG^T = 0$ are satisfied after each execution of the loop 8 the equation $QF^T = RXF^T = 0$ is satisfied, too. This means, whenever Q is not a zero matrix in step 8, problem 1 is already solved. Therefore, we change step 8 into

8. *while* $P \neq \emptyset$ and $Q = 0$ *do*

In this form the algorithm is called algorithm 2a. It is used for calculations in Lie fields, i.e. for transforming right fractions into left fractions.

Clearly, algorithm 2a is much more time efficient than algorithm 2. In order to calculate only one non-zero syzygy also only a part of the Gröbner basis must be generated in most cases.

4. Implementation

The algorithms 2 and 2a have been realised in the computer algebra system AMP (version 6.4) (Drouffe, 1981). The programs SYZYGY (algorithm 2) and LIEFIELD (algorithm 2a) have been tested on an ESER computer EC 1040.

5. Examples

Let x_1, x_2, x_3 be the generators of the Lie algebra so(3), with commutation rules

$$[x_1, x_2] = x_3, \quad [x_2, x_3] = x_1, \quad [x_1, x_3] = -x_2.$$

EXAMPLE 1. Consider the right fraction $(x_1 x_2)x_1^{-1}$. We start with $F = (x_1 x_2, x_1)$. SYZYGY computes the Gröbner basis $G = (x_1 x_2, x_1, x_3, x_2)$, the transformation matrix X ($G^T = XF^T$)

$$X = \begin{pmatrix} 1 & 0 \\ 0 & 1 \\ 1 & -x_2 \\ -x_1 & x_1 x_2 + x_3 \end{pmatrix},$$

the matrix R the rows of which generate the module of syzygies of the Gröbner basis G

$$R = \begin{pmatrix} 1 & -x_2 & -1 & 0 \\ 0 & x_3 & -x_1 & -1 \\ 0 & -1 & x_2 & -x_3 \\ 0 & x_2 & 1 & -x_1 \\ 1 & 0 & 0 & -x_1 \\ x_3 & x_1 & -x_1 x_2 & -x_2 \end{pmatrix},$$

and the matrix Q whose rows generate the module of syzygies of F

$$Q = \begin{pmatrix} 0 & 0 \\ 0 & 0 \\ x_1 x_3 + 2x_2 & -x_1 x_2 x_3 + x_1^2 - 2x_2^2 - x_3^2 - 1 \\ x_1^2 + 1 & -x_1^2 x_2 - x_1 x_3 \\ x_1^2 + 1 & -x_1^2 x_2 - x_1 x_3 \\ 0 & 0 \end{pmatrix}.$$

Therefore, the left fraction equivalent to the right fraction above has the following form:

$$(a_1(x_1x_3+2x_2)+a_2(x_1^2+1))^{-1}(a_1(x_1x_2x_3-x_1^2+2x_2^2+x_3^2+1)+a_2(x_1^2x_2+x_1x_3)),$$

where a_1 and a_2 are arbitrary elements of $U(\mathrm{so}(3))$ such that the denominator is not zero. Note that this is a general solution of the problem, but in the practical use for calculations in Lie fields it is enough to have one non-zero syzygy.

LIEFIELD gives:

$$(x_1x_3+2x_2)^{-1}(x_1x_2x_3-x_1^2+2x_2^2+x_3^2+1) = (x_1x_2)x_1^{-1}.$$

Consider the Weyl algebra A_2 with the basis elements $q_1^{i_1}q_2^{i_2}p_1^{j_1}p_2^{j_2}$ and the only non-vanishing commutators $[p_i, q_i] = 1$ between the generators $q_i, p_i, 1$ $(i = 1, 2)$.

EXAMPLE 2. Consider the right fraction $(3p_2^3-p_1)(1/23q_2p_1+p_2)^{-1}$. The input is $F = (3p_2^3-p_1, 1/23q_2p_1+p_2)$. SYZYGY computes the left Gröbner basis

$$G = (p_2^3-1/3p_1, q_2p_1+23p_2, p_1p_2^2, p_1^2, p_1p_2, p_2^2+1/23p_1, p_1, p_2).$$

The matrix Q contains 28 rows where 22 among them are non-zero. The degree of the first component of the non-trivial syzygies is in the range from 5 up to 9. LIEFIELD gives only the shorter basis

$$G = (p_2^3-1/3p_1, q_2p_1+23p_2, p_1p_2^2, p_1^2)$$

(not yet a left Gröbner basis). The first non-zero syzygy is the third computed one. The returned left fraction is

$$(1/3q_2p_1p_2^3+23/3p_2^4+2p_1p_2^2-1/9q_2p_1^2-23/9p_1p_2)^{-1}(23p_2^6-46/3p_1p_2^3+23/9p_1^2)$$
$$= (3p_2^3p_1)(1/23q_2p_1+p_2)^{-1}.$$

In the first example the relation between the running times of SYZYGY and LIEFIELD is 19·5 sec : 10·3 sec. In the second example it is 61 sec : 12 sec. We presented examples from computation in Lie fields. Therefore, F had had only two elements. Of course, SYZYGY computes the module of the syzygies of $f_1, \ldots, f_m$ also for $m > 2$.

Due to the excessive use of non-commutative multiplications the running time strongly depends on a fast multiplication procedure. The multiplication is realised using LET-commands suitable for any enveloping algebra. For the Weyl algebra twisted products (Lassner, 1980) are three times faster.

6. Concluding Remarks

The algorithm presented is especially aimed to calculations in Lie fields. Instead of left ideals, left Gröbner bases etc., we could consider right ones. In this paper we did not deal with two-sided ideals, since they are not interesting for the computation of syzygies. Further extensions analogously to the commutative case are imaginable. It is possible to introduce reduced Gröbner bases (Buchberger, 1985) and to extend the method to left vector modules (Moeller & Mora, 1986) over enveloping algebras. That makes it possible to compute free resolutions of left (right) ideals in enveloping algebras. Note, however, that not everything transfers from the commutative to the considered non-commutative case, as the following examples show.

- The "purely lexicographical ordering" (Buchberger, 1985) cannot be used in $U(L)$ since property (iii) does not hold.

—Criterion 2 (Buchberger, 1985) does not hold. For instance, $F = (p, q) \subset A_1$ would be a left Gröbner basis according to such a criterion, but this is wrong since $SE(p, q) = -1$ does not have zero as left irreducible part.

We would like to thank Professor B. Buchberger from the Johannes Kepler University, Linz for valuable discussions and hints about his algorithm.

References

Beckmann, P. (1982). Umwandlung von Rechts- in Linksquotienten über nichtkommutativen Ringen, *Wiss. Z. KMU Leipzig. Math.-Nat. R.* **31,** 11–23.

Božek, P., Havliček, M., Navrátil, O. (1985). A new relationship between Lie algebras of Poincaré and de Sitter groups. Preprint, Charles University, Prague.

Buchberger, B. (1965). An algorithm for finding a basis for the residue class ring of a zero-dimensional polynomial ideal. PhD Thesis, Univ. Innsbruck (in German).

Buchberger, B. (1985). Gröbner bases: An algorithmic method in polynomial ideal theory. In: (Bose, N. K., ed.) *Recent Trends in Multidimensional Systems Theory*. D. Reidel.

Cordero, P., Chirardi, G. C. (1972). Realization of Lie algebras and the algebraic treatment of quantum problems. *Fortschr. Phys.* **20,** 105–133.

Dickson, L. E. (1913). Finiteness of the odd perfect and primitive abundant numbers with *n* distinct prime factors. *Am. J. Math.* **35,** 413–426.

Dixmier, J. (1974). *Algèbres Enveloppantes*. Paris: Bordas (Gauthier-Villars).

Drouffe, J. M. (1981). *AMP Language Reference Manual* (version 6). Gif-sur-Yvette CEN Saclay.

Galigo, A. (1985). Some algorithmic questions on ideals of differential operators. *Proc. EUROCAL '85*, L.N.C.S. 204.

Gröbner, W. (1949). *Moderne Algebraische Geometrie*. Wien–Innsbruck: Springer.

Kirillov, A. A. (1972). *Elements of Representation Theory* (in Russian). Moscow: Nauk.

Lassner, W. (1980). Noncommutative algebras prepared for computer calculations. In: *Proc. Int. Conf. System and Techniques of Analytical Computing and Their Appl. in Theor. Phys.*, JINR, D11–80–13, pp. 58–74, Dubna.

Moeller, H. M., Mora, F. (1986). New constructive methods in classical ideal theory. *J. Algebra* **100,** 138–178.

Mora, F. (1985). Gröbner and standard bases for noncommutative polynomial rings. Private communication about a contribution to the conference AAECC-3, Grenoble.

Zacharias, G. (1978). Generalized Gröbner bases in commutative polynomial rings. Bachelor thesis, M.I.T., Dept. Comp. Scie.

On Recognisable Properties of Associative Algebras

TATIANA GATEVA-IVANOVA† AND VICTOR LATYSHEV‡

† *Institute of Mathematics, Bulgarian Academy of Sciences, P.O.B.* 373, 1000 *Sofia, Bulgaria*

‡ *Department of Mechanics and Mathematics, Moscow State University,* 117 234 *Moscow, U.S.S.R.*

(*Received* 5 *February* 1986; *Revised* 24 *October* 1986)

The paper considers computer algebra in a non-commutative setting. So far, such investigations have been centred on the use of algorithms for equality and of universal properties of algebras. Here, the foundation of all computations is the presentation of the algebra under investigation by a finite number of generators subject to a finite number of defining relations, which satisfy the additional property of forming a Gröbner (or standard) basis. Such algebras are called s.f.p.—for standard finite presentation. It is shown that various algebraic properties, such as being finite-dimensional, nilpotent, nil, algebraic are algorithmically recognisable. When the defining relations are words in the generators, this is also shown to be the case for the properties of being semi-simple, prime, semi-prime, etc.

Introduction

In this paper K denotes a fixed field of arbitrary characteristic, and the term K-algebra is used to denote an associative algebra over K. We are interested in the existence of algorithms for recognising certain properties of a K-algebra A. As usual, the starting point of such an investigation is the presentation of A in the form $A = K\langle X\rangle/I$, where $X = \{x_1, \ldots, x_n\}$ is a set of indeterminates, $K\langle X\rangle$ is the free K-algebra on it, and I is an ideal. That, without further restriction, there is no hope of obtaining reasonable statements, is a consequence (at least) of the unsolvability of the word problem in the class of finitely-presented algebras. Thus, we are led to impose some kind of conditions on I. The ones best suited to our purposes turn out to be those considered by Buchberger (1965) for commutative algebras and generalised by Bergman (1978) to the case of associative algebras, in terms of specific properties of some set of generators of I. We now turn to their description, following Bergman (1978).

Let $\langle X\rangle$ denote the free semigroup generated by X; we consider the empty word as belonging to $\langle X\rangle$, and denote it by 1. The elements of $\langle X\rangle$ are ordered as follows: (a) $x_1 < \ldots < x_n$; (b) if $u, v \in \langle X\rangle$ are of the same degree, then $<$ refers to the lexicographic order; (c) if $u, v \in \langle X\rangle$ are of degree m, n respectively, and $m < n$, then $u < v$. Clearly, this order is compatible with the product in $\langle X\rangle$.

We consider only *finitely generated ideals* I (i.e. only finitely-presented algebras A). Let $G = \{g_1, \ldots, g_t\}$ be a generating set for I, and for $i = 1, \ldots, t$ write g_i in the form $w_i - h_i$, where $w_i \in \langle X\rangle$ is the highest term of g_i. In order to avoid trivial cases, we furthermore assume $\deg w_i \geqq 2$ for all i.

Consider the K-linear operators r_{uiv} called *reductions*, defined on the underlying vector space of $K\langle X\rangle$ by the formulas $r_{uiv}(uw_iv) = uh_iv$ $(u, v \in \langle X\rangle)$, and $r_{uiv}(w) = w$ for all other monomials w.

A monomial is called normal (mod I) if it does not contain as a subword any of the w_i's. We say a polynomial f is normal (mod I), if either $f = 0$, or if

$$f = \sum_{i=1}^{s} \alpha_i f_i$$

with $\alpha_i \in K\backslash\{0\}$, f_i is a normal monomial for $i = 1, \ldots, s$, and $f_i < f_j$ when $i < j$.

Since the order on $\langle X\rangle$ satisfies the decreasing chain condition, it follows that for any polynomial $f \in K\langle X\rangle$, one can find by means of a finite number of reductions a normal polynomial $\bar{f}$ such that $\bar{f} \equiv f$ (mod I). Such a polynomial $\bar{f}$ is called a normal form of f. However, $\bar{f}$ is not determined uniquely by f in general. In case every $f \in K\langle X\rangle$ has a unique normal form, the basis G is called *Gröbner* (or *standard*). This is clearly equivalent to the condition that $g \in I$ (if and) only if it can be reduced to 0. In this case $A = K\langle X\rangle/I$ is said to be a *s.f.p.* (=*standard finitely presented*) algebra. The importance of this notation comes from the fact that the normal (mod I) monomials in $K\langle X\rangle$ project to a K-basis of A, cf. Bergman (1978). In particular, if A is s.f.p., it has a solvable word problem. Note that any finitely generated commutative K-algebra is s.f.p. However, in general, whether a given algebra A is s.f.p., is a highly non-trivial problem. Even if this is known to be the case for A, then the actual exhibition of a standard presentation may still be a difficult task, since it depends not only on I and its generators, but also on the algebra generators of A, and the order on $\langle X\rangle$.

In this paper, our starting point is an algebra A which is given via a standard presentation. Our main results show that in this case many significant algebraic properties are *recognisable*, i.e. can be established algorithmically.

THEOREM 1. *Let A be an s.f.p. algebra. Then it is a recognisable property that:*

(i) *A has a growth of a given type, in particular: A is finite dimensional;*
(ii) *A is algebraic over K;*
(iii) *A satisfies the polynomial identity $[X_1, \ldots, X_r] = 0$ for a fixed integer r.*

For algebras without unit, the properties of being nil or nil-potent are recognisable.

The proof is given in remark 9 *for* (i), (ii), *nil and nilpotency. It follows from remark* 12 *for* (iii).

In case A is defined by monomial relations, i.e. $g_i = w_i$ for $i = 1, \ldots, t$, more information is available.

THEOREM 2. *Let A be a monomial algebra. Then it is a recognisable property that:*

(iv) *A has no nilpotent elements;*
(v) *A is semi-simple (in the sense of Jacobson);*
(vi) *A is prime;*
(vii) *A is semi-prime.*

This is an immediate consequence of the more precise result contained in theorem 18.

Related to the question of recognising some property of the *algebra* A, is the problem of the existence of an algorithm for checking if a given *element* of A has some specified property. When A is monomial we answer it positively in the proposition 13 and theorem

15 for the property of belonging to the Jacobson radical. For the properties of being a zero-divisor, nilpotent, or quasi-invertible, we provide algorithms only for elements of a specific type, which are in some sense "external". In particular, they include all homogeneous elements.

We should like to thank Luchezar Avramov for several useful discussions and for suggesting improvements of the exposition. We also want to thank Ragnar-Olaf Buchweitz and Jim Carlson for help with the material in section 5.

1. Ufnarovskij's Graph and Growth of Algebras

This section introduces some techniques which will be used in the rest of the paper, and gives some applications of Ufnarovskij's work on growth of monomial algebras.

For an algebra C, generated over K by $c_1, \ldots, c_n$, denote by C_s the K-linear span of all monomials of degree $\leqq s$ in the c_i's. Recall that C is said to have exponential growth (resp. polynomial growth of degree d) if there exist real numbers $1 \leqq D < E$ such that $D^s \leqq \dim_K C_s < E^s$ (resp.: if d is the smallest non-negative integer such that there exists a degree d real polynomial $p(x)$ with $\dim_K C_s \leqq p(s)$) for all $s \geqq 0$.

It is well known (Borho & Kraft, 1976), that these conditions do not depend on the choice of the (finite set of) generators of C.

A finitely presented algebra $B = K\langle x_1, \ldots, x_n\rangle / J$ is called a monomial algebra, if the ideal J is generated by a finite number of monomials in the x_i's. We shall always assume a monomial algebra comes with a presentation in which the generators x_i and the monomials $W = \{w_1, \ldots, w_t\}$ which generate J are fixed; furthermore, we assume each w_i is of degree $\geqq 2$. It is easily verified that any such W is a Gröbner basis of J.

Following Ufnarovskij (1982), the monomial algebra B determines an oriented graph $U = U(B)$ by the following procedure. The vertices of U are the degree $m-1$ ($m = \max \deg w_i$) monomials in $K\langle X\rangle$ which are not in J. If u and v are such monomials, then there is an edge $\overrightarrow{uv}$ from u to v if and only if there exist variables $x, y \in X$, such that $ux = yv \notin J$. We shall refer to $U(B)$ as the Ufnarovskij graph of B.

A route of length d in the graph U is a sequence of edges $(\overrightarrow{v_0 v_1}, \overrightarrow{v_1 v_2}, \ldots, \overrightarrow{v_{d-1} v_d})$. A route is called simple if it contains no edge twice. A cycle is a subgraph of U such that its vertices $v_0, \ldots, v_d$ satisfy $v_0 = v_d$ and $(\overrightarrow{v_0 v_1}, \ldots, \overrightarrow{v_{d-1} v_d})$ is a simple route. A vertex is called multiple if it enters more than one distinct cycle; if v is not multiple it is dubbed simple. A route is called cyclic if the underlying graph is a cycle.

REMARK 1. Ufnarovskij (1982). There is a one-to-one correspondence between the non-zero words of length $\geqq m-1$ in B and the routes in $U(B)$: the word $x_{i_1} \ldots x_{i_s}$ is mapped to the route $(\overrightarrow{v_0 v_1}, \ldots, \overrightarrow{v_{d-1} v_d})$, where $d = s-(m-1)$, and $v_j = x_{i_{j+1}} x_{i_{j+2}} \ldots x_{i_{j+m-1}}$ for $j = 0, \ldots, d$. Under this correspondence, a cyclic route beginning at v is represented by a word of the form $vp = qv$ for suitable p and q.

We now return to the notation of the Introduction. For a given standard basis $G = \{g_1, \ldots, g_t\}$ we denote by $W = W(G) = \{w_1, \ldots, w_t\}$ the set of the leading monomials of $g_1, \ldots, g_t$. To the algebra $A = K\langle X\rangle / I$ one associates the monomial algebra $\tilde{A} = K\langle X\rangle / J$, where J is generated by the set $W(G)$. We extend the notation of Ufnarovskij graph to algebras with Gröbner basis by simply setting $U(A) = U(\tilde{A})$. We write $L(A)$ for the maximal length of simple routes in $U(A)$ which do not contain any cyclic subroute and call it the *diameter of A*. Let $l = l(A)$ be the number of vertices in $U(A)$. (Obviously $L(A) < l(A) \leqq n^{m-1}$.)

PROPOSITION 2. *Let A be an s.f.p. algebra. Then either:*

(*i*) *A has exponential growth, and this happens if and only if $U(A)$ has a multiple vertex; or*

(*ii*) *A has polynomial growth of degree d, and this happens if and only if $U(A)$ has only simple vertices, and d is the largest number of cycles occurring in a connected component of $U(A)$.*

PROOF. When A is a monomial algebra, this is precisely the main theorem of Ufnarovskij (1982). The general case reduces to the previous one by the observation that the sets of normal monomials in $K\langle X\rangle$ with respect to I and J coincide, hence $A_s \cong \tilde{A}_s$ as K-vector spaces for any $s \geqq 0$.

REMARK 3. Since $\dim_K A < \infty$ means that A has polynomial growth of degree zero, one sees A is infinite-dimensional if and only if $U(A)$ contains a cycle.

REMARK 4. The following elementary fact is used in many computations throughout the paper, and is recorded for further use.

Let a', u, a'' be monomials, such that $deg\, u \geqq m-1$. If $a'u$ and ua'' are normal with respect to I, then so is $a'ua''$.

A particular case is worth special mention.

LEMMA 5. *Let a and u be monomials in $K\langle X\rangle$, such that $deg\, u = m-1$. If uau is normal with respect to I, then so are the monomials $(ua)^i$ for $i \geq 0$. In particular, their images in A are linearly independent.*

As an immediate consequence we note:

COROLLARY 6. *An s.f.p. algebra A is algebraic* (*if and*) *only if it is finite-dimensional.*

For the next corollary and the remark following it, A denotes an algebra without unit. In this case the definition of s.f.p. algebra is modified by replacing the free K-algebra $K\langle X\rangle$ by its subalgebra consisting of polynomials without constant term.

COROLLARY 7. *An s.f.p. algebra is nilpotent if* (*and only if*) *it is nil.*

Furthermore, these properties imply A is finite-dimensional.

PROOF. If A is nil, it is finite-dimensional by corollary 6, and has a K-basis of nilpotent elements by hypothesis.

It is well known that such an algebra is nilpotent.

REMARK 8.

(1) It is an open problem in ring theory whether a finitely-presented nil-algebra is nilpotent; Dnestr Notebook (1982). The preceding corollary shows it has a positive solution in the subclass of s.f.p. algebras.

(2) Since a graded finite-dimensional algebra is nilpotent, the last two statements show that when A is s.f.p. and graded, each one of the properties: A is finite-dimensional, A is algebraic, A is nil, A is nilpotent, implies the remaining three.

REMARK 9. Since $U(A)$ is a finite graph, any property of the algebra A which can be expressed in terms of some property of $U(A)$, is recognisable. According to proposition 2 the character of the growth of A (in particular, whether or not A is finite-dimensional) is a recognisable property. According to corollary 6, this is also true for the property of being algebraic.

If one wants to check nilpotency of A, one first verifies the necessary (according to corollary 6) condition that $\dim_K A$ is finite. Then, for a (normal) monomial u of length $L(A)+m-1$ one checks whether u^p reduces to 0 for

$$p = \sum_{i=0}^{L(A)+m-1} n^i.$$

This condition is necessary and sufficient for nilpotency, since by remark 3 $U(A)$ has no cycles, hence by remark 1 the monomials of length $\leq L(A)+m-1$ generate A as K vector space.

2. Existence of a Polynomial Identity

It is known that in general there exists no algorithm for checking whether a given polynomial identity (= p.i.) is satisfied in some finitely presented algebra (Matiyasevich, 1984). It is not clear whether such an algorithm exists for s.f.p. algebras.

One can ask the weaker question if there exists an algorithm which recognises whether the algebra satisfies any p.i. For monomial algebras the answer has been obtained by Borisenko (1985). He proves that the answer is positive if the corresponding Ufnarovskij graph contains no multiple vertices. For such algebras the existence of a p.i. is equivalent to representability by matrices over some field.

In this section we prove that it is a recognisable property whether an s.f.p. algebra satisfies the identity of Lie nilpotency of a fixed degree.

Furthermore, for monomial algebras we prove the corresponding results for Lie nilpotency of arbitrary degree.

Results of Markov (1979) imply that the T-ideal, generated by the polynomial $[X_1, \ldots, X_r]$ of Lie nilpotency is finitely generated as an ideal. This allows one to recognise algorithmically whether an s.f.p. algebra satisfies the identity of Lie nilpotency of a given degree r. We give a new (and algorithmic) proof of the following result which can also be deduced from Markov (1979).

THEOREM 10. *Let $K\langle X_\infty\rangle$ be the countably generated associative algebra and let T denote its T-ideal, generated by the polynomials $[X_1, \ldots, X_r]$ and $[X_1, X_2] \ldots [X_{2q-1}, X_{2q}]$. Then T is generated as an ideal by its elements of degree $\leq D = \max\{q(r-1), r\}$.*

PROOF. The T-ideal T is generated by the subspaces T_j $(j = 1, 2, \ldots)$ where T_j is the set of polylinear elements of T having degree j and depending on the generators $x_1, \ldots, x_j$. Fix in T_j a Specht basis (Specht, 1950) by numbering the variables according to their indices. For conciseness, the product of generators occurring at the end of an element of the Specht basis will be called its "tail", while the remaining part—which is a product of commutators—will be termed the "head".

When an arbitrary $f \in T_j$ is expressed in terms of the Specht basis, the following types of summands occur:

(i) elements of the Specht basis, whose head contains some commutator of degree $\geq r$; all such elements lie in the ideal, generated by the commutator of length r;

(ii) elements of the Specht basis, whose head contains at least q commutators of length $\leqq r-1$; all such elements are in the right ideal, generated by elements of the type $[X_{i_1}, \ldots, X_{i_{s_1}}] \ldots [X_{k_1}, \ldots, X_{k_{s_q}}]$ and the degrees of these generators do not exceed $q(r-1)$.

(iii) the remaining elements of the Specht basis, which occur in the decomposition of f; their heads are products which involve not more than $q-1$ commutators of length $\leqq r-1$, hence are of degree $\leqq (q-1)(r-1)$.

Note that the heads of elements of type (i) and (ii) are in T, and that this need not be true for elements of type (iii). Write f in the form

$$\sum + \sum_{i=1}^{p} f_i v_i,$$

where $\sum$ incorporates the summands of the first two types, and $v_1, \ldots, v_p$ are all the different monomials occurring as tails of elements of type (iii). Clearly, $\deg f_i \leqq (q-1)(r-1)$. In order to show that $f_i \in T$ for $i=1, \ldots, p$, consider a tail v_{i_0} of maximal length. After replacing by 1 all the variables which enter v_{i_0}, one sees $f_{i_0} \in T$. In order to finish the proof of the theorem we repeat the same argument to $f - f_{i_0} v_{i_0}$, and so on.

Corollary 11. *In the finitely-presented free associative algebra $K\langle K\rangle$, the T-ideal generated by $[X_1, \ldots, X_r]$ is finitely-generated as an ideal.*

Proof. Let $|X| = n$. Then it follows from the results of Latishev (1973) that the quotient algebra $K\langle X\rangle / T$ satisfies the polynomial identity $[X_1, X_2] \ldots [X_{2q-1}, X_{2q}]$ for $q = (r-2)(n+1)$.

Theorem 10 now implies that T is generated as an ideal by its elements of degree $\leqq (r-1)(r-2)(n+1)$.

Remark 12. In order to check that $[X_1, \ldots, X_r] = 0$ is an identity it suffices to show that all elements of T_j can be reduced to zero when $j \leqq (r-1)(r-2)(n+1)$. To this aim it is sufficient to consider only the elements of a basis of this vector space, which is finite-dimensional.

Corollary 13. *Let A be a monomial algebra. If A satisfies an identity of Lie nilpotency, then it satisfies the identity $[X_1, \ldots, X_N] = 0$ with*

$$N = \sum_{i=0}^{2n^{(m-1)}+m-1} n^i.$$

Proof. It has been shown by Borisenko (1985) that such an A can be embedded in an algebra B, of dimension

$$N = \sum_{i=1}^{2n^{(m-1)}+(m-1)} n^i$$

over an extension field $\bar{K}$ of K.

Let $\bar{A}$ be the $\bar{K}$-linear span of A in B. Then the identities with coefficients in K are the same for A and $\bar{A}$. The nilpotency class of the Lie algebra is bounded above by its dimension. Hence, if A is Lie nilpotent, its nilpotency class does not exceed N, and we can apply the previous corollary.

3. Radical Properties of Monomial Algebras

This section proves the algebraic statements which imply that the following properties of a monomial algebra A are recognisable: semi-simplicity (in the sense of Jacobson), primeness, and semi-primeness.

In this section A denotes the monomial algebra $K\langle X\rangle/I$, where I is generated by the monomials $w_1, \ldots, w_t$ of degree $\geqq 2$, with $m = \max \deg w_i$. Furthermore, $L = L(A)$ denotes the diameter of $U(A)$, as defined in section 1.

The inverse image in $K\langle X\rangle$ of the upper nilradical of $A = K\langle X\rangle/I$ is denoted Nil(I).

PROPOSITION 14. *The normal (mod I) polynomial f without constant term does not belong to Nil(I) if and only if there exist monomials u and a such that* $\deg u = m-1$ *and* $ufau \notin I$.

Furthermore, a can be chosen to be of degree $\leqq L+m-1$.

PROOF. Assume a and u exist such that $ufau \notin I$. The highest term of this polynomial is of the form $uf'au$ for some monomial f' from f. By lemma 5, $uf'a$ is not nilpotent modulo I. However, $(uf'a)^i$ is the highest term of $(ufa)^i$, hence ufa is not nilpotent (mod I) as well.

Assume now $f \notin \text{Nil}(I)$, and f has no constant term. One can then find monomials u_i, v_i and constants $\alpha_i \in K$, such that

$$g = \sum_{i=1}^{s} \alpha_i u_i f v_i$$

is not nilpotent (mod I). Replacing if necessary g by its mth power, one can assume $\deg u_i \geqq m-1$ for $i = 1, \ldots, s$. Next, we note that g^{s+1} is a linear combination of products of the form $bu_i fcu_i fd$ with appropriate $b, c, d \in \langle X\rangle$. Since $g^{s+1} \notin I$, there exists such a product which is not in I. Denote by u the word formed by the last $m-1$ letters of u_i, and define a from the equation $cu_i = au$. Clearly, $ufau \notin I$.

The last claim is immediate from the following:

LEMMA 15. *Let* $u \in \langle X\rangle$ *be a monomial of degree* $m-1$, *and let h be an element of* $K\langle X\rangle$. *Assume there exists a monomial* $a \in \langle X\rangle$ *such that* $hau \notin I$ *and* $\deg hau \geqq \deg a + 2m - 2$. *If* $\deg a > L+m-1$, *then there exists a monomial* $b \in \langle X\rangle$ *such that* $hbu \notin I$ *and* $\deg b \leqq L+m-1$.

PROOF. We can assume h is in normal form (mod I). Let h' be the monomial of h for which $h'au$ is the highest term of hau. By assumption, $\deg h' \geqq m-1$.

Now, we need the following fact: if c, f, g are monomials of degree $\geqq m-1$, such that $fcg \notin I$ and c corresponds to a cyclic route beginning at the vertex w of $U(U)$, then $fwg \notin I$. Indeed, in this case $c = wp = qw$ for some monomials p and q. Obviously, $fw \notin I$ and $wg \notin I$, hence, by remark 4, $fwg \notin I$.

Thus, one can substitute the route corresponding to a by a new one of length $\leqq L$: to obtain it just replace any cyclic subroute of a by its initial vertex. Let b be the word corresponding to that new route. It is clear that $\deg b \leqq L+m-1$ and $h'bu \notin I$, hence $hbu \notin I$ as claimed.

The next result describes the Jacobson radical of monomial algebras.

THEOREM 16. *If A is a monomial algebra, its Jacobson radical J coincides with its upper nilradical N.*

PROOF. What has to be shown is that $\mathrm{Nil}(I)$ contains the inverse image $J(I)\subset K\langle X\rangle$ of the Jacobson radical J. So let $f\in J(I)$ be an element in normal form (mod I). Denote its constant term by α. If $\alpha\neq 0$, then $1-\alpha^{-1}f$ is both invertible modulo I, and without constant term: in the graded algebra A this clearly is impossible. Hence, f has no constant term.

Assume $f\notin \mathrm{Nil}(I)$. By proposition 13 there exist $u, a\in\langle X\rangle$ with $\deg u=m-1$, such that $ufau\notin I$. Let uf_iau be the highest monomial in the normal form of $ufau$. Then $uf_iau\notin I$. Noting that $(uf_ia)^2$ is the highest monomial of $(ufau)^2$, and that $(uf_ia)^2\notin I$, by remark 4, one sees that $ufauf_ia\notin I$. By the choice of f_i and by another application of remark 4, one deduces from here that $ufauf_iau\notin I$. Hence, replacing if necessary a by auf_ia, we shall assume that $\deg a\geqq m-1$. A second run of the same argument shows that one also has

$$(ufa)^2\notin I. \tag{1}$$

Since ufa is a non-zero element of $J(I)$, it has a non-zero right quasi-inverse mod I, say $g: ufa+g+ufag\equiv 0$ (mod I). Multiplying on the left by ufa, one further obtains for $h=(ufa)^2+ufag+(ufa)^2g$:

$$h\equiv 0 \text{ (mod } I). \tag{2}$$

If $ag\equiv 0$ (mod I), then by (2) $(ufa)^2\in I$, which contradicts (1). If $ag\not\equiv 0$ (mod I), let ag_j be the highest term in the normal form of ag. It is easily checked that the monomial $(uf_ia)^2g_j\notin I$ is the highest term of the normal form of h, which again contradicts (2).

We note an immediate consequence of remark 4:

LEMMA 17. *If $v\in\langle X\rangle$ is nilpotent modulo I, and $\deg v\geqq m-1$, then $v^2\in I$.*

THEOREM 18. *Let A be a monomial algebra. Then:*

(i) *If A contains a non-trivial nilpotent element, it contains a non-zero monomial v of degree $\leqq L+3m-3$, such that $v^{2m-2}=0$.*
(ii) *The algebra A is semi-simple (in the sense of Jacobson) if and only if for any $v\in\langle X\rangle\backslash I$ of degree $\leqq L+3m-3$ there exist monomials u' and a' with $\deg u'=m-1$, $\deg a'\leqq L+m-1$ such that $u'va'u'\notin I$.*
(iii) *The algebra A is prime if and only if for any $w', w''\in\langle X\rangle\backslash I$ of degree $\leqq m-1$ there exists a monomial $c\in\langle X\rangle$ of degree $\leqq L+3m-3$ such that $w'cw''\notin I$.*
(iv) *The algebra A is semi-prime if and only if for any $v\in\langle X\rangle\backslash I$ of degree $\leqq L+3m-3$ there exists a monomial $c\in\langle X\rangle$ of degree $\leqq L+3m-3$ such that $vcv\notin I$.*

PROOF. (i) We first note that if $c\in\langle X\rangle$ is nilpotent modulo I, then $c^{2m-2}\in I: c^{m-1}$ is either in I, or normal of degree $\geqq m-1$, in which case lemma 17 applies.

Assume $f\in K\langle X\rangle\backslash I$ is nilpotent modulo I. Then the highest term c in its normal form is nilpotent modulo I as well. If $\deg c\leqq L+3m-3$, we are done with $v=c$. Otherwise, write c in the form hau with $h, u\in\langle X\rangle$, both of degree $m-1$. By lemma 17, $(hau)^2\in I$, hence $uh\in I$ from remark 4. The degree of a being $>L+m-1$, there exists by lemma 15 $b\in\langle X\rangle$, such that $v=hbu\notin I$, and $\deg b\leqq L+m-1$. Now $v^2\in I$, and $\deg v\leqq L+3m-2$ as claimed.

(ii) If the Jacobson radical of A is zero, then so is its upper nilradical, and proposition 14 provides the necessary monomials a' and u'.

Assume, conversely, the condition in (ii) holds. By theorem 16 it suffices to prove $\mathrm{Nil}(I)=I$. Assume this is violated, and take a normal $f\in \mathrm{Nil}(I)\backslash I$. By proposition 14 for every monomial u' of degree $m-1$ and every monomial a' of degree $\leqq L+m-1$, $u'fa'u'\in I$. This implies the highest term of f, say w, satisfies $u'wa'u'\in I$ (u' and a' as above), hence by proposition 14 $w\in \mathrm{Nil}(I)\backslash I$. By our hypothesis, $\deg w > L+3m-3$, hence $w=hau$ with $h, u\in\langle X\rangle$, both of degree $m-1$. Applying lemma 15, we find $b\in\langle X\rangle$ with $\deg b\leqq L+m-1$, and $v=hbu\notin I$. Since by remark 4, $u'haua'u'\in I$ implies one of the inclusions $u'h\in I$, $ua'u'\in I$, we also get $u'va'u'\in I$. However, this contradicts the hypothesis, since $\deg v\leqq L+3m-3$.

(iii) Let w' and w'' be normal monomials of degree $\leqq m-1$. If A is prime, there exists an $f\in K\langle X\rangle$ such that $w'fw''\notin I$. Denoting by $w'a'w''$ the highest term of $w'fw''$, one has $w'a'w''\notin I$. If $\deg a' > L+3m-3$, then writing it in the form $a'=u'av'$ with

$$\deg u'+\deg w' = \deg v'+\deg w'' = m-1,$$

we have for $u=w'u'$, $v=v'w''$: $uav\notin I$. By lemma 15 there exists $b\in\langle X\rangle$ of degree $\leqq L+m-1$, such that $w'u'bv'w''=ubv\notin I$.

Conversely, assume condition (iii) holds, and let h and g be normal (mod I) elements of $K\langle X\rangle$, with highest terms v' and v'' respectively. If v' (resp. v'') is of degree $>m-1$, denote by w' (resp. w'') its right-hand (resp. left-hand) segment of length $m-1$. If v' (resp. v'') is of degree $\leqq m-1$, then set $w'=v'$ (resp. $w''=v''$). By hypothesis, there exists a monomial c such that $w'cw''\notin I$. By remark 4, $v'cv''\notin I$, and since this is the highest term of hcg, one has $hcg\notin I$. This proves A is prime.

(iv) If A is semi-prime, one argues as in the first part of the previous proof in order to find c. Conversely, assume condition (iv) holds. As above, it suffices to check that for any monomial $w\in\langle X\rangle\backslash I$ of degree $>L+3m-3$ one can find $c\in\langle X\rangle$ with $wcw\notin I$. Write w in the form uau' with $u, u'\in\langle X\rangle$ and $\deg u=\deg u'=m-1$. Since $w\notin I$, lemma 15 provides $b\in\langle X\rangle$ of degree $\leqq L+m-1$ such that $v=ubu'\notin I$. By hypothesis, there is a $c\in\langle X\rangle$ such that $vcv\notin I$. Remark 4 shows that $wcw\notin I$, as claimed.

4. Zero-divisors, Nilpotents, Quasi-invertibles in Monomial Algebras

In this section A continues to denote a monomial algebra, for which we suppose the defining set of monomials W is minimal, i.e. no $w_i\in W$ divides some $w_j\in W$ when $i\neq j$. Given an element $f\in A$, it is in general a difficult question to determine whether f belongs to any of the classes listed in the title. Below, we give an answer for two particular types of elements.

We now consider the partial order on $\langle X\rangle$, in which a precedes b when a is a left segment of b. A subset of monomials is called *discrete* (resp. *rooted*) if no two of its elements are comparable in this order (resp. it has a least element).

The element

$$f=\sum_{i=1}^{k}\alpha_i f_i\in K\langle X\rangle (f_i\in\langle X\rangle,\ \alpha_i\in K\backslash\{0\})$$

is called discrete (resp. rooted) if it is in normal form (mod I), and $\{f_1,\ldots,f_k\}$ has the corresponding property.

Note that any homogeneous element is discrete, and that a normal polynomial is uniquely a sum of rooted elements.

LEMMA 19. *Let $f \in K\langle X\rangle$ be a discrete element. If $g_1, \ldots, g_l \in \langle X\rangle$ are such that*

$$f\left(\sum_{j=1}^{l} \beta_j g_j\right) \equiv 0 \ (mod\ I)$$

with $\beta_j \in K\backslash\{0\}$, then $f_i g_j \in I$ for any pair of indices (i, j).

PROOF. Under our assumption syzygies of the type

$$\sum_{i,j} \alpha_i \beta_j f_i g_j = 0$$

are impossible. Consequently, the monomials $f_i g_j$ are pairwise different and each one of them has to be reducible to zero.

We say the monomial h "covers" the monomials $f_1, \ldots, f_k$ if $f_i h \equiv 0 \pmod I$, $i = 1, \ldots, k$. For instance, the monomials g_j of the preceding lemma cover the monomials f_i.

Note that the monomials which annihilate f on the right form a right ideal, generated by monomials h, which are right divisors of the basis monomials w_i and cover $f_1, \ldots, f_k$. Clearly, one can suppose the h's form a discrete set. In particular, the right annihilator of the polynomial f mod I is a finitely-generated right ideal which is a free right $K\langle X\rangle$-module of rank $\leqq t(m-1)$.

PROPOSITION 20. *For a discrete element*

$$f = \sum_{i=1}^{k} \alpha_i f_i,$$

the property of being a (left) zero-divisor is recognisable.

In fact, it suffices to check whether there exists a right divisor of the monomial relations $w_1, \ldots, w_t$ which covers all the monomials $f_1, \ldots, f_k$.

It is also worth remarking that lemma 19, on which the proof is based, does not hold in general without further assumptions on the polynomial f. In fact, one has:

EXAMPLE 21. Let I be the ideal of $K\langle X\rangle$ generated by $w = yxzya^2x$, where $x, y, z \in X$, $a \in \langle X\rangle$;

$$f = yxzy - yxzya, \qquad g = xzya^2x + axzya^2x.$$

It is easily checked that $fg \equiv 0 \pmod I$ and that the claim of the lemma does not hold. Note that in this case no right divisor of w covers all the monomials of f.

We turn to the investigation of nilpotency of elements.

LEMMA 22. *Let*

$$f = \sum_{i=1}^{k} \alpha_i f_i$$

be a discrete polynomial with $\deg f_i \geqq m-1$. Then f is nilpotent mod I if and only if for any $r \geqq 0$, $i, j_1, \ldots, j_r \in \{1, \ldots, k\}$ are satisfied the congruences:

$$f_i f_{j_1} \cdots f_{j_r} f_i \equiv 0 \pmod I. \tag{3}$$

Furthermore, if f is nilpotent, then $f^{k+1} \equiv 0$ (mod I).

PROOF. We first note that if f is nilpotent mod I, then $f_i^2 \equiv 0 \pmod{I}$ for $i = 1, \ldots, k$. Indeed, let $f^d \equiv 0 \pmod{I}$. Then the congruences $f_{i_1} \ldots f_{i_d} \equiv 0 \pmod{I}$ hold since f is discrete. In particular, $f_i^d \equiv 0 \pmod{I}$, which by lemma 17 yields the claim. The congruences (3) imply $f^{k+1} \equiv 0 \pmod{I}$. Let us show that conversely, the nilpotency of f implies (3). Assume $f_i f_{j_1} \ldots f_{j_2} \not\equiv 0 \pmod{I}$. Then the monomial $f_i f_{j_1} \ldots f_{j_2}$ enters the decomposition of f^{r+1} which is nilpotent mod I. As seen above,

$$(f_i f_{j_1} \ldots f_{j_r}) \cdot (f_i f_{j_1} \ldots f_{j_r}) \equiv 0 \pmod{I},$$

which by remark 4 implies the congruence (3).

THEOREM 23. *Let*

$$f = \sum_{i=1}^{k} \alpha_i f_i$$

be a discrete element. If f is nilpotent modulo I, then

$$f^{(m-1)(t(m-1)+1)} \equiv 0 \pmod{I}, \tag{4}$$

where t is the number of monomials w_i which generate I and $m = \max \deg w_i$.

PROOF. If f is discrete, so is f^j for any $j \geqq 1$. Replacing if necessary f by f^{m-1}, we can assume $\deg f_i \geqq m-1$ for $i = 1, \ldots, k$. As seen in the preceding proof, $f_i = u_i h_i$ for $i = 1, \ldots, k$, with u_i a right divisor of at least one of the monomials w_j. Choose, for each of the f_i's, such a divisor of maximal possible degree, and denote it by $\hat{f}_i$. The number of right divisors of the monomials w_j, hence also the number of different $\hat{f}_i$'s, does not exceed $t(m-1)$. Let us establish the validity of (4). Note that f^d, $d = t(m-1)+1$ is a linear combination of products $f_{i_1} \ldots f_{i_d}$, $i_j \in \{1, \ldots, k\}$ for $j = 1, \ldots, d$, and we shall show each one of these monomials lies in I.

In fact, such a product contains at least two monomials f_{i_p}, f_{i_q} with $\hat{f}_{i_p} = \hat{f}_{i_q}$, i.e. it has the form $\ldots f_{i_p} \ldots f_{i_q} \ldots$. According to lemma 22, $f_{i_p} \ldots f_{i_p} \equiv 0 \pmod{I}$. The choice of $\hat{f}_{i_p}$ implies $f_{i_p} \ldots \hat{f}_{i_p} \equiv \pmod{I}$ and since $f_{i_q} = \hat{f}_{i_p} h_{i_q}$, one has $\ldots f_{i_p} \ldots f_{i_q} \ldots \equiv 0 \pmod{I}$.

COROLLARY 24. *For discrete elements, being nilpotent is a recognisable property.*

THEOREM 25. *Assume*

$$f = \sum_{i=1}^{k} \alpha_i f_i$$

is discrete, with $\deg f_i \geqq m-1$. Then if f is (right) quasi-invertible mod I, it is nilpotent mod I.

In particular, quasi-invertibility mod I is a recognisable property of such elements.

PROOF. Let the right quasi-inverse of f, written in normal form be

$$g = \sum_{j=1}^{l} \beta_j g_j.$$

One has

$$f + g + fg \equiv 0 \pmod{I}. \tag{5}$$

All monomials f_i and $f_j g_h$ are pairwise distinct, this follows from the inequality

$\deg g_g \geqq m-1$ $(h=1,\ldots,l)$ which is implied by (5). Thus, with a convenient renumbering of the monomials, one can write

$$g = -f + \sum_{j=1}^{l-k} \beta_j g_j.$$

The equality (5) can be rewritten in the form

$$\sum_{j=1}^{l-k} \beta_j g_j - \sum_{i,j} \alpha_i \alpha_j f_i f_j + \sum_{r=1}^{k} \sum_{s=1}^{l-k} \alpha_r \beta_s f_r g_s \equiv 0 \ (\text{mod } I). \tag{6}$$

The claim $f^{k+1} \in I$ follows from the congruences

$$f_i a f_i \equiv 0 \ (\text{mod } I) \quad \text{for } i = 1, \ldots, k, \tag{7}$$

and a a product (possibly empty) of f_j's which we establish next.

Let $a = f_{j_q} \ldots f_{j_1}$, and assume $f_i a f_i \notin I$. Then $f_{j_1} f_i \notin I$, and since all of the monomials

$$\{f_i f_j, f_r g_s \mid 1 \leqq i, j, r \leqq k,\ 1 \leqq s \leqq l-k\}$$

are pairwise distinct, the relation (6) implies that there exists an index s_1 such that $g_{s_1} = f_{j_1} f_i$ and $\beta_{s_1} = \alpha_{j_1} \alpha_i$.

The left-hand side of (6) contains, by (7), a monomial $g_{s_2} = f_{j_2} g_{s_1}$. Iteration on this procedure yields a chain of monomials $g_{s_1} < g_{s_2} < \ldots < g_{s_q}$, all of them not in I, such that $g_{s_z} = f_{j_z} g_{s_{z-1}}$ for $z = 2, \ldots, q$. Since $g_{s_q} = f_{j_q} \ldots f_{j_1} f_i = a f_i$, from (5) and (7) one concludes that the process of construction of new monomials g_{s_z} can be continued indefinitely. This is absurd, hence (7) holds.

Before turning to rooted elements, we state a result valid in a more general situation.

Proposition 26. *Let $f \in K\langle X\rangle$ be normal of the form*

$$f = \alpha_1 ab_1 + \ldots + \alpha_k ab_k,$$

with $\alpha_i \in K \backslash \{0\}$, $b_i \in \langle X\rangle$, and a is a monomial of degree $\geqq m-1$.

If f is right quasi-invertible mod I, then $f^2 \equiv 0$ (mod I).

Proof. We can assume $ab_1 < ab_2 < \ldots < ab_k$ for the order on $\langle X\rangle$ described in the introduction. Let $g = \beta_1 g_1 + \ldots + \beta_l g_l$ be its right quasi-inverse mod I, written in normal form with $\beta_i \in K \backslash \{0\}$ and $g_1 < \ldots < g_1$. Then

$$f + g + fg \equiv 0 \ (\text{mod } I) \tag{8}$$

which implies that for $j = 1, \ldots, l$, $g_j = au_j$ for some monomial u_j. Note that the monomial $ab_k g_l$ is the highest one on the left-hand side of (8), hence $ab_k g_l \in I$. Since $g_l = au_l$ and $\deg a \geqq m-1$, remark 4 shows $ab_k a \in I$. By induction, we can now assume $ab_i a \in I$ for $i = k, k-1, \ldots, j+1$. If $ab_j a \notin I$, then $ab_j g_l = ab_j au_l \notin I$. However, the congruence

$$f + g + fg \equiv \alpha_j \beta_l ab_j g_l + h \ (\text{mod } I),$$

where h is a linear combination of monomials which are strictly lower than $ab_j g_l$, contradicts (8).

We now have $ab_i a \in I$ for $i = 1, \ldots, k$, which implies $f^2 \equiv 0$ (mod I).

Corollary 27. *When f is as in proposition 26 (in particular, when f is rooted), the properties of being right quasi-invertible or nilpotent are equivalent and recognisable.*

REMARK 28. In general, a quasi-invertible element need not be nilpotent. For instance, assuming $\mathrm{char}(K) \neq 2$ and I is the ideal in $K\langle X\rangle$, generated by $a^2 \neq 1$ for a monomial $a \in \langle X\rangle$, the element $g = -1/2 - 1/4a$ is a right quasi-inverse mod I of $f = 1 - a$. However, f obviously is not nilpotent.

5. Some Programs

NOTATION

K	a field
$X = \{x1, \ldots, xn\}$	set of indeterminates
$K\langle X\rangle$	free associative K-algebra generated by X

The following types of variables will be used:

Variables	*Type of variables*	*Remarks*
$G = \{g1, \ldots, gt\}$	set of polynomials in $K\langle X\rangle$	G is a Gröbner basis of the ideal $\langle G\rangle$ (cf. Introduction)
$W = \{w1, \ldots, wt\}$	array of strings	W is the set of all leading monomials of $g1, \ldots, gt$
m	integer	$m = \max \deg gi$
k	integer	
$N(k)$	array of strings	$N(k)$ is the set of all normal mod $\langle G\rangle$ monomials of length k
l	integer	$l = \mathrm{card}\, N(m-1)$ is the number of vertices of the Ufnarovskij graph $U(G)$ of the s.f.p. algebra $A = K\langle X\rangle/\langle G\rangle$ (cf. section 1)
$M = M(G) = M(U(G))$	matrix of integers size $l \times l$	M is the incidence matrix of the graph $U(G)$ $M(i,j) = \begin{cases} 1 & \text{if there is an edge from vertex } i \text{ to vertex } j \text{ in } U(G); \\ 0 & \text{otherwise.} \end{cases}$ The kth power of M, M^k has the property that $M^k(i,j)$ is the number of routes of length k from vertex i to vertex j in $U(G)$.

PROBLEM 1

Given: W, T.
Find: $N(1), \ldots, N(T)$.

PROGRAM 1

Input: W, T.
Output: $N(1), \ldots, N(T)$.

Variables	*Type of variables*	*Remarks*
W	array of strings	
T, n	integers	$n =$ cardinality of X
p, i, j, k, s	integers	
X	array of strings of length 1	

$N(i)$	array of strings	$N(i)$ is the set of normal monomials of length i
a, b	strings	

Function

$$\mathrm{Nor}(a, x) = \begin{cases} ax & \text{if } ax \text{ is a normal monomial} \\ \emptyset & \text{otherwise.} \end{cases}$$

```
BEGIN
  input T;
  input W;
  N(1) := X
  p := 0
  While p < T  do
    begin
      p := p+1
      N(p) := ∅
    end
  i := 0
  While i < T-1  do
    begin
      i := i+1
      if N(i) = ∅  then             If N(i) = ∅ then there are no
        begin                       normal monomials of length
          exit                      bigger than i, so N(i+1), . . .,
        end                         N(T) = ∅, end of the program
      k := card N(i)                cardinality of N(i)
      j := 0
      While j < k  do
        begin
          j := j+1
          a := N(i)(j)
          s := 0
          While s < n  do
            begin
              s := s+1
              x := X(s)
              b := Nor(a, x)
              if b ≠ ∅  then        ax is normal
                begin
                  N(i+1) := N(i+1) ∪ {b}
                end
          end {While}
      end {While}                   We have obtained N(i+1)
  end {While}
END
```

PROBLEM 2

Given: G (Therefore, Given: W).

Find: $M = M(U(G))$.

PROGRAM 2
Input: W.
Output: M.

Variables	*Type of variables*	*Remarks*
m	integer	$m = \max\{\deg w \mid w \in W\}$
i, j	integers	
$N(m-1)$	array of strings	$N(m-1)$ is the set of all normal monomials of length $m-1$; they are the vertices of the graph $U(G)$
l	integer	$l = \text{card } N(m-1)$
M	matrix of integers size $l \times l$	M is the incidence matrix of the graph $U(G)$ $M(i,j) = \begin{cases} 1 & \text{if } N(m-1)(i)x = yN(m-1)(j) \not\equiv 0 \pmod{G} \text{ for some } x, y \in X \\ 0 & \text{otherwise} \end{cases}$
a	string	normal monomial
x, y	strings of length 1	$x, y \in X$
Function Nor(a, x)		$\text{Nor}(a,x) = \begin{cases} ax & \text{if } ax \text{ is normal} \\ \emptyset & \text{otherwise} \end{cases}$

```
BEGIN
  Find: N(m-1); l
  M :≠ 0                                          zero matrix
  i := 0
  While i < l  do
    begin
      i := i+1
      j := 0
      While j < l  do
        begin
          x := right(N(m-1)(j), 1)                x = the right segment of
                                                  N(m-1)(j) of length 1
          y := left(N(m-1)(i), 1)                 y = left segment of N(m-1)(i)
                                                  of length 1
          if (Nor(N(m-1)(i), x) ≠ 0) and          N(m-1)(i)x = yN(m-1)(j) is
             (N(m-1)(i)+x = y+N(m-1)(j))  then    a normal monomial, i.e. there
                                                  is an edge from N(m-1)(i) to
                                                  N(m-1)(j) in U(G)
             begin
               M(i, j) := 1
             end
        end {While j}
  end {While i}
END
```

PROBLEM 3
Given: G.
Find: The type of growth of $A = \langle X \rangle / G$.

PROGRAM 3
Input: $M = M(G)$.
Output: R.

Variables	*Type of variables*	*Remarks*
l	integer	number of vertices in the graph $U(G)$
d	integer	degree of polynomial growth
$C, C(1), \ldots, C(l), D$	sets of integers	sets of vertices of $U(G)$
M, N, L	matrices of integers size $l \times l$	M is the incidence matrix of $U(G)$
P, Q	matrices of integers size at most $l \times l$	
R	string	type of growth
i, j, k	integers	auxiliary variables

```
BEGIN
  N := E                          E = identity matrix of size l×l
  L := 0                          0 = zero matrix of size l×l
  C := 0                          0 = the empty set
  j := 0                          j will denote the power of M under
                                  investigation

  While j < l  do
    begin
      j := j+1                    increment j, the power of the incidence matrix
      N := N × M                  store jth power of M in N
      D := i: N(i, i) ≠ 0         D is the set of indices of diagonal elements of
                                  M^j which are non-zero. This is the set of all
                                  vertices lying on a cyclic route of length
                                  precisely j
      If (∃ i∈D: N(i, i) > 1) or  If either condition is satisfied, there exists a
      (∃ k: 1 < k < j and         vertex lying on two different cycles, hence
          k∤j: D∩C(k) ≠ ∅)
      then
        begin
          R := "exponential"      the growth is exponential (cf. proposition 2)
          exit                    and we are done
        end
    C(j) := D\C                   C(j) is the set of vertices which lie on a cycle
                                  of length j but no shorter one
    C := C∪C(j)                   This is the set of all vertices on a cyclic route
                                  of length ≦j
    L := L∨N                      L∨N(i, k) = L(i, k)∨N(i, k) where
                                  a∨b = 0 iff a = b = 0,
                                  a∨b = 1 else.
                                  —L(i, k) = 1 iff there is a route from vertex i
                                  to vertex k of length at most j
  end {While}                     There can be no route of length ≧l which
                                  does not contain a cycle.
```

	$L(i, k) = 1$ if there is a route from vertex i to vertex k.
	We know now that the algebra A has polynomial growth
if $C = 0$ then	If no vertex lies on a cycle . . .
begin	(0 = empty set)
$R :=$ polynomial growth of degree 0	—The algebra A is finite-dimensional (cf. proposition 2)
exit	
end	
$d := 1$	
$P := M \backslash C$	$M \backslash C$ denotes the matrix obtained by deleting the ith row and ith column whenever $i \notin C$, i.e. the vertex i does not lie on a cycle
$P(j, j) := 0$	Set all diagonal entries (which were 1) equal to zero
If $(\exists\, i < j : P(i, j) \,.\, P(j, i) = 1)$	—i.e. the vertices i and j lie on the same cycle . . .
then delete jth row and jth column	—We obtain the incidence matrix of $\hat{U}$, the graph with all cycles in U collapsed to a single vertex. This graph is oriented without cycles. There is an edge between two vertices iff the corresponding cycles in the original graph U could be connected by a route
$Q := P$	
While $Q \neq 0$ do	
begin	
$d := d + 1$	
$Q := Q \times P$	The index of nilpotency of P-min $(i \mid P^i = 0)$ is equal to $1 +$ the extremal length of a route in the graph $\hat{U}$, i.e. to the maximal number of different cycles occurring in the same connected component of U
end While	—d is the index of nilpotency of P, hence, by proposition 2, d is the degree of the polynomial growth
begin	
$R :=$ polynomial growth of degree d end	
END	

References

Bergman, G. (1978). The diamond lemma for ring theory. *Adv. Math.* **29,** 178–218.

Borisenko, V. (1985). On matrix representations of finitely presented algebras defined by a finite number of words. *Vestnik Mosc. Univ. Ser. 1. Matematika. Mehanika*, No. 4, 75–77 (in Russian).

Borho, W., Kraft, H. (1976). Über die Gelfand–Kirilov dimensions. *Math. Ann.* **220,** 1–24.

Buchberger, B. (1965). An algorithm for finding a basis for the residue class ring of a zero-dimensional polynomial ideal (German). PhD Thesis, Univ. of Innsbruck (Austria), Math. Inst., 1965, see also *Aequationes Mathematicae* **4/3,** 374–383, 1970.

Latyshev, V. (1973). On some varieties of associative algebras. *Izvestija Akad. Nauk SSSR, Ser. Mat.* **37,** 1010–1037 (in Russian); English transl.: *Math. USSR, Izvestija* **7** (1973), 1011–1038 (1975).

Markov, V. (1979). Systems of generators of T-ideals of finitely generated free algebras. *Algebra Logika* **18,** 587–589 (in Russian); English transl.: *Algebra Logic* **18,** 371–378 (1980).
Matiyasevich, Y. (1984). On the investigations concerning some algorithmic problems of algebra and number theory. *Trudy Mat. Inst. AN SSSR* **168,** 218–236 (in Russian). (Proceedings of the Steklov Inst.).
Specht, W. (1950). Gesetze in Ringen I. *Math. Z.B.* **52,** 557–589.
Ufnarovskij, V. (1982). A growth criterion for graphs and algebras defined by words. *Mat. Zametki* **31,** 465–472 (in Russian); English transl.: *Math. Notes* **37,** 238–241 (1982).